THIRD EDITION

ATLAS OF
FUNCTIONAL
NEUROANATOMY

THIRD EDITION

ATLAS OF
FUNCTIONAL
NEUROANATOMY

WALTER J. HENDELMAN, M.D., C.M.

University of Ottawa
Ottawa, Ontario, Canada

CRC Press
Taylor & Francis Group
Boca Raton London New York

CRC Press is an imprint of the
Taylor & Francis Group, an **informa** business

CRC Press
Taylor & Francis Group
6000 Broken Sound Parkway NW, Suite 300
Boca Raton, FL 33487-2742

© 2016 by Taylor & Francis Group, LLC
CRC Press is an imprint of Taylor & Francis Group, an Informa business

No claim to original U.S. Government works

Printed on acid-free paper
Version Date: 20150512

International Standard Book Number-13: 978-1-4665-8534-8 (Pack - Book and Ebook)

This book contains information obtained from authentic and highly regarded sources. While all reasonable efforts have been made to publish reliable data and information, neither the author[s] nor the publisher can accept any legal responsibility or liability for any errors or omissions that may be made. The publishers wish to make clear that any views or opinions expressed in this book by individual editors, authors or contributors are personal to them and do not necessarily reflect the views/opinions of the publishers. The information or guidance contained in this book is intended for use by medical, scientific or health-care professionals and is provided strictly as a supplement to the medical or other professional's own judgement, their knowledge of the patient's medical history, relevant manufacturer's instructions and the appropriate best practice guidelines. Because of the rapid advances in medical science, any information or advice on dosages, procedures or diagnoses should be independently verified. The reader is strongly urged to consult the relevant national drug formulary and the drug companies' and device or material manufacturers' printed instructions, and their websites, before administering or utilizing any of the drugs, devices or materials mentioned in this book. This book does not indicate whether a particular treatment is appropriate or suitable for a particular individual. Ultimately it is the sole responsibility of the medical professional to make his or her own professional judgements, so as to advise and treat patients appropriately. The authors and publishers have also attempted to trace the copyright holders of all material reproduced in this publication and apologize to copyright holders if permission to publish in this form has not been obtained. If any copyright material has not been acknowledged please write and let us know so we may rectify in any future reprint.

Visit the Taylor & Francis Web site at
http://www.taylorandfrancis.com

and the CRC Press Web site at
http://www.crcpress.com

This book is dedicated to

my family

my teachers

my colleagues

and to

all who seek to understand the brain

CONTENTS

Section 4   The Limbic System   200

PREFACE TO THE THIRD EDITION

Why a new edition of the *Atlas of Functional Neuroanatomy*? After all, the structures of the brain have not changed since the last edition. However, our knowledge concerning the functional aspects of the brain has certainly increased, and this has led to a revised understanding of the brain and its structures.

The atlas is primarily a teaching book—for medical students, non-neurology residents, and all the other health professionals who need a solid understanding of the brain—in order to help patients who have neurological problems. As a result of our better understanding of brain function, the emphasis on what is needed in practice has certainly changed. At the same time we have witnessed a decrease in the amount of time devoted to this subject matter in the curriculum. And, above all, the way students now approach learning has undergone a transformation, if not a total revolution.

This new edition has given us the opportunity to reorganize the material so that the learner can progress in a more sequential manner to acquire the knowledge of the subject matter. New material was added to explain the visual system, the meninges, and the venous system, as well as the limbic system. Many illustrations have been enhanced and improved. In all, there has been an increase of about 10% in the number of illustrations. There has also been a substantial increase in the number of neuroradiological images because this is the way neuroanatomy is "seen" in the clinical setting.

The format is the same as before, with each illustration accompanied by explanatory text. As before, there is a selective labeling of structures, with the objective of understanding the functional brain, guided by the targeted audience for this atlas. The text accompanying each illustration has been re-thought and refreshed and kept as brief as warranted. There is extensive cross-referencing to other illustrations in the atlas. The Annotated Bibliography has been updated, and the Glossary has been retained. A new component has been added, clinical cases, that requires a knowledge of the neuroanatomy.

In addition to visual enhancement of almost all the illustrations, there are two very significant enrichments for this edition of the atlas, a Web site with interactive features, and narrated video demonstrations of the brain which are now accessed on the Web site. The atlas Web site (www.atlasbrain.com) includes all the illustrations, with rollover labeling of the structures, animation of all the pathways and some connections. The added bonus is video presentations, narrated laboratory demonstrations using actual brain material that were produced by the author at an earlier phase of the use of technology for learning.

I am also the co-author of an affiliated book, *The Integrated Nervous System: A Systematic Diagnostic Approach* (2010, CRC Press), with two practicing neurologists, Drs. Humphreys (pediatric) and Skinner (adult). This is a case-based teaching text about neurological problem-solving through the presentation of neurological diseases, specifically how to go about the localization of the lesion (where) and how to resolve the likely etiology (what). Accompanying the text is a Web site with more than 40 additional cases, each with a history, physical examination, test results, and an explanation of the diagnosis. This is an additional clinical resource which is often referenced in the atlas.

All this is to say that this revised edition of the atlas presents a fresh view of neuroanatomy, with additional learning resources. Hopefully, those who require a knowledge of neuroanatomy—students in medicine and the allied health professions and others studying the brain, as well as those involved with neurological patients—now have a suitable resource to help them to comprehend this complex and fascinating subject matter—the brain.

Walter J. Hendelman, M.D., C.M.
Professor Emeritus
Faculty of Medicine
University of Ottawa
Ottawa, Canada

PREFACE TO THE SECOND EDITION

This atlas grew out of the seeds of discontent of a teacher attempting to enable medical students to understand the neuroanatomical framework of the human brain, the central nervous system. As a teacher, it is my conviction that each slide or picture that is shown to students should be accompanied by an explanation; these explanations formed the basis of an atlas. Diagrams were created to help students understand the structures and pathways of the nervous system and each illustration was accompanied by explanatory text, so that the student could study both together.

The pedagogical perspective has not changed over the various editions of the atlas as it expanded in content, but the illustrations have evolved markedly. They changed from simple artwork to computer-based graphics, from no color to two colors, to the present edition in full color. The illustrations now include digital photographs, using carefully selected and dissected specimens.

Most of the diagrams in the atlas were created by medical students, with artistic and/or technological ability, who could visualize the structural aspects of the nervous system. These students, who had completed the basic neuroanatomy course, collaborated with the author to create the diagrams intended to assist the next generation of students to learn the material more easily and with better understanding. I sincerely thank each of them for their effort and dedication and for their frequent, intense discussions about the material (please see the acknowledgments). They helped decide which aspects should be included in an atlas intended for use by students early in their career with limited time allotted for this course of study during their medical studies.

This atlas has benefited from the help of colleagues and staff in the department of which I have been a member for over 30 years and from professional colleagues who have contributed histological and radiological enhancements, as well as advice. Their assistance is sincerely appreciated.

The previous edition of this atlas included a CD-ROM containing all the images in full color. At that time, few texts had such a learning companion. It is to the credit of CRC Press that they were willing to accept the idea of this visual enhancement as an aid to student learning. The CD-ROM accompanying this new edition of the atlas, thanks to another student, employs newer software that allows the creative use of "rollover" labeling and also adds animation to some of the illustrations (please see the User's Guide).

A final comment about the word "functional" in the title is appropriate. The central nervous system, the CNS, is a vast, continually active set of connections, ever changing and capable of alteration throughout life. The orientation of the written text is to describe both the structural aspects of the CNS and the connections between the parts, and to explain the way those structures of the brain operate as a functional unit. In addition, there are clinically relevant comments included in the descriptive text, where there is a clear relation between the structures being described and neurological disease.

No book could be completed without the support and encouragement of the people who are part of the process of transforming a manuscript to a published work, from the publisher and the project editor, to the technical staff that handles the illustrations, to the proofreaders and copyeditors who work to improve and clarify the text. Each individual is an important contributor to the final product, and I wish to thank them all.

I sincerely hope that you, the learner, enjoy studying from the *Atlas of Functional Neuroanatomy* and its accompanying CD-ROM, and that the text and illustrations, along with the dynamic images, help you to gain a firm understanding of this fascinating, complex organ—the brain.

Walter J. Hendelman, M.D., C.M.
Ottawa, Canada

PREFACE TO THE FIRST EDITION

The instructional goal of this *Atlas* is to assist the student of the brain to achieve an understanding and a three-dimensional visualization of the human central nervous system (CNS).

The *Atlas of Functional Neuroanatomy* is written for medical students who are studying the CNS for the first time, students in allied health fields, and professionals-in-training (physicians, nurses, physical and occupational therapists) who require a visual reference to the structures of the CNS, as well as students in certain undergraduate courses, particularly neuroscience and psychology. Regardless of the student, the challenge to the teaching faculty is the same—how could we improve, enhance, and facilitate the learning process? We, as teachers, must see the learning task from the perspective of the student—how can he/she learn, understand, and assimilate this very complex subject matter, particularly with the volume of material to be learned and the short period of time typical of new curricula?

Clearly the challenge to any author is to try to organize and reduce the information load and present core material with adequate explanation. The *Atlas of Functional Neuroanatomy* contains diagrams and text, an optimum way of guiding the student through the complexity of the structure and function of the CNS, with some clinical references to make the information relevant to the real world of people with diseases. The illustrations include diagrams and photographs, each labeled to the degree necessary for a student learning the material for the first time, or for a professional requiring a resource review of the CNS. The focus is on the illustrations, each of which is accompanied by explanatory text on the facing page, supplemented by a brief introduction to various sections (e.g., brainstem, motor systems, limbic).

This *Atlas* is built upon three previous editions (titled *Student's Atlas of Neuroanatomy*), which began with some illustrations, then photographic material, and subsequently added air brush diagrams on the basal ganglia, thalamus, and limbic system. After use in our course here and much feedback from students, the material of the *Atlas* has been significantly reorganized and rethought, and much of the text rewritten. The approach of this revised *Atlas* is to integrate the structure and function so that a student can understand the neurological approach to disease of the nervous system, with a focus on *where* the information has been interrupted.

The *Atlas* starts with an **Orientation** to the various parts of the nervous system, presented from the spinal cord upward to the brain. Radiographic material has been added, since this is the way the CNS will be viewed and investigated by all our students in clinics. The second section, **Functional Systems**, presents the sensory and motor pathways as they traverse the nervous system. The addition of color to these diagrams contributes substantially to their visual impact.

The third section, **Neurological Neuroanatomy**, has both an anatomical and neurological orientation. Sufficient information is given to allow the student to work through the neurological question—*where* is the disease process occurring (i.e., neurological localization)? The emphasis in this section is on the brainstem. To assist in this goal, a select series of cross sections of the human brainstem is included. In addition, new illustrations have been added on the blood supply to the brain, using color and graphic overlays, since vascular lesions are still most common and relate closely to the functional neuroanatomy. This section is supplemented with cross sections of the human brainstem.

The section on the **Limbic System** has been completely revised and much reduced in content from the previous edition. It is placed as the last section of the *Atlas* because it can be taught at various points in the medical curriculum, e.g., as part of "mind" or psychiatry. Other courses might not include this specialized topic.

Consistent with the computer/digital revolution, many of the illustrations were converted into computer graphics, with tones of shading. *Color* has been added to facilitate the visualization of the CNS pathways, in both system-based (Section B) and cross-sectional diagrams (Section C). Some students might still want to add color to the illustrations; coloring the illustrations has assisted many students by adding an active component to the learning process. (A guide to color coding is included after the list of illustrations.)

Much of the subject matter's difficulty is terminology—complex, difficult to spell, sometimes inconsistent, with a Latin base, and sometimes with names of individuals (used often by neurologists, neurosurgeons, and neuroradiologists). A *Glossary* of terms is appended to help the student through this task.

Students might wish to consult more complete texts on the anatomy and physiology of the nervous system and certainly some neurology books. A guide to this reference material is included in the *Annotated Bibliography*. Added to this are suggestions for material available on CD-ROMs, as well as the Internet. Students are encouraged to seek out additional resources of this nature.

The digital revolution has led to the expectation of a visual presentation with clear graphics on the screen! Therefore, this edition of the *Atlas* is being published with an accompanying *CD-ROM,* allowing the use of full-color illustrations, where relevant. As in the book, each graphic has a brief explanatory text. We hope that students will have easy access to view this CD and that this additional resource will enhance learning! The author is grateful to CRC Press for agreeing to publish the *Atlas* with the CD.

Many individuals have contributed to the *Atlas*. Their efforts are deeply appreciated. We have worked collaboratively to try to present a clear understandable view of the structure and function of the CNS. All have worked under my direction and therefore the ultimate value of the *Atlas,* whether favorable or unfavorable, rests on the shoulders of the author. You, the learner, will be our best judge.

Special thanks are extended to the members of CRC Press LLC without whose help this project would not have been completed.

Walter J. Hendelman, M.D., C.M.

ACKNOWLEDGMENTS

This atlas has been a cumulative work in progress, adding and altering, and deleting material over time. The illustrations have been created by talented and dedicated individuals—artists, photographers, and students, and with the help of staff and colleagues—whom the author has had the pleasure of working with over these many years.

SPECIAL ACKNOWLEDGMENT TO DR. TIM WILLETT

Dr. Willett began working with me on the second edition of the Atlas (published by CRC Press in 2006) when he was still a medical student, one of my students in the course on the nervous system. Together we did the dissection and digital photography of the brain specimens that remain in the present atlas. Using Photoshop, he undertook to redo many of the illustrations in the atlas.

Tim is not only the primary illustrator once again of this third edition of the atlas; he is also a partner in researching and understanding the complex neuroanatomical literature, thus making sure that the illustrations are consistent with the information available in various texts. In addition, he has honed his creative skill in designing the optimum presentation of the neuroanatomy, suitable for the primary learning audience of this atlas.

Tim has created several new illustrations and redid many others for the present edition, as well as taking on the responsibility of selecting the neuroradiology to go along with the illustrations.

PREVIOUS EDITIONS

The atlas was originally published with the title *Student's Atlas of Neuroanatomy* by the University of Ottawa Press (1987 and 1988) and then by Saunders (1994). The diagrams in the first editions were created by Mr. Jean-Pierre Morrissey, a medical student at the time he did the work. To these were added photographs of brain specimens taken by Mr. Stanley Klosevych, who was then the director of the Health Sciences Communication Services, University of Ottawa. Dr. Andrei Rosen subsequently created the airbrush diagrams (note particularly the various pathway summaries and the limbic system illustrations) and expanded the pool of illustrations. The efforts of the staff of the University of Ottawa Press and of Saunders who published these editions is very much appreciated and acknowledged.

The first edition of the atlas, published in 2000 by CRC Press under its present title, *Atlas of Functional Neuroanatomy,* included new computer-generated diagrams done by Mr. Gordon Wright, a medical illustrator, that replaced many of the earlier illustrations. Mr. Wright also put together the CD-ROM for that edition that contained all the illustrations in the atlas.

CD-ROM FOR THE SECOND EDITION

Mr. Patrick O'Byrne, a doctoral candidate in the nursing program at the Faculty of Health Sciences, University of Ottawa, put together the CD-ROM for the second edition, by using Macromedia Flash software to create rollover labeling and animated illustrations.

The material on the CD-ROM is now available on the atlas Web site (www.atlasbrain.com) for the second edition of the atlas using the original files from the CD-ROM.

PRESENT EDITION

ILLUSTRATIONS

Mr. Perry Ng, medical illustrator, is the primary illustrator for *The Integrated Nervous System*. CRC Press as the publisher of both books has permitted illustrations from that text, modified, to be used in the atlas (those on the meninges, Figure 7.2 and Figure 7.3, and the cerebrospinal fluid, Figure 7.8). Mr. Ng also created new illustrations of the venous system (Figure 7.4, Figure 7.5, and Figure 7.6) and the limbic system (Figure 10.7 and Figure 10.8).

SPECIMENS

The brain specimens and *in situ* specimens were made available by Dr. M. Hincke, Professor and Head, Division of Clinical and Functional Anatomy, Department of Cellular and Molecular Medicine, Faculty of Medicine, University of Ottawa. Special thanks to Ms. Shannon Goodwin, prosector, for assistance in this task. Ms. Goodwin also provided the *in situ* specimens of the vertebral column (Figure 1.2 and Figure 1.10) and those used to create the venous sinuses (Figure 7.4, Figure 7.5, and Figure 7.6).

RADIOGRAPHS

Dr. Michael Kingstone has been our consultant neuroradiologist for the present (and previous) edition, with contributions from colleagues at the Ottawa Hospital. Most radiographs have been replaced from the previous edition with new images, by using the upgraded capability of the newer machines and accompanying software. Special thanks to Dr. Santanu Chakraborty for the Diffusion Tensor Image (DTI) of the projection fibers (Figure 2.4).

HISTOLOGICAL SECTIONS

Colleagues and staff of the Department of Pathology, Children's Hospital of Eastern Ontario, Ottawa, are responsible for preparing the histological sections of the human brainstem, added to in the previous edition by sections of the human spinal cord.

SPINAL CORD MACRO SPECIMENS

Specimens of the human spinal cord (Figure 3.9) were provided by Mr. Alain Tremblay and Dr. John Woulfe, Eastern Ontario Regional Laboratory Association, Ottawa, Canada.

ANNOTATED BIBLIOGRAPHY

Michelle Leblanc of the Health Sciences Library of the University of Ottawa provided the assistance to update the Annotated Bibliography.

WEB SITE

www.atlasbrain.com

All the illustrations in the atlas are on the Web site, with roll-over (mouse over) labeling and animation of all the pathways and some connections. This work was done, using the appropriate software, by Shannon Goodwin, prosector, Division of Clinical and Functional Anatomy, Department of Cellular and Molecular Medicine, Faculty of Medicine, University of Ottawa. Ms. Goodwin also did the same for the second edition of the atlas (also available on the same Web site) based upon the material on the CD-ROM (see above). The author gratefully acknowledges her effort and expertise in this task.

VIDEOS

Mr. Klosevych (mentioned earlier) and I collaborated to create the video demonstrations of the brain by using ¾-inch VHF tape, which we edited frame by frame.

Ms. Mariane Tremblay, Web, Multimedia and Learning Technologies Designer and Technician with Medtech—Information Management Services of the Faculty of Medicine at the University of Ottawa—ably resuscitated the video files, which are now available on the Web site.

SUPPORT

The previous editions were supported financially, in part, by grants from Teaching Resources Services of the University of Ottawa. The present (and previous) edition received support from CRC Press.

The support of my home department at the Faculty of Medicine of the University of Ottawa (initially the Department of Anatomy and now called the Department of Cellular and Molecular Medicine) including colleagues, secretaries, and other staff in the gross anatomy laboratory, is gratefully acknowledged. Computer support has consistently been available from Medtech—Information Management Services of the Faculty of Medicine at the University of Ottawa.

Clinical colleagues in the Division of Neurology (Department of Medicine) of the Ottawa Hospital readily assisted with information on the clinical aspects of diseases and the relationship with neuroanatomy. Special thanks to my co-authors of the *Integrated Nervous System,* Dr. Chris Skinner and Dr. Peter Humphreys, for their input on various clinically related topics.

CRC PRESS

I am grateful to Ms. Barbara Norwitz for her advice and guidance as my executive editor for the first and second editions of this atlas. Mr. Lance Wobus, senior editor, has ably guided me through this edition. My thanks also to Ms. Jill Jurgensen, and Ms. Charlene Counsellor, and all the staff of the press for their care and attention in publishing this edition and previous editions.

Finally, to the many classes of students, who have provided inspiration, as well as comments, suggestions and feedback, with thanks to all.

Walter J. Hendelman, M.D., C.M.
April 2015

AUTHOR BIOGRAPHY

Walter Hendelman, M.D., C.M., is a Canadian, born and raised in Montreal. He did his undergraduate studies at McGill University in science with honors in psychology. As part of his courses in physiological psychology, he assisted in an experimental study of rats with lesions of the hippocampus, which was then a little-known area of the brain. At that time, Professor Donald Hebb was the chair of the Psychology Department and was gaining prominence for his theory known as "cell assembly," explaining how the brain functions.

Dr. Hendelman proceeded to do his medical studies at McGill. The medical building was situated in the shadow of the world-famous Montreal Neurological Institute (MNI), where Dr. Wilder Penfield and colleagues were forging a new frontier in the understanding of the brain. Subsequently, Dr. Hendelman completed an internship and a year of pediatric medicine, both in Montreal.

Having chosen the brain as his life-long field of study and work, his next decision involved the choice of either clinical neurology or brain research—Dr. Hendelman chose the latter, with the help of Dr. Francis McNaughton, a senior neurologist at the MNI. Postgraduate studies continued for 4 years in the United States, in the emerging field of developmental neuroscience, using the "new" techniques of nerve tissue culture and electron microscopy. Dr. Richard Bunge was his research mentor at Columbia University Medical Center in New York City, and his neuroanatomy mentor was Dr. Malcolm Carpenter, author of the well-known textbook, *Human Neuroanatomy.*

Dr. Hendelman returned to Canada and has made Ottawa his home for his academic career at the Faculty of Medicine of the University of Ottawa, in the Department of Anatomy, now merged with Physiology and Pharmacology into the Department of Cellular and Molecular Medicine. He began his teaching in Gross Anatomy and Neuroanatomy and in recent years has focused on the latter. His research continued, with support from Canadian granting agencies, using nerve tissue culture to examine the development of the cerebellum; later on he became involved in studies on the development of the cerebral cortex. Several investigations were carried out in collaboration with summer and graduate students and with other scientists. He has been a member of various neuroscience and anatomy professional organizations, has attended and presented at their meetings, and has numerous publications on his research findings.

In addition to research and teaching and the usual academic "duties," Dr. Hendelman was involved with the Faculty of Medicine and university community, including a committee on research ethics. He has also been very active in curriculum planning and teaching matters in the Faculty of Medicine. During the 1990s, when digital technology became available, Dr. Hendelman recognized its potential to assist student learning, particularly in the anatomical subjects, and helped bring technology into the learning environment of the Faculty. He also organized a teaching symposium for the Canadian Association of Anatomy, Neurobiology and Cell Biology on the use of technology for learning the anatomical sciences.

In 2002, Dr. Hendelman completed a program in medical education and received a Master's degree in Education from the Ontario Institute of Studies in Education, affiliated with the University of Toronto. In the same year, following retirement, he began a new stage of his career, with the responsibility for the development of a Professionalism Program for medical students at the University of Ottawa.

As a student of the brain, Dr. Hendelman has been deeply engaged as a teacher of the subject throughout his career. Dedicated to assisting those who wish to learn functional neuroanatomy, he produced several teaching videotapes using anatomical specimens and five previous editions of this atlas (under various titles). As part of this commitment he has collaborated in the creation of two computer-based learning modules, one on the spinal cord based upon the disease syringomyelia and the other on voluntary motor pathways; both contain original graphics to assist in the learning of the challenging and fascinating subject matter, the human brain.

Most recently, to keep up with the use of technology, Dr. Hendelman guided the development of a Web site for the second edition of this atlas with interactive features. In addition, the second edition has been translated into Italian and now into French, and the French edition also has a Web site. Finally, Dr. Hendelman is a co-author of *The Integrated Nervous System: A Systematic Diagnostic Approach,* published by CRC Press (2010).

In his non-professional life, Walter Hendelman is a husband, a father, an active member of the community, a lover of music, sometimes a choir member and now a member of a community band (trumpet), a commuter cyclist, and an avid skier and skater.

LIST OF ILLUSTRATIONS

Appendix: Neurological Neuroanatomy

INTRODUCTION TO THE ATLAS

We are about to embark on an interesting and challenging journey—an exploration of the human brain. The complexity of the brain has not yet been adequately described in words. The analogies to switchboards or computers, although in some way appropriate to describe some aspect of brain function, do not do the least bit of justice to the totality. The brain functioning as a whole is infinitely more than its parts. Our brains encompass and create a vast universe.

In the past decade, we have come to appreciate that our brains are in a dynamic state of change in all stages of life. We knew that brain function was developing throughout childhood, and this has been extended into the teen years and even into early adulthood. We now are beginning to understand that the brain has the potential to change throughout life, in reaction to the way we live and our personal experiences in this world. The generic term for this is plasticity, and the changes may alter extensively the connections of the brain and its pattern of processing information, whether from the external world, from our internal environment, or from the brain itself as it generates "thoughts" and "feelings."

PEDAGOGICAL PLAN

An understanding of the **central nervous system (CNS)** and how it functions requires knowing its component parts and their specialized operations and the contribution of each of the parts to the function of the whole.

- *Section 1* of the atlas introduces the student to the CNS from an anatomical viewpoint, by looking at the CNS from the outside and then examining its internal structures, first the cerebral hemispheres and then the brainstem, cerebellum and spinal cord.
- The subsequent section, *Section 2*, uses these components to build the various functional systems, such as the sensory and motor systems.
- The coverings of the brain, the meninges, the venous sinuses, and the cerebrospinal fluid as well as the blood supply to the brain and spinal cord are discussed in *Section 3*.
- The parts of the brain involved with emotional behavior, the limbic system, are discussed in *Section 4*.
- Detailed histological organization is found in the *Appendix*.

The student is also provided with a *Glossary* of terms to help navigate the complex terminology of neuroanatomy and neurology. In addition, there is an *Annotated Bibliography* to guide the student/learner to additional printed and some Web-based resources.

LEARNING RESOURCES

WEB SITE: www.atlasbrain.com

The Web site adds another dimension to the learning process. Ideally, the student is advised to read the text, using *both* the text illustration and the illustration on the Web site. All the labeling on the Web site uses rollover (mouse-over) technology, adding to the learning process by providing the student with an opportunity to identify structures, with immediate feedback. In addition, animation has been added to certain illustrations, such as the pathways, where understanding and seeing the tract that is being described, along with the relays and crossing (decussation), can hopefully assist the student in developing a three-dimensional understanding of the nervous system. Some complex connections are also animated, such as those of the basal ganglia.

The Web site also has the *Glossary* of terms (as in the printed atlas). Please note: This Web site also includes the web material for the second edition of the Atlas and for its French translation.

NOTICE TO USERS

All the material on the Web site is copyright of the author. Please note that use of the Web site is free! The illustrations on the Web site may be used for educational purposes **only in an educational setting**, with proper attribution to the author including the title of the atlas and the publisher.

VIDEO LESSONS

The Web site also has laboratory demonstrations, narrated by the author, of the brain and skull. These are often useful before beginning a section of study and probably for review purposes. For those who have limited or no access to actual brain material, these videos should provide a valuable learning opportunity.

	Color	Tracts	Thalamus	Cortex
Sensory	Blue (light)	DC-ML	VPL	Postcentral gyrus (upper), Parietal lobe
	Blue (dark)	ALS	VPL	
	Cyan	Trigeminal	VPM	Postcentral gyrus (lower)
	Fuchsia	Audition	MGB	Heschl, Temporal lobe
	Magenta	Vision	LGB	Calcarine, Occipital lobe
	Purple		LP, Pul	Sensory association areas
Sensory & Motor	Olive	Vestibular, MLF		
	Yellow	Reticular	IL, CM, Mid	
Motor	Orange (yellowish)	Voluntary motor		Precentral gyrus, Frontal lobe generally
	Orange (reddish)	Parasympathetic		
	Aqua	Basal ganglia	VA/VL	
	Green	Cerebellum	VA/VL	Premotor
Limbic	Red & Pink	Limbic	DM	Prefrontal
	Brown	Limbic	AN, LD	Cingulate
Special Nuclei	Black	Substantia nigra		
	Red	Red nucleus, Amygdala		

USER'S GUIDE

COLOR CODING

Color adds a significant beneficial dimension to the learning of all anatomical subjects and particularly neuroanatomy. The colors in this atlas have been assigned a functional role in that they are used consistently for the presentation of sensory, motor and other structures. The following is the color coding used in this atlas, as shown on the opposite page:

SENSORY: nuclei and tracts
    Dorsal column (DC)—medial lemniscus   LIGHT BLUE
    (ML)
    Anterolateral system (ALS)   DEEP BLUE
    Trigeminal pathways   CYAN
  AUDITION   FUCHSIA
  VISION   MAGENTA
  ASSOCIATION AREAS (sensory)   PURPLE
SENSORY and MOTOR
    VESTIBULAR (medial longitudinal   OLIVE
    fasciculus [MLF])
    RETICULAR FORMATION   YELLOW
MOTOR: nuclei and tracts
    Voluntary   ORANGE (yellowish)
    Parasympathetic   ORANGE (reddish)
  BASAL GANGLIA   AQUA
  CEREBELLUM   GREEN
LIMBIC
    Prefrontal cortex   RED AND PINK
    Cingulate gyrus, amygdala   BROWN
SPECIAL NUCLEI
    Substantia nigra   BLACK
    Red nucleus (and tract)   RED

Thalamic abbreviations in this illustration are as follows (see Figure 4.3):

- AN, anterior nuclei
- CM, centromedian
- DM, dorsomedial
- IL, intralaminar
- LD, lateral dorsal
- LGB, lateral geniculate body
- LP, lateral posterior
- MGB, medial geniculate body
- Mid, midline
- Pul, pulvinar
- VA, ventral anterior
- VL, ventral lateral
- VPL, ventral posterolateral
- VPM, ventral posteromedial

For students who enjoy a different learning approach, a black and white photocopy of the illustration can be made and then the color added, promoting active learning.

Some students may wish to add more color to some of the airbrush diagrams, including the basal ganglia, thalamus and limbic system.

REFERENCE TO OTHER FIGURES

Reference is made throughout the atlas to other illustrations that contain material relevant to the subject matter or structure being discussed. Although this may be somewhat distracting to the learner reading a page of text, the author recommends looking at the illustration and the accompanying text being referenced, to clarify or enhance the learning of the subject matter or structure.

CLINICAL ASPECT

Various clinical entities are mentioned in the text, where there is a clear connection between the structures being discussed and a clinical disease, for example, Parkinson's disease and the substantia nigra. In Section 3, the vascular territories are discussed, and the deficits associated with occlusion of these vessels are reviewed. Textbooks of neurology should be consulted for a detailed review of clinical diseases (see the Annotated Bibliography). Management of the disease and specific drug therapies are not part of the subject matter of this atlas.

ADDITIONAL DETAIL

On occasion, a structure is described that has some importance but may be beyond what is necessary, at this stage, for an understanding of the system or pathway under discussion. In other cases, a structure is labeled in an illustration but is discussed at another point in the atlas.

DEVELOPMENTAL ASPECT

For certain parts of the nervous system, knowledge of its development contributes to an understanding of the structure seen in the adult. This is particularly so for the spinal cord, as well as for the ventricular system. Knowledge of development is also relevant for the cerebral hemispheres and for the limbic system (i.e., the hippocampal formation).

NOTE TO THE LEARNER

This notation is added at certain points in the text when, in the author's opinion, it may be beneficial for a student learning the matter to review a certain topic, or consult the videos; in other cases, there is a recommendation to return to the section at a later stage. Sometimes, consulting other texts is suggested. Of course, this is advice only, and each student will approach the learning task in his or her own way.

CLINICAL CASES

A new section has been added at the end of the book to introduce the learner to clinical conditions. The emphasis here is on neurological problems for which knowledge of neuroanatomy is critical. Some of these patients may be seen early in their disease in the office of their family physicians and by non-neurologists. Answers to these problems and additional clinical cases will be added on the Web site.

ADDITIONAL RESOURCE

The Integrated Nervous System: A Systematic Diagnostic Approach, by the author (WH) and two neurologists Dr. P. Humphreys (pediatric) and Dr. C. Skinner (adult) (published by CRC Press, 2010), provides a clinical resource and extension for students. This is a neurologically oriented book which is based on clinical problems and has the same learning approach as the atlas; the Web site accompanying this text has additional neuroanatomical resources and an extensive collection of clinical case-based material (see Annotated Bibliography).

Reference is made in the atlas to material in the Integrated Text and to its Web site that amplifies or supplements the structure or topic or clinical entity being discussed. Illustrations from this textbook (done by Perry Ng) have been used in the present atlas with permission of the co-authors and the publisher.

INVITATION TO LEARNERS

The author invites readers to submit questions, comments and corrections at feedback@atlasbrain.com. The questions and their answers may be posted on the Web site, with no identities revealed.

CENTRAL NERVOUS SYSTEM ORGANIZATION

ORIENTATION
FUNCTIONAL NEUROHISTOLOGY

A brief review of the histology of the nervous system is needed before beginning a description of its neuroanatomy. The foremost cell of the central nervous system (CNS) is the **neuron**, and the human nervous system has billions of neurons. A neuron has a cell body (also called soma, or **perikaryon**), **dendrites**, which extend a short distance from the soma, and an **axon**, which connects one neuron with others. Neuronal membranes are specialized for electro-chemical events, which allow these cells to receive and transmit messages to other neurons. The dendrites and cell bodies of the neurons receive information that causes a neuron to "fire" or to change its firing pattern (increase or decrease). The axons transmit the firing pattern of the cell to the other neurons via synapses (see later). Generally, each neuron receives synaptic input—afferents—from hundreds or perhaps thousands of neurons, and its axon distributes this information—efferents—via collaterals (branches) to hundreds of neurons, either nearby or at a distance.

Within the CNS, neurons that share a common function are usually grouped together; such groupings are called **nuclei** (singular is **nucleus**, which is somewhat confusing because it does not refer to the part of a cell). In other parts of the brain, the neurons are grouped at the surface, to form a **cortex**. In a cortical organization, neurons are arranged in layers, and the neurons in each layer are functionally alike and different from those in other layers. Older cortical areas have three layers (e.g., the cerebellum, the hippocampus); more recently evolved cortices have six layers (the cerebral cortex) and sometimes sublayers.

Some neurons in the nervous system are directly linked either to incoming sensory information or to motor functions. In the CNS of more advanced organisms, the vast majority of neurons interconnect between the sensory and motor aspects (i.e., they form circuits that participate in the processing of information). These neurons are called **interneurons**, and more complex information processing such as occurs in the human brain is correlated with the dramatic increase in the number of interneurons in our brains.

Much of the substance of the brain consists of **axons**, also called **fibers**, which connect one part of the brain with other areas. These fibers function so that the various parts of the brain communicate with each other. Some axons go a short distance and link neurons locally, and others travel a long distance and connect different areas of the brain and spinal cord. Many of the axons are myelinated, forming a so-called insulation, which serves to increase the speed of axonal conduction; the thicker the **myelin sheath**, the faster the conduction. Axons originating from one area (cortex or nucleus) and destined for another area usually group together and form a **tract**, also called a **pathway** (or fasciculus).

The communication between neurons occurs almost exclusively at specialized junctions known as **synapses**, using biological molecules called **neurotransmitters**. These neurotransmitters modify ion movements across the neuronal membranes of the synapse and alter neurotransmission—they may be excitatory or inhibitory in their action, or they may modulate synaptic excitability. The action of neurotransmitters depends also on the specific receptor type; there is an ever increasing number of receptor subtypes allowing for even more complexity of information processing within the CNS. The post-synaptic neuron modifies its firing pattern depending on the summative effect of all the synapses acting on it at any moment in time. Drugs are being designed to act on these receptors for therapeutic purposes.

The other major cells of the CNS are **glia**, so-called supporting cells; there are more glia than neurons. There are two types of glial cells:

- **Astrocytes**, which are involved in supportive structural and metabolic events.
- **Oligodendrocytes**, which are responsible for the formation and maintenance of the myelin that ensheathes the axons.

Some of the maturation involving motor capabilities and language that we see in infants and children can be accounted for by the progressive myelination of the various pathways within the CNS throughout childhood.

FUNCTIONAL NEUROANATOMY

One approach to an understanding of the nervous system is to conceptualize that it is composed of a number of functional modules, starting with simpler ones and evolving in higher primates and humans to a more complex organizational network of cells and connections. The function of each part is dependent on and linked to the function of all the modules acting in concert.

SPINAL CORD

The first of the functional units is the **spinal cord** (see Figure 1.1, Figure 1.2, Figure 1.10, and Figure 1.11), which connects the CNS with the skin and muscles of the body via the peripheral nervous system (**PNS**). Simple and complex reflex circuits are located within the spinal cord. It receives sensory information (**afferents**) from the skin and body wall, which is then transmitted to higher centers of the brain. Movement (motor) instructions from higher centers are sent to the spinal cord, and these motor commands (**efferents**) are then delivered to the muscles. Certain motor patterns are organized in the spinal cord, and these are under the influence of motor areas in the brainstem and cerebral cortex. Also located within the spinal cord are neurons of the autonomic nervous system (**ANS**), which innervates the intestines and the internal organs and the glands.

BRAINSTEM

As the functional systems of the brain have become more complex, new control "centers" have evolved; these are often spoken of as the higher centers. The first set of these centers is located in the **brainstem**, which is situated above the spinal cord and within the skull (in humans). The brainstem includes three distinct areas—**medulla**, **pons**, and **midbrain** (see Figure 1.1, Figure 1.2, Figure 1.6, and Figure 1.8). Some nuclei within the brainstem reticular formation (see Figure 3.6A and Figure 3.6B) are concerned with essential functions such as the control of the pulse and respiration and the regulation of blood pressure. Other nuclei within the brainstem reticular formation are involved in setting our level of arousal, and they play an important role in maintaining our state of consciousness. Special nuclei in the brainstem are responsible for some basic types of movements in response to gravity (i.e., vestibular). In addition, almost all the **cranial nerves** and their nuclei,

which supply the structures of the head, are anchored in the brainstem (see Figure 1.8, Figure 3.4, and Figure 3.5).

CEREBELLUM

The **cerebellum**, which is situated behind the brainstem (inside the skull) in humans (see Figure 1.1, Figure 1.2, Figure 1.9, Figure 3.7, and Figure 3.8), has strong connections with the brainstem. The cerebellum has a simpler form of cortex that consists of only three layers. Parts of the cerebellum are quite old in the evolutionary sense, and parts are relatively newer. This "little brain" is involved in motor coordination and also in the planning of movements. How this is accomplished will be understood once the input-output connections of the various parts of the cerebellum are studied (in Section 2).

DIENCEPHALON

Next in the hierarchy of the development of the CNS is the area of the brain called the **diencephalon** (see Figure 1.7, Figure 1.8, and Figure 2.6). Its largest part, the **thalamus**, develops in conjunction with the cerebral hemispheres and acts as the gateway to the cerebral cortex. The thalamus consists of several nuclei, each of which projects to a different part of the cerebral cortex and receives reciprocal connections from the cortex. The **hypothalamus**, a much smaller part of the diencephalon, serves mostly to control the neuroendocrine system via the pituitary gland and also organizes the activity of the ANS. Parts of the hypothalamus are intimately connected with the expression of basic drives (e.g., hunger, thirst, and reproduction), with the regulation of water in our bodies, and with the manifestations of "emotional" behavior as part of the limbic system (see later).

CEREBRAL HEMISPHERES

With the continued evolution of the brain, the part of the brain called the forebrain undergoes increased development, a process called encephalization. This has culminated in the development of the **cerebral hemispheres**, which dominate the brains of higher mammals, and reaching their zenith (so we think) in humans. Most neurons of the cerebral hemispheres are found at the surface and form the **cerebral cortex** (see Figure 1.1, Figure 1.3, and Figure 1.7), most of which is six layered (also called the **neocortex**). In the human, the cerebral cortex is thrown into ridges (gyri, singular **gyrus**) and valleys (sulci, singular **sulcus**). The enormous expansion of the cerebral cortex in the human, both in terms of size and complexity, has caused this part of the brain to become the dominant controller of the CNS. The cerebral cortex is capable, so it seems, of over-riding most of the other regulatory systems. We need our cerebral cortex for almost all interpretations and actions related to the functioning of the sensory and motor systems, for consciousness, language, and thinking.

BASAL GANGLIA

Buried within the cerebral hemispheres are the **basal ganglia**, large collections of neurons (see Figure 2.5A and Figure 2.5B), which are involved mainly in the initiation and organization of motor movements. These neurons affect motor activity and other cortical functions through their influence on the cerebral cortex.

LIMBIC SYSTEM

Several areas of the brain are involved in behavior, which is characterized by the reaction of the animal (or person) to situations it encounters. This reaction in humans consists of both a psychological component and physiological changes that are termed "emotional." Various parts of the brain are involved with these activities, and collectively they have been named the **limbic system**. This network includes cortex, various subcortical areas, parts of the basal ganglia, the hypothalamus, and parts of the brainstem. (The limbic system is described in Section 4.)

In summary, the nervous system has evolved so that its various parts have assigned tasks. For the nervous system to function properly, there must be communication among the various parts. Some of these links are the major sensory and motor pathways, called **tracts** (or fascicles). Much of the mass of tissue in our hemispheres is made up of these **pathways** (e.g., see Figure 2.2B and Figure 2.3).

Within all parts of the CNS are the remnants of the neural tube from which the brain developed; these spaces are filled with **cerebrospinal fluid (CSF)**. The spaces in the cerebral hemispheres are actually quite large and are called **ventricles** (see Figure 2.1A and Figure 2.1B). The formation, flow, and reabsorption of CSF are discussed in Section 3 (see Figure 7.8).

The CNS is laced with blood vessels because neurons depend on a continuous supply of oxygen and glucose. This aspect is discussed further in Section 3, on vasculature.

STUDY OF THE CENTRAL NERVOUS SYSTEM

Early studies of the normal brain were generally descriptive. Brain tissue does not have a firm consistency, and the brain needs to be fixed for gross and microscopic examination. One of the most common fixatives used to preserve the brain for study is formalin. After fixation with formalin, the brain can be handled and sectioned. Areas containing predominantly neuronal cell bodies (and their dendrites and synapses) become grayish after formalin fixation, and these areas are traditionally called **gray matter** (see Figure 2.9A). Tracts containing myelinated axons become white after formalin fixation, and such areas are simply called the **white matter** (see Figure 2.2B).

We have learned much about the normal function of the human CNS through diseases and injuries to the nervous system. Diseases of the nervous system can involve the neurons, either directly (e.g., metabolic disease) or by reducing the blood supply that is critical for the viability of nerve cells. Some degenerative diseases affect a particular group of neurons. Other diseases can affect the cells supporting the myelin sheath and thereby disrupt neurotransmission. Biochemical disturbances may disrupt the balance of neurotransmitters and cause functional disease states.

The introduction of imaging of the nervous system— computed tomography (CT) and magnetic resonance imaging (MRI), particularly functional MRI (fMRI)—is revealing fascinating information about the organization and functional aspects of the CNS. We are slowly beginning to piece together an understanding of what is considered by many as the last and most important frontier of human knowledge—an understanding of the **brain**.

CLINICAL ASPECT

Certain aspects of clinical neurology are included in this book, both to amplify the text and to indicate the importance of knowing the functional anatomy of the CNS. Knowing where a lesion is located (the localization) often indicates the nature of the disease (the diagnosis), thereby leading to treatment and allowing the physician to discuss the prognosis with the patient.

Chapter 1

Overview and External Views

FIGURE 1.1—CENTRAL NERVOUS SYSTEM OVERVIEW A

ANTERIOR PERSPECTIVE (PHOTOGRAPH—COMPOSITE)

The first task in developing an understanding of the brain is delineating its component parts. This figure is a photographic view of the entire central nervous system (CNS)—the brain hemispheres (the cerebrum, also called the cerebral hemispheres), the brainstem, and the spinal cord—from an anterior perspective.

THE CEREBRAL HEMISPHERES

The large cerebrum, divided (so it seems) into two separate portions, the cerebral hemispheres, is by far the most impressive structure of the human CNS and the one that most are referring to when speaking about "the brain." In fact the "two" cerebral hemispheres are connected across the midline (see Figure 2.2A). The cerebral hemispheres occupy most of the interior of the skull, the cranial cavity.

THE BRAINSTEM

The brainstem is also seen from this perspective. It consists of three parts, from above downward the **midbrain**, **pons**, and **medulla**. The midbrain portion is usually obscured by the cerebral hemispheres in this view (see Figure 1.8 and Figure 3.2).

The *cranial nerves,* which supply the structures of the head and neck, are attached to the brainstem. The brainstem and cranial nerves are considered in Chapter 3 (see Figure 3.4 and Figure 3.5).

Part of the **cerebellum**, the "little brain," can also be seen from this perspective. The cerebellum is introduced in Chapter 3 (see Figure 3.7 and Figure 3.8) and is further discussed with the motor system in Section 2 (see Figure 5.15, Figure 5.16, and Figure 5.17).

THE SPINAL CORD

This long extension of the CNS continues down from the medulla and is found in the vertebral canal (see Figure 1.2 and Figure 1.10). The meninges, the connective tissue coverings of the spinal cord, have been opened, thereby showing the attached nerve roots (motor and sensory). The spinal cord is discussed with Figure 1.10, Figure 1.11, and Figure 3.9. (The meninges, consisting of dura, arachnoid and pia, are discussed in Section 3.)

Note to the Learner: For safe handling of brain tissue, current guidelines recommend the use of disposable gloves when handling any brain tissue, to avoid possible contamination with infectious agents, particularly the so-called slow viruses. In addition, formalin is a harsh fixative and can cause irritation of the skin. Many individuals react to the smell of the formalin and may develop an asthmatic reaction. People who handle formalin-fixed tissue must take extra precautions to avoid these problems. In most laboratories, the brains are soaked in water before being put out for study.

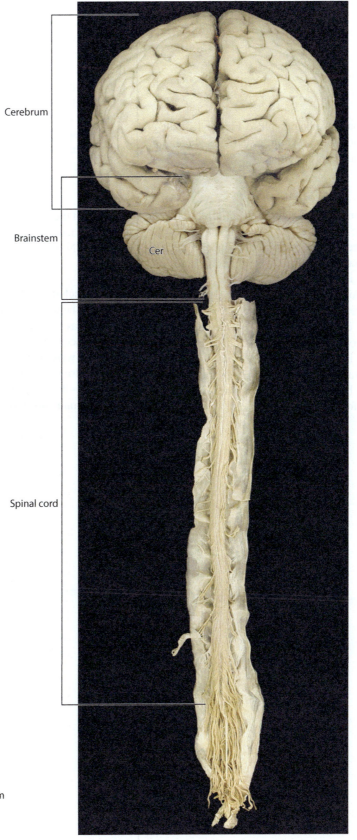

Cerebrum

Brainstem

Cer

Spinal cord

Cer = Cerebellum

FIGURE 1.1: Central Nervous System Overview A—Anterior Perspective (photograph—composite)

FIGURE 1.2—CENTRAL NERVOUS SYSTEM OVERVIEW B

LATERAL PERSPECTIVE (PHOTOGRAPH—*IN SITU*)

This is the companion photograph to Figure 1.1 that is provided to assist the learner in visualizing the brain, brainstem, and spinal cord *in situ*.

THE CEREBRAL HEMISPHERES

The skull and brain have been cut in the midline, called a mid-sagittal view; therefore, one is looking at half of the brain. One can see that the cerebral hemispheres occupy most of the cranial cavity. The cerebral hemispheres are further discussed in Chapter 2.

Proceeding downward is a narrowed part of the central nervous system, the brainstem, which occupies the lower part of the cranial cavity; the pontine bulge is the most noticeable portion of the brainstem. The brainstem is further discussed in Chapter 3.

Behind the brainstem is the cerebellum, the so-called little brain. The brainstem and cerebellum occupy the posterior cranial fossa of the skull.

Additional features of the meninges of the brain and the way that the cranial cavity is partitioned are considered in Section 3.

The central nervous system continues downward as the spinal cord, situated in the vertebral canal. The spinal cord, in the adult human, terminates at the level of the upper lumbar vertebral level, whereas the vertebral canal continues (further explained with Figure 1.10). The spinal cord and spinal canal are further discussed in this chapter and also in Chapters 2 and 3 (as well as in Section 2).

EXTERNAL VIEWS

The rest of the illustrations in this chapter are more detailed external views of the various parts of the central nervous system (CNS).

Note to the Learner: The video of the cerebral hemispheres includes a demonstration of the exterior views of the brain hemispheres from the various perspectives (this chapter), as well as structures in the interior of the hemispheres (see Chapter 2). This video can be accessed on the Web site (www.atlasbrain.com).

It is also instructive at this stage to view the video on the Web site—the Interior of the Skull—which demonstrates how the brain "fits into" our skull.

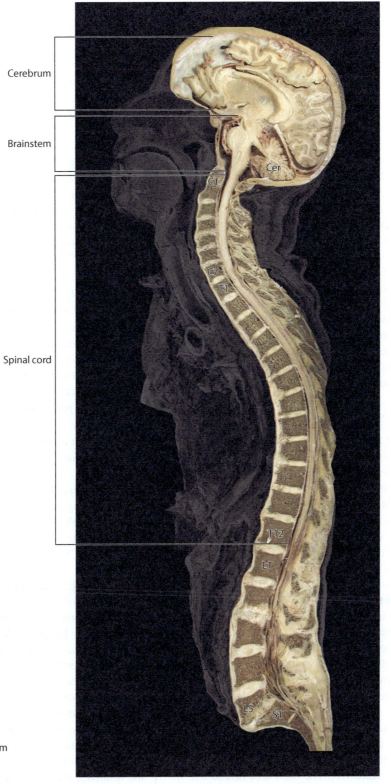

Cerebrum

Brainstem

Spinal cord

Cer = Cerebellum

FIGURE 1.2: Central Nervous System Overview B—Lateral Perspective (photograph—*in situ*)

FIGURE 1.3—CEREBRAL HEMISPHERES 1

CEREBRAL CORTEX: LATERAL VIEW AND LOBES (PHOTOGRAPHS)

When people talk about "the brain," they are generally referring to the cerebral hemispheres, also called the **cerebrum**. The brain of higher apes and humans is dominated by the cerebral hemispheres. The outer layer, the **cerebral cortex** with its billions of neurons and its vast interconnections, is responsible for sensory perceptions, movements, language, thinking, memory, consciousness, and certain aspects of emotion. In short, the intact cerebral hemispheres are needed for all aspects of higher levels of function and to adapt to our constantly changing circumstances including socially and emotionally.

Most of the cerebral cortex is organized in six layers, known collectively as the **neocortex**, with the neurons of each layer having a different function. In formalin-fixed material, the neuronal cortex takes on a grayish appearance and is often referred to as the gray matter (see Figure 2.9A and Figure 2.10A).

UPPER ILLUSTRATION

The surface of the hemispheres in humans (and in some other mammals) is thrown into irregular folds. This infolding allows an extensive expansion of the cerebral cortex to be contained within a restricted volume; otherwise, our skulls would have to be enormous!

The ridges are called **gyri** (singular, **gyrus**), and the intervening crevices are called **sulci** (singular, **sulcus**). A very deep sulcus is called a **fissure**. Three of these are indicated in the figure—the central fissure, the lateral fissure and the parieto-occipital fissure (better seen on the medial view of the hemispheres in Figure 1.7); these tend to be constant in all human brains. The whole surface of the hemispheres, the **cerebral cortex**, is composed of neurons and their connections.

The basic division of each of the hemispheres is into **four lobes**—frontal, parietal, temporal, and occipital (which are more or less in congruity with the bones of the skull overlying these parts of the brain). The **central fissure** divides the area anterior to it, which is the frontal lobe, from the area posterior to it, the parietal lobe. The parietal lobe extends posteriorly to the parieto-occipital fissure (see also Figure 1.7). The brain area behind that fissure is the occipital lobe. The **lateral fissure** separates the frontal and parietal lobes from the temporal lobe below. (The lateral fissure is seen more clearly in Figure 1.4.)

LOWER ILLUSTRATION

The lower illustration uses color to differentiate the various lobes of the brain. In very general terms, the various lobes of the brain are described as having the following functions:

- The **frontal lobes** (in humans) are generally thought of as having "executive" functions, the decision-making part of the brain for so-called voluntary conscious actions, both in the present and in planning for the future. Motor activities are associated with the frontal lobes. Parts of the frontal lobes are connected with the "emotional" limbic system (discussed in Section 4).
- The **parietal lobes** are considered to be both sensory and visuo-spatial in function, incorporating multiple sensory inputs.
- The **temporal lobes** are closely associated with the auditory system (hearing) and language functions (on the "dominant" side). Their medial aspect (not seen on this perspective) plays a significant role in memory functions, as well as being part of the limbic system.
- The **occipital lobes**, better seen on the medial aspect of the brain (see Figure 1.7), are closely associated with the visual system.

Various areas of the human brain are highly specialized for language functions (discussed further with Figure 4.5). Most of the cerebral cortex in humans is not connected directly to a sensory or motor function and is known as an "**association**" area, a term that can perhaps be explained functionally as inter-relating the various activities in the different parts of the brain. Most important, the homologous association areas of the two sides, which are connected across the midline (via the corpus callosum, see Figure 2.2A), may share similar functions but may not participate in exactly the same manner in carrying out its tasks.

Part of the cerebral cortex has been "buried" from view—the insula—and is shown in Figure 1.4. The surface of the cerebral hemispheres can also be visualized from a number of other directions—from below (inferior view, see Figure 1.5 and Figure 1.6); and after dividing the two hemispheres along the inter-hemispheric fissure (in the midline), the hemispheres are seen to have a medial surface as well (see Figure 1.7).

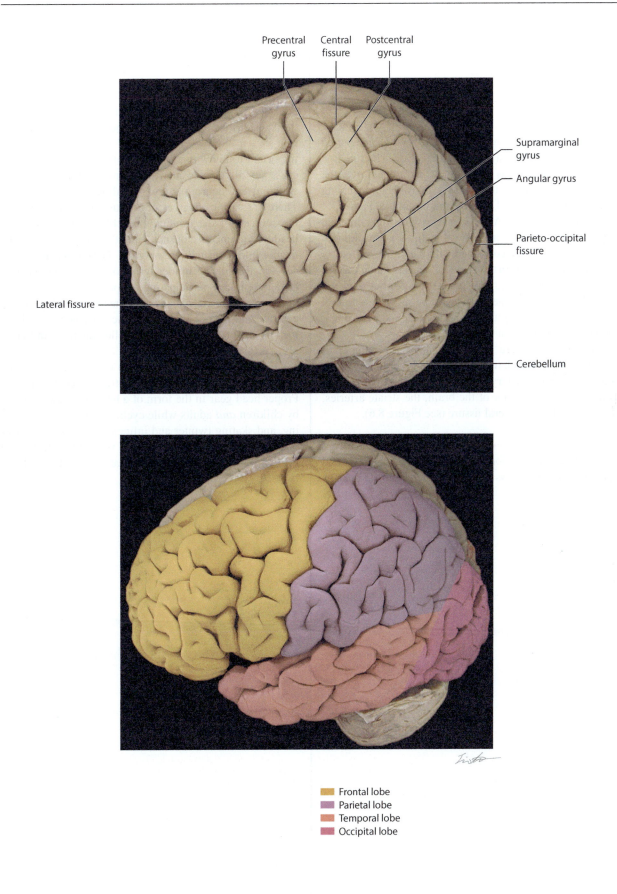

FIGURE 1.3: Cerebral Hemispheres 1—Cerebral Cortex: Lateral View and Lobes (photographs)

FIGURE 1.4—CEREBRAL HEMISPHERES 2

THE INSULA (PHOTOGRAPH)

The lateral fissure has been "opened" to reveal some buried cortical tissue; this area is called the insula. The function of this cortical area has been somewhat in doubt over the years. It seems that this is the area responsible for receiving taste sensations, relayed from the brainstem (see Figure 3.4 and Appendix Figure A.8). Sensations from our internal organs may reach the cortical level in this area. In addition, parts of the insula are connected with the limbic system.

The specialized cortical gyri for hearing (audition) are also to be found within the lateral fissure, but they are actually part of the upper surface of the superior temporal gyrus (as shown in Figure 6.1, Figure 6.2, and Figure 6.3) which is located within the lateral fissure.

The lateral fissure has within it a large number of blood vessels, which have been removed—branches of the middle cerebral artery (discussed with Figure 8.4). Branches to the interior of the brain, the striate arteries, are given off in the lateral fissure (see Figure 8.6).

Note to the Learner: The insular cortex can be recognized on a coronal section of the brain (see Figure 2.9A) and also on horizontal views of the brain (see Figure 2.10A), as well as with brain imaging (computed tomography and magnetic resonance imaging; see Figure 2.9B and Figure 2.10B).

CLINICAL ASPECT

A closed head injury that affects the brain is one of the most serious forms of accident. The general term for this is a concussion, a "bruising" of the brain. It is most important to know that this type of brain injury does not cause any discernible changes when the brain is examined with current neuroimaging modalities (CT and MRI). There are various degrees of concussion depending upon the severity of the trauma. The effects vary from mild headache to unconsciousness, and they may include some memory loss, usually temporary. New tests are being developed to detect milder forms of concussion, particularly in children participating in sports. The after-effects of concussion include headaches, the loss of ability to concentrate and the need for sleep. There are now guidelines for a period of "brain rest" before allowing these individuals to return to normal activities.

Everything possible should be done to avoid a brain injury, particularly when participating in sport activities. Proper head gear in the form of a helmet should be worn by children *and* adults while cycling, skiing, snowboarding, and skating (winter and inline). Closed head injuries occur most frequently with motor vehicle accidents, and the use of seatbelts and of proper seats for children reduces the risk.

Central
fissure

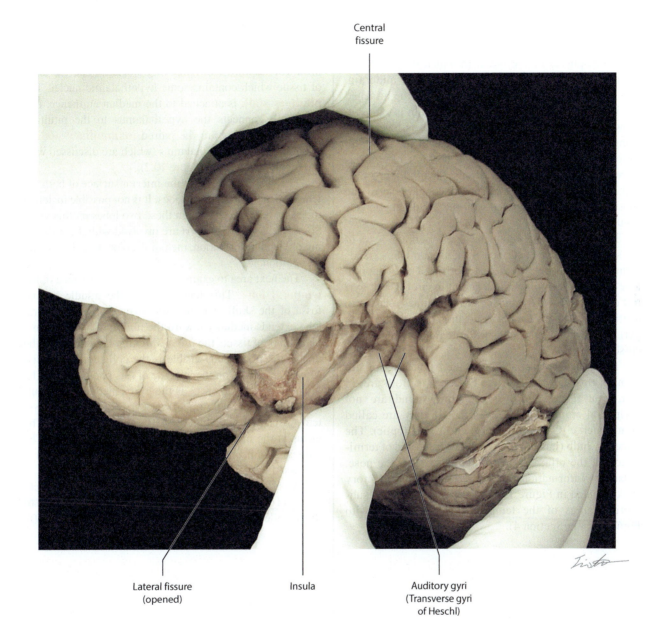

Lateral fissure
(opened)

Insula

Auditory gyri
(Transverse gyri
of Heschl)

FIGURE 1.4: Cerebral Hemispheres 2—The Insula (photograph)

FIGURE 1.5—CEREBRAL HEMISPHERES 3

CEREBRAL CORTEX: INFERIOR VIEW WITH MIDBRAIN CUT (PHOTOGRAPH)

This figure is a view of the inferior surface of the cerebral hemispheres of the brain, in which the brainstem has been sectioned through at the level of the midbrain, thus removing most of the brainstem and the attached cerebellum. The cut surface of the midbrain is exposed, showing a linear area of brain tissue that is black; this elongated cluster of cells is the nucleus of the midbrain called the substantia nigra, and it consists of neurons with pigment inside the cells (discussed with Figure 4.2C and Appendix Figure A.3). The functional role of the substantia nigra is discussed with the basal ganglia (see Figure 5.14).

The frontal lobes occupy the anterior cranial fossa of the skull. The inferior surface of the frontal lobe extends from the frontal pole to the anterior tip of the temporal lobe (and the beginning of the lateral fissure). These gyri rest on the roof of the orbit and are sometimes referred to as the **orbital gyri**. This is association cortex, and these gyri have strong connections with the limbic system (discussed in Section 4).

The olfactory tract and optic nerve (and chiasm) are seen on this view (and in the Figure 1.6). Both are, in fact, central nervous system pathways and are not peripheral cranial nerves, even though they are called cranial nerve (CN) I (olfactory) and CN II (optic). The olfactory bulb (labeled in Figure 1.6) is the site of termination of the olfactory nerve filaments from the nose. Olfactory information is then carried in the olfactory tract (labeled in Figure 1.6) to various cortical and subcortical areas of the temporal lobe (discussed with Figure 10.4 in Section 4).

The optic nerves (CN II, cut) exit from the orbit and continue to the optic chiasm, where there is a partial crossing of visual fibers (see Figure 6.4). The regrouped visual pathway continues, now called the optic tract (see Figure 6.4 and Figure 6.6).

Posterior to the chiasm is the area of the hypothalamus, part of the diencephalon. Behind the optic chiasm are the median eminence and then the mammillary (nuclei) bodies, both of which belong to the hypothalamus. The **median eminence** (not labeled) is an elevation of tissue which contains some hypothalamic nuclei. The **pituitary stalk** is attached to the median eminence, and this stalk connects the hypothalamus to the pituitary gland. Behind this are the paired **mammillary bodies**, two nuclei of the hypothalamus (which are discussed with the limbic system; see Figure 10.2).

This dissection reveals the inferior surface of both the temporal and the occipital lobes. It is not possible to define the exact boundary between these two lobes on this view. Some of these inferior gyri are involved with the processing of visual information, including color, as well as facial recognition.

The next area to examine is the inferior surface of the **temporal lobe**. This lobe occupies the middle cranial fossa of the skull. The inferior surface of the temporal lobe extends medially toward the midbrain and ends in a blunt knob of tissue, the **uncus**. Moving laterally from the uncus, the first sulcus visible is the collateral sulcus/fissure (seen clearly on the left side of this photograph). The **parahippocampal gyrus** is the gyrus medial to this sulcus; it is an extremely important gyrus of the limbic system (discussed with Figure 9.3B and Figure 9.5A). The uncus is the most medial protrusion of this gyrus. (The clinical significance of the uncus and uncal herniation is discussed with Figure 1.6.)

Also visible on this specimen is the posterior thickened end of the corpus callosum (discussed with Figure 1.6) that is called the splenium (see Figure 1.7 and Figure 2.2B).

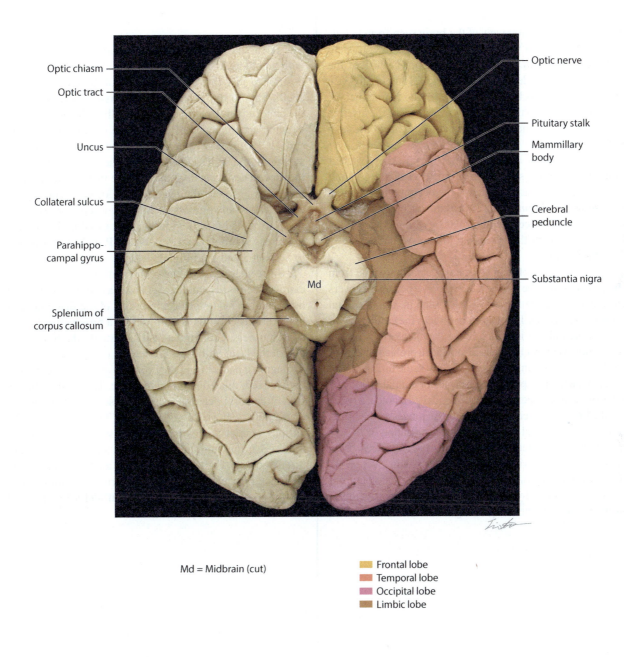

Optic chiasm

Optic tract

Uncus

Collateral sulcus

Parahippo-
campal gyrus

Splenium of
corpus callosum

Optic nerve

Pituitary stalk

Mammillary
body

Cerebral
peduncle

Substantia nigra

Md

Md = Midbrain (cut)

Frontal lobe
Temporal lobe
Occipital lobe
Limbic lobe

FIGURE 1.5: Cerebral Hemispheres 3—Cerebral Cortex: Inferior View with Midbrain Cut (photograph)

FIGURE 1.6—CEREBRAL HEMISPHERES 4

CEREBRAL CORTEX: INFERIOR VIEW WITH BRAINSTEM (PHOTOGRAPH)

This is a photographic view of the brain seen from below, the inferior view, a view that now includes the brainstem and the cerebellum. The medulla and pons, parts of the brainstem, can be identified (see Figure 1.8 and Figure 3.2), but the midbrain is hidden from view. The cranial nerves are still attached to the brainstem, and some of the arteries (although bluish, these are arteries) to the brain are also present.

The brainstem and cerebellum occupy the posterior cranial fossa of the skull. In fact, in this view, the cerebellum obscures the visualization of the occipital lobe (which was shown in Figure 1.5, after removal of most of the brainstem and the cerebellum). Various cranial nerves can be identified with the brainstem (discussed subsequently with Figure 1.8).

A thick sheath of dura separates the occipital lobe from the cerebellum—the **tentorium cerebelli** (because it covers the cerebellum; see Figure 7.6; also seen in a mid-sagittal view in Figure 1.7). The tentorium divides the cranial cavity into an area above it, which is the supratentorial space, a term often used by clinicians to indicate a problem in any of the lobes of the brain, and an area below it, the infratentorial space, which corresponds to the posterior cranial fossa. The tentorium has an opening, sometimes called a notch, a space for the brainstem (see Figure 7.5; the meninges of the brain, dura, arachnoid, and pia are discussed in Section 3).

Part of the arterial system is also seen in this brain specimen (the arterial supply is discussed in Section 3). The basilar artery (which is blue), which is situated in front of the pons, ends by dividing into the posterior cerebral arteries to supply the occipital regions of the brain. The cut end of the internal carotid artery is seen, but the remainder of the arterial circle of Willis is not dissected on this specimen (see Figure 8.1); the arterial supply to the cerebral hemispheres is fully described in Section 3).

The olfactory tract and optic nerve (and chiasm) are again seen on this view.

CLINICAL ASPECT

The uncus has been clearly identified in the specimens, with its blunted tip pointed medially. The uncus is in fact positioned just above the free edge of the tentorium cerebelli. Should the volume of brain tissue increase above the tentorium as a result of brain swelling, hemorrhage, or a tumor, accompanied by an increase in intracranial pressure (ICP), the hemispheres would be forced out of their supratentorial space. The only avenue to be displaced is in a downward direction, through the tentorial notch, and the uncus becomes the leading edge of this pathological event. The whole process is clinically referred to as **uncal herniation**. (This will be further discussed in the context of Increased Intracranial Pressure, ICP, in the Introduction to Section 3 and with Figure 7.1.)

Since the edges of the tentorium are very rigid, the extra tissue in this small area causes a compression of the brain matter, leading to compression of the thalamus and upper brainstem; this is followed by a progressive loss of consciousness. Cranial nerve III is usually the first structure at the level of the midbrain to be compressed as well, damaging it and causing the eye to be abducted, depressed with a fixed, dilated pupil on that side. This is an ominous sign of cerebral decompensation caused by a mass lesion in the supratentorial compartment. (The function of the various cranial nerves is discussed with Figure 3.4 and Figure 3.5; the pupillary light reflex and its pathway are discussed with Figure 6.7.) This is a medical emergency! Continued herniation will lead to further compression of the lower structures of the brainstem and a loss of vital functions, followed by rapid death.

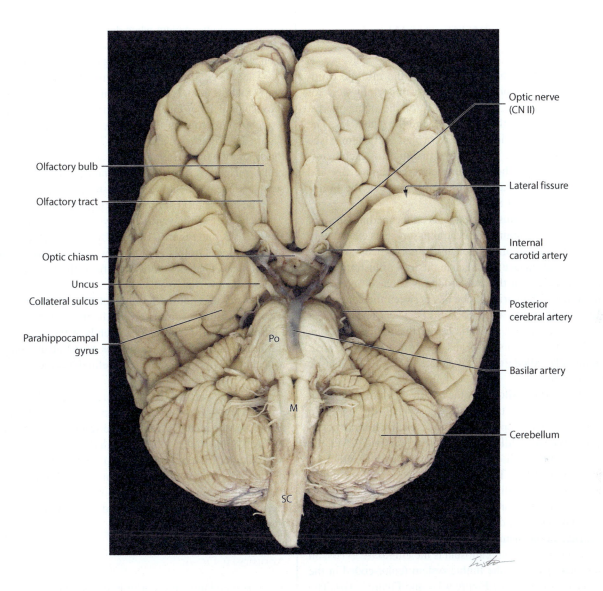

Olfactory bulb

Olfactory tract

Optic chiasm

Uncus

Collateral sulcus

Parahippocampal
gyrus

Optic nerve
(CN II)

Lateral fissure

Internal
carotid artery

Posterior
cerebral artery

Basilar artery

Cerebellum

Po

M

SC

Po = Pons
M = Medulla
SC = Spinal cord

FIGURE 1.6: Cerebral Hemispheres 4—Cerebral Cortex: Inferior View with Brainstem (photograph)

FIGURE 1.7—CEREBRAL HEMISPHERES 5

CEREBRAL HEMISPHERES: MEDIAL VIEW AND LOBES (PHOTOGRAPHS)

This view of the brain sectioned in the midline (mid-sagittal plane) is probably the most important view for understanding the gross anatomy of the hemispheres, the diencephalon, the brainstem, and the ventricles. The section has divided the corpus callosum (see Figure 2.2A), gone between the thalamus of each hemisphere (through the 3rd ventricle, see Figure 2.8 and Figure 3.1), and passes through all parts of the brainstem (see Figure 1.8).

The focus in this illustration is on the medial aspect of the lobes of the brain. The central fissure does extend onto this part of the brain (although it is not as deep as on the dorsolateral surface). As seen in the lower illustration, the medial surface of the frontal lobe is situated anterior to the fissure. The parietal lobe lies between the central fissure and the deep **parieto-occipital fissure**. The **occipital lobe** is now visible; the main fissure that divides this lobe is the **calcarine fissure**, and the primary visual area, commonly called **area 17**, is situated along its banks (see Figure 6.6).

The corpus callosum—the bundle of white matter that interconnects the two hemispheres—has the expected "white matter" appearance. Inside each cerebral hemisphere is a space filled with cerebrospinal fluid (CSF), the lateral ventricle (discussed in Chapter 2 with Figure 2.1A and Figure 2.1B). The **septum pellucidum**, a membranous septum that divides the anterior portions of the lateral ventricles of one hemisphere from that of the other side, has been torn during dissection, revealing the lateral ventricle of one hemisphere behind it.

Above the corpus callosum is the **cingulate gyrus**, an important gyrus of the limbic system (color-coded in the lower illustration; see Figure 9.1A and Figure 9.1B). The fornix, a fiber tract of the limbic system, is located in the lower edge of the septum (discussed with Figure 9.3A and Figure 9.3B).

The cerebellum lies behind the brainstem (and the 4th ventricle). It has been sectioned through its midline portion, the **vermis** (see Figure 3.7.) Although it is not necessary to name all of its various parts, it is useful to know two of them—the lingula and the nodulus. (The reason for this will become evident when describing the cerebellum; see Figure 3.7.) The tonsil of the cerebellum can also be seen in this view (see Figure 1.8, Figure 3.2 and Figure 3.7).

The cut edge of the tentorium cerebelli, one of the major folds of dura, is seen separating the cerebellum from the occipital lobe. One of the dural venous sinuses, the straight sinus, runs in the midline of the tentorium (see Figure 7.4, Figure 7.5, and Figure 7.6). This view clarifies the separation of the supratentorial space, namely the cerebral hemispheres, from the infratentorial space, the brainstem, and the cerebellum in the posterior cranial fossa.

The mid-sagittal section goes through the midline 3rd ventricle, part of the ventricular system (see Figure 2.9A, Figure 2.9B, and Figure 7.8), thereby revealing the diencephalic region. On this medial view, the thalamic portion of the diencephalon is separated from the hypothalamic part by a groove, the **hypothalamic sulcus** (see Figure 3.2). This sulcus starts at the foramen of Monro (the interventricular foramen, discussed with the ventricles; see Figure 2.1A and Figure 7.8) and ends at the aqueduct of the midbrain. The optic chiasm is found at the anterior aspect of the hypothalamus, and behind it is the mammillary body (see Figure 1.5).

The three parts of the brainstem can be distinguished on this view—the midbrain, the pons with its bulge anteriorly, and the medulla (refer to the ventral view shown in Figure 1.8 and Figure 3.1). Through the midbrain is a narrow channel for cerebrospinal fluid (CSF), the aqueduct of the midbrain (see the discussion of the CSF in Section 3). The midbrain behind the aqueduct includes the superior and inferior colliculi, referred to as the tectum (see Figure 1.9 and Figure 6.7).

ADDITIONAL DETAIL

There is a thalamus present in both cerebral hemispheres (see Figure 2.6). Often, a bundle of fibers, the interthalamic adhesion, connects the thalami across the midline. (This anatomical detail will be referred to with Figure 2.7.)

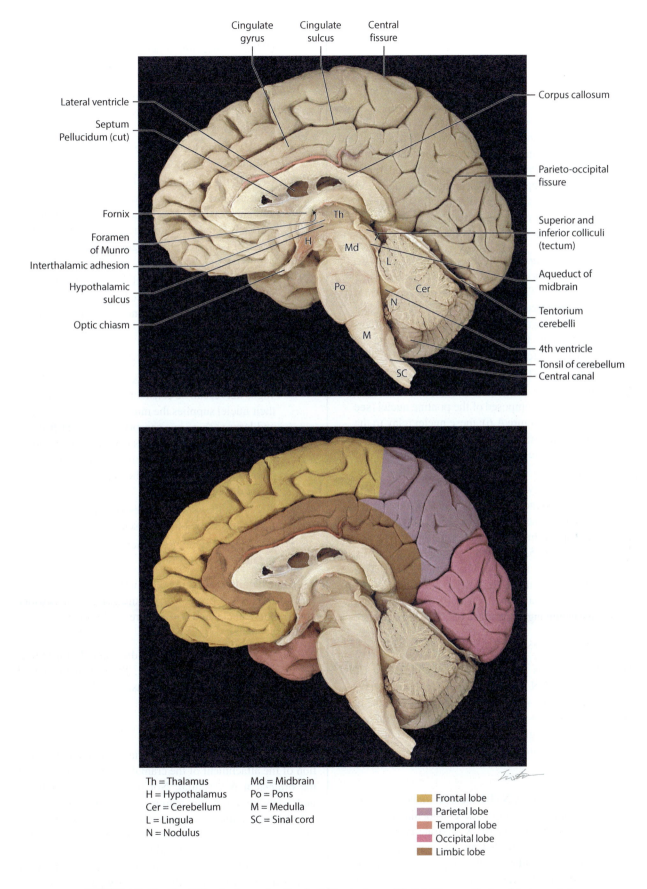

Cingulate gyrus

Cingulate sulcus

Central fissure

Lateral ventricle

Septum Pellucidum (cut)

Corpus callosum

Parieto-occipital fissure

Fornix

Foramen of Munro

Interthalamic adhesion

Hypothalamic sulcus

Optic chiasm

Th

H

Md

Po

L

Cer

N

M

SC

Superior and inferior colliculi (tectum)

Aqueduct of midbrain

Tentorium cerebelli

4th ventricle

Tonsil of cerebellum

Central canal

Th = Thalamus
H = Hypothalamus
Cer = Cerebellum
L = Lingula
N = Nodulus

Md = Midbrain
Po = Pons
M = Medulla
SC = Sinal cord

Frontal lobe
Parietal lobe
Temporal lobe
Occipital lobe
Limbic lobe

FIGURE 1.7: Cerebral Hemispheres 5—Cerebral Hemispheres: Medial View and Lobes (photographs)

FIGURE 1.8—BRAINSTEM A

BRAINSTEM AND CRANIAL NERVES: VENTRAL VIEW (PHOTOGRAPH)

This specimen has been obtained by dissecting the brainstem and cerebellum (along with the diencephalon and parts of the basal ganglia) from the remainder of the brain. This has been done by cutting the fibers of the internal capsule, fibers ascending to and descending from the cerebral cortex (called projection fibers; discussed with Figure 2.4).

The three parts of the brainstem can be differentiated on this ventral view (from above downward):

- *The midbrain:* The midbrain region has the two large "pillars" anteriorly, called the **cerebral peduncles**. These contain fibers descending from the cerebral cortex to the brainstem (cortico-bulbar tract; see Figure 5.10); to the pontine nuclei (cortico-pontine fibers; see Figure 5.15); and to the spinal cord (cortico-spinal tract; see Figure 5.9).
- *The Pons:* The pontine portion is distinguished by its bulge anteriorly, the **pons proper**, an area that is composed of the pontine nuclei (see Appendix Figure A.6); these nuclei relay to the cerebellum (see Figure 5.15). The cortico-spinal fibers are dispersed amongst these nuclei.
- *The medulla:* The medulla is distinguished by the **pyramids**, two distinct elevations on either side of the midline. The direct voluntary motor pathway from the cortex to the spinal cord, the cortico-spinal tract, actually forms these pyramids (see Figure 5.9, Appendix Figure A.8, Appendix Figure A.9, and Appendix Figure A.10). This tract crosses the midline as the **pyramidal decussation**, demarcating the end of the medulla and the beginning of the spinal cord (see Figure 6.12).

CRANIAL NERVE FUNCTIONS

The cranial nerves (CN) III—XII, not including CN I (olfactory) and CN II (optic), are peripheral nerves—sensory and motor—that supply the structures of the head and neck, including autonomic (parasympathetic) nerves to various salivary glands and special muscles of the eyeball. Each nerve is unique functionally.

Midbrain Level

- Cranial nerve (CN) III, the oculomotor nerve—It supplies several of the extraocular muscles that move the eyeball. A separate part, called the Edinger-Westphal nucleus, provides parasympathetic fibers to the pupil and the muscle controlling the lens.
- CN IV, the trochlear nerve—This supplies one extraocular muscle.

Pontine Level

- CN V, the trigeminal nerve—Its major nucleus subserves an extensive sensory function for the face and scalp, and structures of the head (e.g., the nasal sinuses and the meninges). A smaller nucleus supplies motor fibers to jaw muscles (mastication).
- CN VI, the abducens nerve—This supplies one extraocular muscle.
- CN VII, the facial nerve—Of its several nuclei, one supplies the muscles of facial expression, and another nucleus is parasympathetic to two salivary glands; a third nucleus subserves the sense of taste from the tongue.
- CN VIII, the vestibulocochlear nerve—This nerve consists of two special senses, the auditory portion for hearing (see Figure 6.1) and the vestibular portion for balance and equilibrium (see Figure 6.8).

Medullary Level

- CN IX, the glossopharyngeal, and CN X, the vagus nerve, can be considered together. One of their nuclei supplies the muscles of the pharynx and larynx; there is a sensory component from the same areas. Nerve IX is parasympathetic to one salivary gland; the vagus nerve is the major parasympathetic nerve to the organs of the thorax and abdomen.
- CN XI, the spinal accessory nerve—This innervates some of the muscles of the neck.
- CN XII, the hypoglossal nerve—It is motor to the muscles of the tongue.

More details concerning the innervation of each of the cranial nerves are given with Figure 3.4 for the sensory cranial nerve nuclei and with Figure 3.5 for the motor cranial nerve nuclei.

Note to the Learner: This dissected brainstem is shown in a schematic in Figure 3.1. The schematic is used in many illustrations of various aspect of the brainstem throughout the atlas.

CLINICAL ASPECT

Knowing the functions of the cranial nerves and the location of the attachment of the cranial nerves to each part of the brainstem is essential. Not only does this assist in understanding the neuroanatomy of this (difficult) region, but this information is critical in determining lesions and the localization of a lesion of the brainstem region (discussed further in Section 3 and Clinical Cases).

A lesion of the brainstem is likely to interrupt one or more sensory or motor pathways as they pass through the brainstem. Because of the close relationship with the cerebellum, there may be cerebellar signs as well.

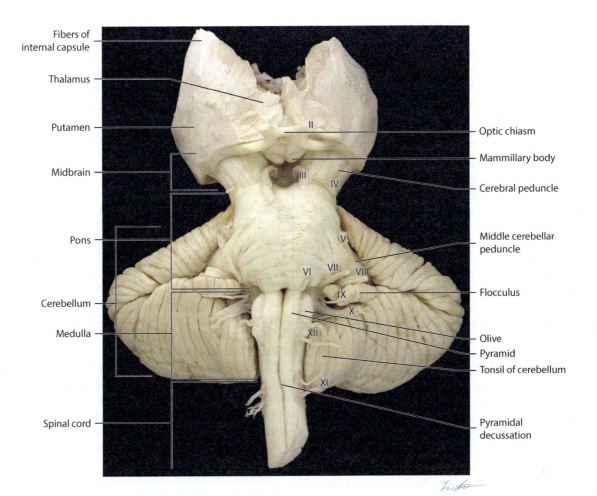

Fibers of internal capsule

Thalamus

Putamen

Midbrain

Pons

Cerebellum

Medulla

Spinal cord

II

III

IV

V

VI VII VIII

IX

X

XII

XI

Optic chiasm

Mammillary body

Cerebral peduncle

Middle cerebellar peduncle

Flocculus

Olive

Pyramid

Tonsil of cerebellum

Pyramidal decussation

II = Optic nerve
III = Oculomotor nerve
IV = Trochlear nerve
V = Trigeminal nerve
VI = Abducens nerve
VII = Facial nerve
VIII = Vestibulocochlear nerve
IX = Glossopharyngeal nerve
X = Vagus nerve
XI = Spinal accessory nerve
XII = Hypoglossal nerve

FIGURE 1.8: Brainstem A—Brainstem and Cranial Nerves: Ventral View (photograph)

FIGURE 1.9—BRAINSTEM B

BRAINSTEM AND CEREBELLUM: DORSAL (POSTERIOR) VIEW (PHOTOGRAPH)

This specimen of the brainstem and diencephalon including the basal ganglia, with the cerebellum attached, is being viewed from the dorsal or posterior perspective in the upper image. The 3rd ventricle, the ventricle of the diencephalon, separates the thalamus of one side from that of the other (see Figure 2.8, Figure 2.9A, and Figure 2.10A; see also Figure 1.7, in which the brain is separated down the midline in the mid-sagittal plane). The diencephalon is discussed with Figure 2.6.

The dorsal part of the midbrain is seen to have four elevations, named the colliculi (see also Figure 3.3). The upper elevations are the **superior colliculi**, and they are functionally part of the visual system, a center for visual reflexes (see Figure 6.7 and Figure 6.9). The lower elevations are the **inferior colliculi**, and these are relay nuclei in the auditory pathway (see Figure 6.1). These colliculi form the so-called **tectum** (see Figure 4.2C), a term that is often used; a less frequently used term for these colliculi is the quadrigeminal plate.

The **pineal**, a glandular structure, hangs down from the back of the diencephalon and sits between the colliculi.

Although not quite in view in this illustration, the trochlear nerves (CN IV) emerge posteriorly at the lower level of the midbrain, below the inferior colliculi (see Figure 3.3).

The posterior aspect of the pons and the medulla are hidden by the cerebellum—some of these structures are seen in the lower illustration (a photographic view), and some are seen in a diagram with the cerebellum removed (see Figure 3.3).

THE CEREBELLUM

The cerebellum, sometimes called the little brain, is easily recognizable by its surface, which is composed of narrow ridges of cortex, called **folia** (singular, **folium**). The cerebellum is located beneath a thick sheath of the meninges, the tentorium cerebelli, inferior to the occipital lobe of the hemispheres (see Figure 1.2, Figure 1.7, and Figure 7.6), in the posterior cranial fossa of the skull.

The cerebellum is involved with motor control and is part of the motor system, influencing posture, gait, and voluntary movements as part of motor modulation (discussed in more detail with the Motor Systems in Section 2). Its function is to facilitate the performance of movements by coordinating the action of the various participating muscle groups. This is often spoken of simply as "smoothing out" motor acts (further discussed with Figure 3.8).

Anatomically, the cerebellum can be described by looking at its appearance in a number of ways. The human cerebellum *in situ* has an upper or superior surface, as seen in this photograph, and a lower or inferior surface (shown in lower photograph). The central portion is known as the **vermis**. The lateral portions are called the **cerebellar hemispheres**.

Sulci separate the folia, and some of the deeper sulci are termed fissures. The **primary fissure** is located on the superior surface of the cerebellum, which is the view seen in this photograph. The **horizontal fissure** is located at the margin between the superior and inferior surfaces. Using these sulci and fissures, the cerebellar cortex has traditionally been divided into a number of different lobes, but many (most) of these do not have a distinctive functional or clinical importance, so only a few are mentioned when the cerebellum is discussed (see Figure 3.7 and Figure 3.8).

BRAINSTEM AND CEREBELLUM: DORSAL (INFERIOR) VIEW (PHOTOGRAPH)

The lower image is a photograph of the same specimen as the upper photograph, but the specimen is tilted to reveal the inferior aspect of the cerebellum and the posterior aspect of the medulla. The posterior aspect of the pons is still obscured by the cerebellum. The posterior aspect of the midbrain can no longer be seen. The upper end of the thalamus is still in view.

The horizontal fissure of the cerebellum is now clearly seen; it is used as an approximate divider between the superior and inferior surfaces of the cerebellum (see Figure 3.7). The vermis of the cerebellum is clearly seen between the hemispheres. Just below the vermis is an opening into a space—the space is the 4th ventricle (described with the ventricular system in Section 3). The opening called the foramen of Magendie is between the ventricle of the "inside" of the brain and the subarachnoid (cerebrospinal fluid) space outside the brain (discussed with Figure 7.8).

The part of the brainstem immediately below the foramen is the medulla, its posterior or dorsal aspect. The most significant structure seen here is a small elevation representing an important sensory relay nucleus, the nucleus gracilis, part of the pathway for discriminative touch sensation, called the gracilis tract (or fasciculus). (The details of this pathway are discussed with Figure 5.2 and Figure 5.5.) The nucleus for the cuneatus tract is not seen in this photograph (see Figure 3.3). These nuclei are discussed with the brainstem cross-sections in the Appendix (see Appendix Figure A.10). The medulla ends and the spinal cord begins where the C1 nerve roots emerge.

The cerebellar lobules adjacent to the medulla are known as the **tonsils** of the cerebellum (see Figure 1.7 and the ventral view of the cerebellum in Figure 1.8). The tonsils are found just above the foramen magnum of the skull (see Figure 3.2). (The clinical aspect is discussed with Figure 3.2 and further discussed in the Introduction to Section 3 and with Figure 7.1.)

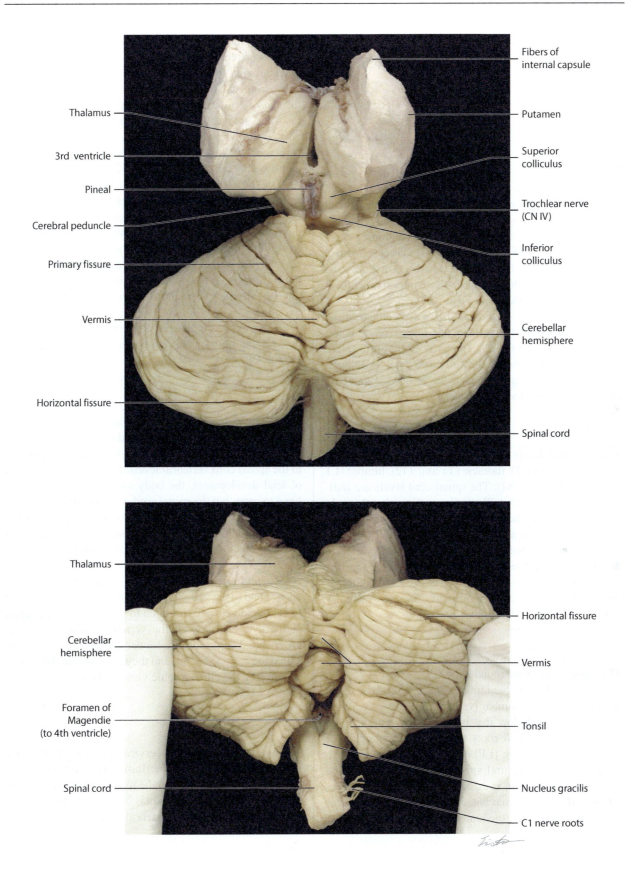

FIGURE 1.9: Brainstem B—Brainstem and Cerebellum: Dorsal posterior and inferior views (photographs)

FIGURE 1.10—SPINAL CORD 1

SPINAL CORD: VERTEBRAL CANAL (PHOTOGRAPH *IN SITU* AND T2 MAGNETIC RESONANCE IMAGING SCAN)

The spinal cord is the extension of the central nervous system (CNS) below the level of the skull. It is an elongated structure that is located within the vertebral canal, covered with the **meninges**—dura, arachnoid, and pia—and surrounded by the **subarachnoid space** containing cerebrospinal fluid (CSF) (see Figure 7.3). There is also a space between the dura and the vertebra known as the **epidural space**. Both these spaces have important clinical implications (discussed with Figure 7.3).

The photograph on the left shows a specimen of the spinal column (with the vertebra and intervertebral discs) and the spinal cord *in situ*. The photograph on the right shows a similar view of the vertebral canal and spinal cord—as a T2-weighted magnetic resonance imaging (MRI)—in the mid-sagittal plane.

SPECIMEN

The vertebral levels are indicated on the illustration—cervical (C1 and C7), thoracic (T1 and T12), lumbar (L1 and L5) and sacral (S1). The spinal cord levels are indicated (between the two illustrations). Note that these do not correspond to the vertebral levels as the spinal cord in the adult ends at about the L1- L2 level.

The spinal cord is situated within the vertebral canal, with little room to spare. It tapers to a cone-like ending in the lumbar region, called the conus medullaris, below which are seen a number of "nerves"—actually nerve roots (within the lumbar cistern).

MAGNETIC RESONANCE IMAGE

The spinal cord is surrounded by CSF, which is seen as white on this T2-weighted image. The spinal cord is again seen to taper in the lumbar (vertebral) region, as with the intact specimen. Below that is the lumbar cistern, filled with CSF and the nerve roots, the location for sampling of CSF (lumbar puncture [LP], spinal tap—discussed with Figure 7.3). The epidural space, the space between the dura and the vertebra, is difficult to discern at the upper levels of the spinal canal but, as seen in the MRI scan, is often filled with fat in the lumbar region (see Figure 7.3).

The spinal cord, notwithstanding its relatively small size compared with the rest of the brain, is absolutely essential for our normal function. It is the connector between the CNS and our body (other than the head). On the sensory (afferent) side, the information arriving from the skin, muscles, and viscera informs the CNS about what is occurring in the periphery; this information then "ascends" to higher centers in the brain. (The sensory pathways are described in Section 2.)

On the motor (efferent) side, the nerves leave the spinal cord to control our muscles. Although the spinal cord has a functional organization within itself, these neurons of the spinal cord receive their "instructions" from higher centers, including the cerebral cortex, via several descending tracts. This enables us to carry out normal movements, including normal walking and voluntary activities. (The motor pathways are described in Section 2.)

The spinal cord also has a motor output to the viscera and glands, part of the autonomic nervous system (see Figure 3.9).

DEVELOPMENTAL PERSPECTIVE

During early development, the spinal cord and the vertebral canal are the same length, and the entering and exiting nerve roots are found at the same level corresponding to the spinal cord vertebral levels. During the second part of fetal development, the body and the bony spine continue to grow, but the spinal cord does not. After birth, the spinal cord fills the vertebral canal only to the level of L2, the second lumbar vertebra (also seen in the MRI scan). The space below the termination of the spinal cord is the **lumbar cistern**, filled with CSF.

Therefore, because the spinal cord segments do not correspond to the vertebral segments, the nerve roots must travel in a downward direction to reach their proper entry and exit levels between the vertebrae, more so for the lower spinal cord roots. These nerve roots are collectively called the **cauda equina**, and they are found in the lumbar cistern (see the photographic view of the cauda equina in Figure 1.11).

CLINICAL ASPECT

The four vertebral levels—cervical, thoracic, lumbar, and sacral—are indicated on the illustration. The spinal cord levels are indicated between the images. One must be very aware of which reference point is being used when discussing spinal cord injuries—vertebral or spinal.

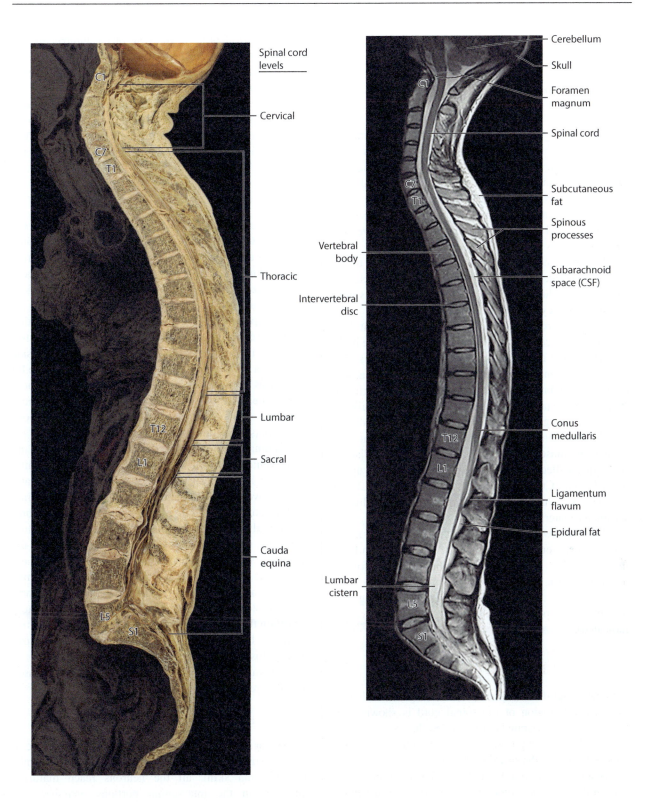

FIGURE 1.10: Spinal Cord 1—Spinal Cord: Vertebral Canal (photograph *in situ* and T2 magnetic resonance imaging scan)

FIGURE 1.11—SPINAL CORD 2

SPINAL CORD: LONGITUDINAL VIEW AND CAUDA EQUINA (PHOTOGRAPHS)

This is a photographic image of the spinal cord removed from the vertebral canal. The dura-arachnoid has been opened, and the anterior aspect of the cord is seen, with the attached spinal roots—ventral (motor) and dorsal (sensory); from this anterior perspective, most of the roots seen are the ventral (i.e., motor roots); The denticulate ligament, a pial extension, is found between the ventral and dorsal roots (see also Figure 8.7).

The spinal cord is divided into parts according to the region innervated: cervical (8 spinal roots); thoracic (12 spinal roots); lumbar (5 spinal roots); sacral (5 spinal roots); and coccygeal (1 root).

The nerve roots attached to the spinal cord that connect the spinal cord with the skin and muscles of the body give the spinal cord a segmented appearance. This segmental organization is reflected onto the body in accordance with embryological development. Areas of skin are supplied by certain nerve segments—each area is called a **dermatome** (e.g., inner aspect of the arm and hand, C8; umbilical region, T10), with overlap from adjacent segments. The muscles are supplied usually by two adjacent segments, called **myotomes** (e.g., biceps of the upper limb, C5 and C6; quadriceps of the lower limb, L3 and L4). This known pattern is very important in the clinical setting (see later).

There are two enlargements of the spinal cord—at the cervical level for the upper limb, the roots of which form the **brachial plexus**, and at the lumbosacral level for the lower limb, the roots of which form the **lumbar and sacral plexuses**. The spinal cord tapers at its ending, and this lowermost portion is called the **conus medullaris** (as indicated).

CAUDA EQUINA (PHOTOGRAPH)

A higher-magnification photographic image of the lowermost (sacral) region of the spinal cord is shown. The tapered end of the spinal cord is called the **conus medullaris**, and this lower portion of the spinal cord corresponds approximately to the sacral segments.

Below the vertebral level of L2 in the adult, inside the vertebral canal, are numerous nerve roots, both ventral and dorsal, collectively called the **cauda equina**. The roots are traveling from the spinal cord levels to exit at their appropriate (embryological) intervertebral level (see Figure 7.3). These nerve roots are found within the **lumbar cistern**, an expansion of the subarachnoid space, a space containing cerebrospinal fluid (CSF) (see Figure 7.8).

The roots are floating in the CSF of the lumbar cistern. The nerve roots are further discussed with Figure 1.12.

The pia mater of the spinal cord gathers at the tip of the conus medullaris into a ligament-like structure, the **filum terminale**, which attaches to the dura-arachnoid at the termination of the vertebral canal, which is the end of the spinal dural sheath, at the level of (vertebral) S2. The three meningeal layers then continue and attach to the coccyx as the coccygeal ligament.

CLINICAL ASPECT

The segmental organization of the spinal cord and the known pattern of innervation to areas of skin and to muscles allow a knowledgeable practitioner, after performing a detailed neurological examination, to construct an accurate localization of the injury or disease (called the lesion) at the spinal cord (segmental) level.

The spinal cord can be affected by tumors, either within the cord (intramedullary) or outside the cord (extramedullary). There is a large plexus of veins on the outside of the dura of the spinal cord, and this is a possible site for metastases from pelvic (including prostate) tumors. These tumors press on the spinal cord as they grow and cause symptoms as they compress and interfere with the various pathways (see Section 2).

Traumatic lesions of the spinal cord occur following motor vehicle, bicycle, and driving accidents. Protruding discs can impinge on the spinal cord. Other traumatic lesions involve gunshot and knife wounds. If the spinal cord is completely transected (i.e., cut through completely), all the tracts are interrupted. For the ascending pathways, this means that sensory information from the periphery is no longer available to the brain. On the motor side, all the motor commands cannot be transmitted to the anterior horn cells, the final common pathway for the motor system. The person therefore is completely cut off on the sensory side and loses all voluntary control below the level of the lesion. Bowel and bladder control are also lost.

The vascular supply to the spinal cord is discussed with Figure 8.7 and Figure 8.8.

DEVELOPMENTAL ASPECT

Embryologically, the spinal cord commences as a tube of uniform size. In those segments that innervate the limbs (muscles and skin), all the neurons reach maturity. However, in the intervening portions, massive programmed cell death occurs during development because there is less peripheral tissue to be supplied. In the adult, therefore, the spinal cord has two "enlargements"—the cervical for the upper limb and the lumbosacral for the lower limb, each giving rise to the nerve plexus for the upper and lower limbs, respectively.

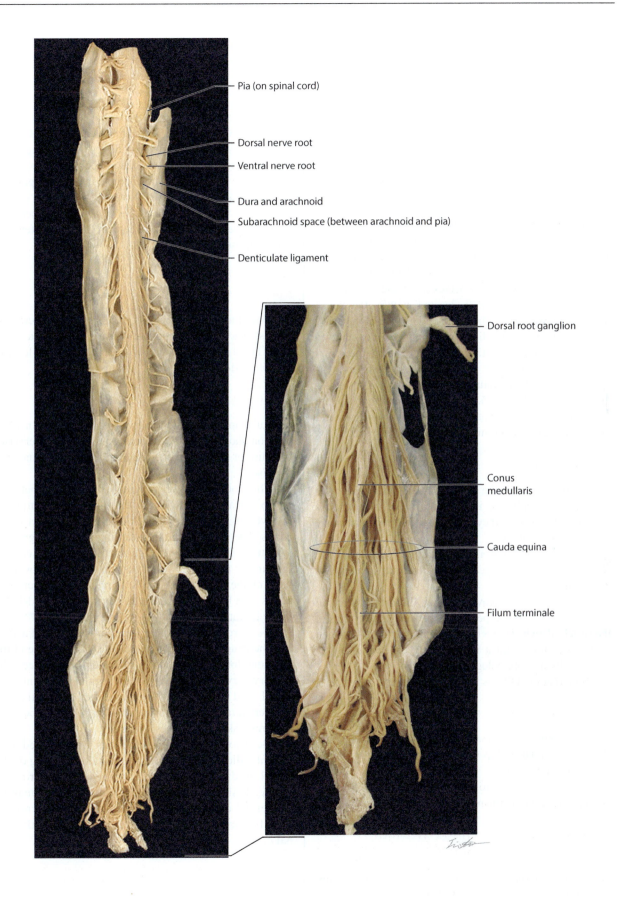

Pia (on spinal cord)

Dorsal nerve root

Ventral nerve root

Dura and arachnoid

Subarachnoid space (between arachnoid and pia)

Denticulate ligament

Dorsal root ganglion

Conus medullaris

Cauda equina

Filum terminale

FIGURE 1.11: Spinal Cord 2—Spinal Cord: Longitudinal View and Cauda Equina (photographs)

FIGURE 1.12—SPINAL CORD 3

NERVE ROOTS (WITH T2 MAGNETIC RESONANCE IMAGING SCANS)

Two sets of nerve roots connect the spinal cord with the periphery. The four illustrations show the locations of the roots and their relationships with the vertebrae.

DORSAL ROOTS

The dorsal roots are the sensory roots that carry information from the periphery into the central nervous system. These fibers enter the spinal cord dorsally and are connected with the dorsal horn (see Figure 4.1 and Figure 5.1).

The cell bodies of these nerves are located in the **dorsal root ganglion (DRG)**, located within the space between the vertebrae. The peripheral portion of the nerve can be very long, extending (in adults) from the spine to the toes. The central portion of the nerve enters the spinal cord (at the appropriate level), where some fibers synapse (pain and temperature—the anterolateral system—discussed with Figure 5.3) and others continue up the spinal cord (discriminative touch, joint position and vibration—dorsal column—discussed with Figure 5.2).

VENTRAL ROOTS

The ventral roots are the axons of the anterior horn cells en route to supply the innervation to the muscles. As shown in Figure 1.11, they leave the spinal cord anteriorly by a series of rootlets.

The dorsal and ventral roots are united just beyond the DRG and are known together as the (mixed) spinal nerve (see Figure 7.3).

The nerve roots leave or enter the vertebral canal between the vertebrae, specifically between the pedicles of the vertebral arch. This is called the intervertebral foramen in anatomy texts, and it is commonly called the **neural foramen** by neuroradiologists.

Note that the DRG is located in the neural foramen.

FIGURE 1.12A (UPPER LEFT—DRAWING)

This is a longitudinal view of the two roots and the DRG, showing the relationship with the vertebra and exiting in the intervertebral space. The nerve roots of the lumbar region are descending to exit the vertebral canal at the appropriate level via the neural foramen.

FIGURE 1.12B (UPPER RIGHT—T2-WEIGHTED MAGNETIC RESONANCE IMAGE)

A longitudinal view of the same level shows the exiting nerve roots and a DRG, as indicated.

FIGURE 1.12C (LOWER LEFT—T2-WEIGHTED MAGNETIC RESONANCE IMAGE)

An axial view of the nerve roots through the neural foramen at the cervical level.

The meningeal layers of the spinal cord are discussed in Section 3 (see Figure 7.3); cerebrospinal fluid (CSF) is white in this T2-weighted magnetic resonance image. The dorsal roots can be seen attached, in their respective positions, to the spinal cord. These extend in the same plane to exit at the appropriate level. Note that the dural "sleeve" extends to the point where the mixed spinal nerve is formed. The subarachnoid space with CSF also extends to this point (see Figure 7.3).

FIGURE 1.12D (LOWER RIGHT—T2-WEIGHTED MAGNETIC RESONANCE IMAGE)

An axial view of the spinal cord at the lumbar vertebral level shows a number of roots of the cauda equina in the CSF. These are the nerves descending to a lower level to exit the vertebral canal (as shown in the drawing).

CLINICAL ASPECT

It is quite apparent that the nerve roots are in a vulnerable position should there be a "protrusion" of the intervertebral disc or degenerative changes at the joints connecting the discs.

The roots to the lower extremity, those exiting between L4 to L5 and L5 to S1 are the ones most commonly involved in the everyday back injuries that affect many adults. The student should be familiar with the signs and symptoms that accompany degenerative disc disease in the lumbar region. Lower back pain with pain down the back of the leg ("sciatica") is a frequent occurrence in adults, and how best to manage it is still heavily debated in the literature.

Nerve roots can be anesthetized by injection of a local anesthetic agent into their immediate vicinity. One of the locations for this injection is in the epidural space. The sensory nerve roots to the perineal region, which enter the spinal cord at the sacral level, are often anesthetized in their epidural location during childbirth. This procedure requires a skilled anesthesiologist.

Occasionally, neurologic deficits seen in a pediatric patient indicate that the filum terminale is pulling on the spinal cord; this is called a tethered spinal cord syndrome. If this is suspected clinically, further imaging studies are done, and in some cases the filum terminale must be surgically cut to relieve the tension on the spinal cord.

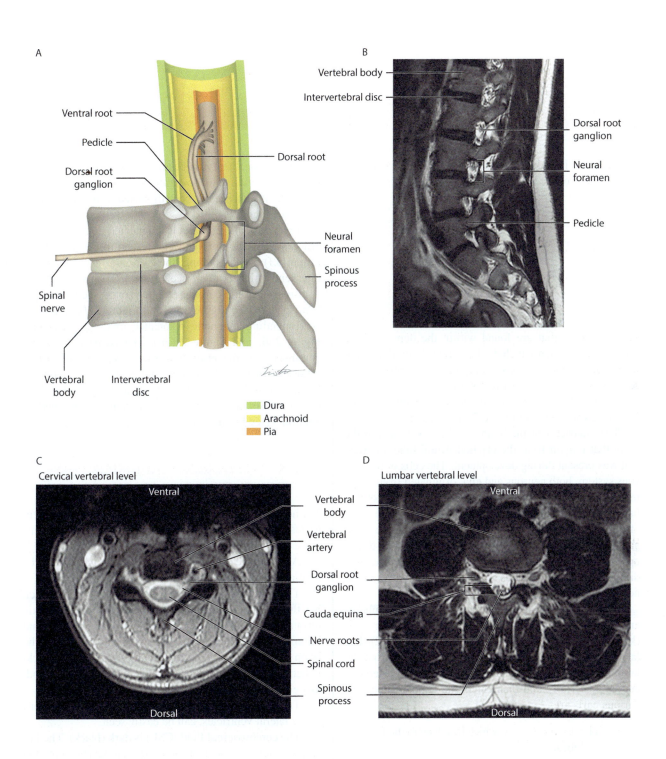

A

Ventral root

Pedicle

Dorsal root ganglion

Dorsal root

Spinal nerve

Neural foramen

Spinous process

Vertebral body

Intervertebral disc

Dura
Arachnoid
Pia

B

Vertebral body

Intervertebral disc

Dorsal root ganglion

Neural foramen

Pedicle

C
Cervical vertebral level

Ventral

Dorsal

D
Lumbar vertebral level

Ventral

Dorsal

Vertebral body

Vertebral artery

Dorsal root ganglion

Cauda equina

Nerve roots

Spinal cord

Spinous process

FIGURE 1.12: Spinal Cord 3—Nerve Roots (with T2 magnetic resonance imaging scans)

Chapter 2

Internal Structures—Cerebral Hemispheres

This chapter introduces the structures found *within* the various parts of the brain.

FIGURE 2.1A—CEREBRAL VENTRICLES 1

VENTRICLES: LATERAL VIEW (WITH AXIAL CT RADIOGRAPHS)

The structures that are found within the depths of the cerebral hemispheres include the cerebral ventricles, the white matter, and the basal ganglia. The ventricles are cavities within the brain filled with **cerebrospinal fluid (CSF)**. The formation, circulation, and locations of the CSF are explained with Figure 7.8 in Section 3.

The ventricles of the brain are the spaces within the brain that remain from the original neural tube, the tube that was present during development. The cells of the nervous system, both neurons and glia, originated from a germinal matrix that was located adjacent to the lining of this tube. The cells multiply and migrate away from the walls of the neural tube to form the nuclei (including the basal ganglia) and cerebral cortex. As the nervous system develops, the mass of tissue grows and the size of the tube diminishes, leaving various spaces in different parts of the nervous system.

The parts of the tube that remain in the hemispheres are the cerebral ventricles, called the **lateral ventricles** (known as ventricles I and II, also ventricles 1 and 2). The lateral ventricle of the hemispheres, shown here from the lateral perspective, is shaped like the letter C in reverse; it curves posteriorly and then enters the temporal lobe. Its various parts are:

- The **anterior horn**, which lies deep to the frontal lobes.
- The central portion or **body**, which lies deep to the parietal lobes.
- The **atrium** or **trigone**, where it widens and curves and then enters the temporal lobe as the **inferior horn**.

In addition, there may be an extension into the occipital lobes, the **occipital** or posterior horn, and its size varies. (The lateral ventricle has been dissected from this perspective—see Figure 9.4.)

Each lateral ventricle is connected to the midline 3rd ventricle (the 3rd ventricle is seen in Figure 1.9 and Figure 3.1) by an opening, the **foramen of Monro**—the interventricular foramen (see Figure 1.7). This connection is also seen in the coronal and horizontal sections of the hemispheres (see Figure 2.9A and Figure 2.10A), and the corresponding magnetic resonance imaging scans (see Figure 2.9B and Figure 2.10B). The ventricular system continues into the brainstem and is represented faintly (ghosted) in this illustration (and is described with the brainstem in Figure 3.1, Figure 3.2, and Figure 3.3). The flow of CSF in the ventricular system is described in Section 3 (see Figure 7.8).

VENTRICLES: HORIZONTAL (AXIAL) VIEW (COMPUTED TOMOGRAPHY SCANS)

The horizontal (axial) view was chosen to visualize the ventricles with a computed tomography (CT) scan because this is the image used most frequently in the clinical setting.

A CT image shows the skull bones (in white) and the relationship of the brain with the skull.

The outer cortical tissue is visible, with gyri and sulci, but there is not as much detail as seen with a magnetic resonance imaging (MRI) scan (shown in Figure 2.9B and Figure 2.10B). The structures seen in the interior of the brain include the white matter, which is a "fuzzy" speckled gray; the basal ganglia and thalamus can be discerned, as well as the internal capsule.

The ventricular spaces, particularly the anterior horn of the lateral ventricles, can be easily seen.

The cerebrospinal fluid (CSF) is dark (black). The lateral ventricles are seen at three different levels—note the different configuration depending on the level of the "cut." The choroid plexus can be seen within the lateral ventricles (see Figure 7.8 and Figure 9.4).

The cerebellum can be recognized, with its folia, but there is no sharp delineation between it and the hemispheres.

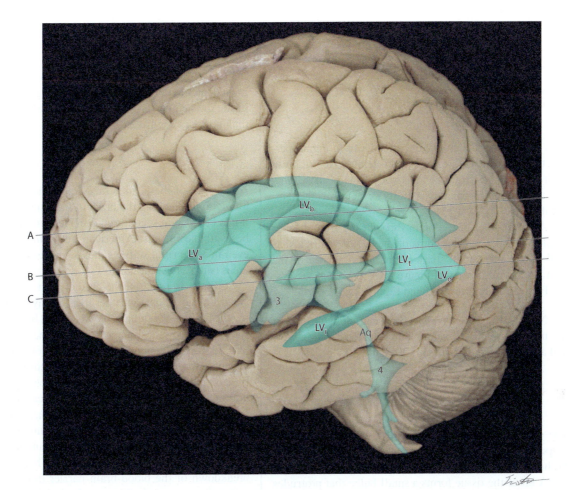

Lateral ventricle:
LV$_a$ = Anterior horn
LV$_b$ = Body
LV$_t$ = Atrium (trigone)
LV$_o$ = Occipital horn
LV$_i$ = Inferior horn

3 = 3rd ventricle
4 = 4th ventricle
Aq = Aqueduct of midbrain
Cp = Choroid plexus

FIGURE 2.1A: Cerebral Ventricles 1—Ventricles: Lateral View (with axial CT radiographs)

FIGURE 2.1B—CEREBRAL VENTRICLES 2

VENTRICLES: ANTERIOR VIEW (WITH CORONAL CT RADIOGRAPHS)

This is a view of the brain from an anterior (coronal) perspective, showing the location of the cerebral ventricles in each hemisphere. Note the reverse C shape of the ventricles and their position more laterally as they enter the temporal lobe.

A coronal cut of the brain through the frontal lobes would show the anterior horns of the lateral ventricles, as well as the very small space of the inferior horn of the lateral ventricle (review of Figure 2.1A may be necessary; see Figure 2.8, Figure 9.5A, and Figure 9.5B).

Note to the Learner: The 3rd ventricle, aqueduct of the midbrain, and 4th ventricle are "faded" in this view and are shown (and discussed) in Figure 3.1.

VENTRICLES: ANTERIOR (CORONAL) VIEW (COMPUTED TOMOGRAPHY SCANS)

The ventricles are again dark with these CT scans. The shape and size of the ventricle changes with the location of the horizontal "cut"—frontal horn, body, or the trigone area.

A bit of the tissue forms a small bulge that protrudes into the anterior horn, normally. This protrusion is formed by the head of the caudate nucleus (see Figure 2.5A and Figure 2.5B); this caudate bulge is also seen on a horizontal view of the brain (see Figure 2.1A and also in Figure 2.2B).

The cut includes the temporal lobe, and the inferior (temporal) horn of the lateral ventricle is seen as a crescent-shaped slit. In this case, it is the hippocampus that protrudes into this part of the ventricle (discussed with Figure 9.4, Figure 9.5A, and Figure 9.5B).

Note the tissue inside the ventricles on the cut through the atrium (the one on the far right). This is some of the choroid plexus of the lateral ventricles, the site of production of the cerebrospinal fluid (see also Figure 9.4, further discussed with Figure 7.8).

The ventricular system of the diencephalon and brainstem is shown ("ghosted") and will be shown and discussed with Figure 3.1, Figure 3.2, Figure 3.3, and Figure 7.8.

Note to the Learner: The same photographic image is used later to develop the understanding of the structures inside the hemispheres, namely, the basal ganglia and the thalamus (see Figure 2.8).

CLINICAL ASPECT

A CT scan of the head is an X-ray done with computed tomography showing the brain in the interior of the skull. It can be done in seconds and this technology would now be available in most American and Canadian hospitals, even in some rural settings, but not necessarily in remote areas (such as the far north).

CT scan is a state-of-the-art diagnostic modality which is used to investigate whether there is any "lesion" affecting the brain. For example, blood (caused by a hemorrhage) shows as a "bright" image with CT and the descriptive term for this is "hyper-density." The loss of blood supply to a region would be seen as a "hypo-dense" region (e.g., because of the blockage of an artery supplying a part of the brain, as discussed in Section 3). CT scan is superior to MRI scanning when investigating lesions with high calcium content such as bony lesions or meningiomas.

CT scans can be "enhanced" using an iodinated compound injected intravenously to see whether there is a breakdown of the blood-brain barrier (for example with various tumours and other vascular lesions lesions such as aneurysms or arteriovenous malformations).

MRI, which uses a powerful magnet and radio waves (and is not an X-ray) provides complementary information to CT scans and produces images to delineate the structures with different water and fat content inside the skull.

ADDITIONAL NOTE

The computed tomography scan (particularly the image on the far right) shows an abnormal feature—cortical atrophy. There is a definite increase in the space between the brain and skull seen in these images, as well as in the sulci separating the gyri of the brain. This type of image would be seen in patients with dementia, and further studies of this patient would be warranted based on these images.

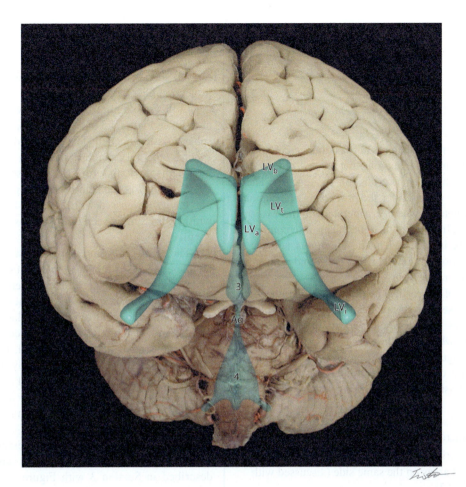

Lateral ventricle:
LV_a = Anterior horn
LV_b = Body
LV_t = Atrium (trigone)
LV_i = Inferior horn

3 = 3rd ventricle
4 = 4th ventricle
Aq = Aqueduct of midbrain
Cp = Choroid plexus

C_h = Caudate n. (head)

FIGURE 2.1B: Cerebral Ventricles 2—Ventricles: Anterior View (with coronal CT radiographs)

FIGURE 2.2A—WHITE MATTER 1

CORPUS CALLOSUM: SUPERIOR VIEW (PHOTOGRAPH)

One of the other major features of the human cerebral cortex is the vast number of neurons that are devoted to communicating with other neurons of the cortex. These interneurons are essential for the processing and elaboration of information, whether generated in the external world or internally by our "thoughts." This intercommunicating network is reflected in the enormous number of axonal connections among cortical areas and with other parts of the central nervous system.

The axons of the cortical neurons and the connections to and from these neurons are located within the depths of the hemispheres. With fixation in formalin, these myelinated axons are white, and these areas are called the **white matter** (see Figure 2.2B and Figure 2.3). In the spinal cord, these are called tracts; in the hemispheres, these bundles are classified in the following way:

- **Commissural** bundles—connecting cortical areas across the midline.
- **Association** bundles—interconnecting the cortical areas on the same side (discussed with Figure 2.3).
- **Projection** fibers—connecting the cerebral cortex with subcortical structures, including the basal ganglia, thalamus, brainstem, and spinal cord (see Figure 2.4); many of these

fibers are located within the internal capsule (see Figure 4.4).

All such connections are bidirectional, including the projection fibers.

In this photograph, the brain is viewed from directly above (see Figure 1.3 and Figure 7.1), with the interhemispheric fissure opened. The dural fold between the hemispheres, the falx cerebri, has been removed from the interhemispheric fissure. This thick sheath of dura keeps the two halves of the hemispheres in place within the cranial cavity (discussed with Figure 7.4 and Figure 7.5). A whitish structure is seen in the depths of the fissure—the **corpus callosum**.

The corpus callosum is the largest of the commissural bundles, as well as the latest in evolution. This is the anatomical structure required for each hemisphere to be kept informed of the activity of the other hemisphere. The axons connect to and from the lower layers of the cerebral cortex, and in most cases the connections are between homologous areas and are reciprocal. In fact, the corpus callosum was already seen previously when viewing the cortical tissue on the medial aspect of the hemispheres, as represented by the frontal, parietal, and occipital lobes (see Figure 1.7). The corpus callosum has been divided in the process.

In this specimen, the blood vessels supplying the medial aspect of the hemispheres are present (fully described in Section 3 with Figure 8.5). Moreover, the cerebral ventricles are located below (i.e., inferior to) the corpus callosum (see Figure 1.7, Figure 2.9A, and Figure 2.9B).

The clinical aspect of the corpus callosum is discussed with Figure 2.2B.

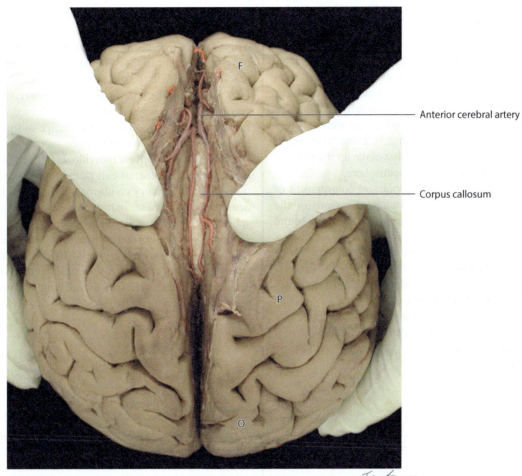

Anterior cerebral artery

Corpus callosum

F = Frontal lobe
P = Parietal lobe
O = Occipital lobe

FIGURE 2.2A: White Matter 1—Corpus Callosum: Superior View (photograph)

FIGURE 2.2B—WHITE MATTER 2

CORPUS CALLOSUM: MEDIAL DISSECTED VIEW (PHOTOGRAPH)

The dissection of this specimen needs some explanation. The brain is again seen from the medial view (as in Figure 1.7; its anterior aspect is on the left side of this photograph) with the corpus callosum exposed. The septum pellucidum is not present, exposing the lateral ventricle, with the head of the caudate nucleus protruding into the anterior horn (as seen previously). Cortical tissue has been removed from this brain by using blunt dissection techniques. If this dissection is done successfully, the fibers of the corpus callosum can be followed, as well as other white matter bundles (see Figure 2.3). These fibers intermingle with other fiber bundles that make up the mass of white matter in the depth of the hemisphere.

The corpus callosum is the massive commissure of the forebrain that connects homologous regions of the two hemispheres of the cortex across the midline. In the midline cut, the thickened anterior aspect of the corpus callosum is called the genu, and the thickened posterior portion is the splenium.

This dissection shows the white matter of the corpus callosum, followed to the cortex.

If one looks closely, looping U-shaped bundles of fibers can be seen connecting adjacent gyri; these are part of the local association fibers.

CLINICAL ASPECT

Even though the connections of the corpus callosum are well described, the function of the corpus callosum under normal conditions is difficult to discern. In rare cases, persons are born without a corpus callosum, a condition called agenesis of the corpus callosum, and these individuals as children and adults usually cannot be distinguished from anatomically normal individuals, unless specific testing is done.

The corpus callosum has been sectioned surgically in certain individuals with intractable epilepsy, which is epilepsy that has not been controllable using multiple anticonvulsant medications. The idea behind this surgical procedure is to stop the spread of the abnormal discharges from one hemisphere to the other. Generally, the surgical procedure has been helpful in well-selected cases, and there is apparently no noticeable change in the person or in his or her level of brain function.

Studies done in these individuals have helped to clarify the role of the corpus callosum in normal brain function. Under laboratory conditions, it has been possible to demonstrate in these individuals how the two hemispheres of the brain function independently, after the sectioning of the corpus callosum. These studies show how each hemisphere responds differently to various stimuli, and they also show the consequences of the failure of information to be transferred from one hemisphere to the other.

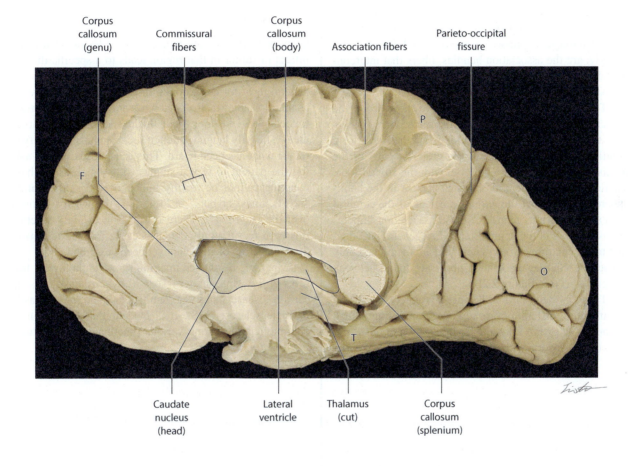

Corpus callosum (genu)

Commissural fibers

Corpus callosum (body)

Association fibers

Parieto-occipital fissure

Caudate nucleus (head)

Lateral ventricle

Thalamus (cut)

Corpus callosum (splenium)

F = Frontal lobe
P = Parietal lobe
T = Temporal lobe
O = Occipital lobe

FIGURE 2.2B: White Matter 2—Corpus Callosum: Medial Dissected View (photograph)

FIGURE 2.3—WHITE MATTER 3

ASSOCIATION FIBERS: LATERAL DISSECTED VIEW (PHOTOGRAPH)

The dorsolateral aspect of the brain is viewed in this photograph (see Figure 1.3). The lateral fissure has been opened, with the temporal lobe below; deep within the lateral fissure is the insula (as in Figure 1.4).

Under the cerebral cortex is the white matter of the brain (see also Figure 9.4). It is possible to dissect various fiber bundles (not easily) by using a blunt instrument (e.g., a wooden tongue depressor). Some of these bundles, functionally, are the association bundles, fibers that interconnect different parts of the cerebral cortex on the same side (classified with Figure 2.2A).

This specimen has been dissected to show two of the association bundles within the hemispheres. The **superior longitudinal fasciculus** (fasciculus is another term for a bundle of axons) interconnects the posterior parts of the hemisphere (e.g., the parietal lobe) with the frontal lobe. There are other association bundles present in the hemispheres that connect the various portions of the cerebral cortex. The various names of these association bundles are usually not of much importance in a general introduction to the central nervous system and are mentioned only if need be. Shorter association fibers are found between adjacent gyri (see Figure 2.2B).

These association bundles are extremely important in informing different brain regions of on-going neuronal processing, thus allowing for integration of our activities (e.g., sensory with motor and limbic). One of the major functions of these association bundles in the human brain seems to be to bring information to the frontal lobes, especially to the prefrontal cortex, which acts as the "executive director" of brain activity (see Figure 1.3 and Figure 6.13).

One of the most important association bundles, the **arcuate bundle**, connects the two language areas. It connects Broca's area anteriorly with Wernicke's area in the superior aspect of the temporal lobe, in the dominant (left) language hemisphere (see Figure 4.5).

CLINICAL ASPECT

Damage to the arcuate bundle from a lesion such as an infarct or tumor in that region leads to a specific disruption of language, called conduction aphasia. **Aphasia** is a general term for a disruption or disorder of language. In conduction aphasia, the person has normal comprehension (intact Wernicke area) and fluent speech (intact Broca area). The only language deficit seems to be an inability to repeat what has been heard. This is usually tested by asking the patient to repeat single words or phrases whose meaning cannot be readily understood (e.g., the phrase "no ifs, ands, or buts" or "the quick brown fox jumped over the lazy dog"). There is some uncertainty whether this is in fact the only deficit because isolated lesions of the arcuate bundle have not yet been described.

Magnetic resonance imaging shows pathological features of the white matter—seen as hyperintense foci—in certain disease states (e.g., multi-focal plaques such as in multiple sclerosis MS). It is evident that disruption of communication among various functional areas disturbs the functioning brain.

Superior longitudinal
fasciculus

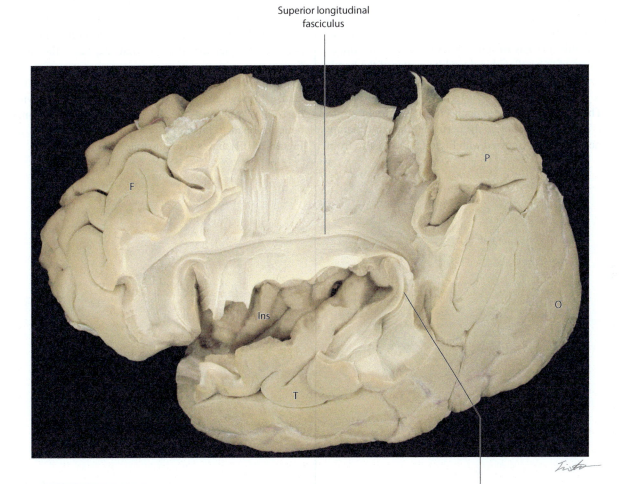

Arcuate bundle

F = Frontal lobe
P = Parietal lobe
T = Temporal lobe
O = Occipital lobe

Ins = Insula

FIGURE 2.3: White Matter 3—Association Fibers: Lateral Dissected View (photograph)

FIGURE 2.4—WHITE MATTER 4

PROJECTION FIBERS (DIFFUSION TENSOR IMAGING)

This illustration was done with *Diffusion Tensor Imaging* (DTI, often erroneously called "tractography"), one of the latest imaging modalities but not one used clinically (at the time this text was written). What is shown is a number of images of "fibers" within the white matter of the hemispheres. These include commissural fibers and projection fibers.

The cerebral cortex is connected to all brain structures. The fibers coming *to* the cortex are sensory, mainly via the thalamus (discussed with Figure 5.5) and also from the basal ganglia (discussed with Figure 5.14). These are sometimes called centripetal fibers. The fibers *from* the cortex to all "lower centers," including motor fibers (cortico-spinal and cortico-bulbar, see Figure 5.9 and Figure 5.10), are often called centrifugal fibers. Collectively, they are classified as projection fibers.

The projection fibers course to or from the cortex (the upper figure), are found within the white matter (the corona radiata, in the middle figure), and collect in a funnel-like manner as the internal capsule (the lower figure).

Many of the fibers to and from the cerebral cortex course through the internal capsule, which is shown in horizontal views of the brain (photograph and magnetic resonance imaging scan—see Figure 2.10A and Figure 2.10B) and is also illustrated in Figure 4.4.

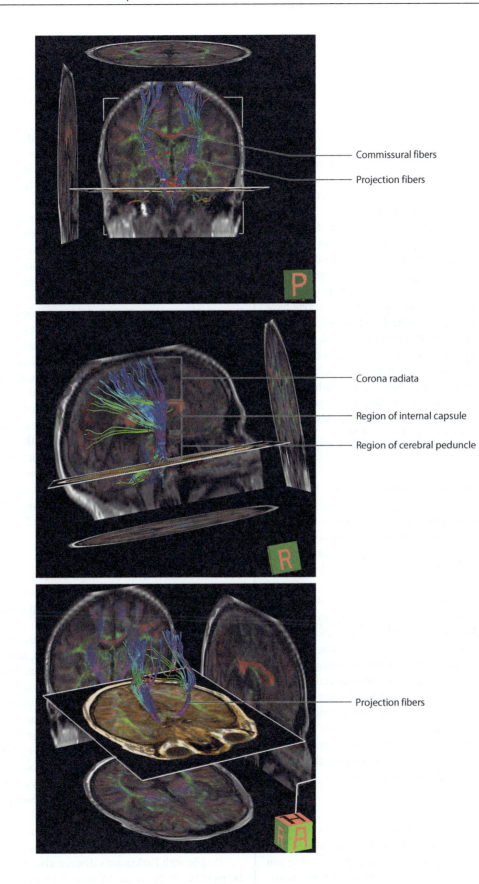

FIGURE 2.4: White Matter 4—Projection Fibers (Diffusion Tensor Imaging)

FIGURE 2.5A—BASAL GANGLIA 1

BASAL GANGLIA: ORIENTATION

The large collections of gray matter within the hemispheres, belonging to the forebrain, in addition to the white matter and the ventricles already described, are collectively called the **basal ganglia**. The term **striatum** is often used for the basal ganglia, but this term is not always used with neuroanatomical precision.

Our understanding of the functional role of the basal ganglia is derived largely from disease states affecting these neurons and their connections. In general, humans with lesions in the basal ganglia have some form of motor dysfunction, a **dyskinesia** (i.e., a movement disorder). However, as discussed in this chapter, these neurons have connections with both neocortical and limbic areas and are definitely involved in other brain functions.

The basal ganglia are described in a series of illustrations. This diagram is for orientation and terminology. Figure 2.5B contains more anatomical details and the functional aspects. The details of the connections and the circuitry involving the basal ganglia are described in Section 2 (see Figure 5.14 and Figure 5.18).

UPPER ILLUSTRATION

From the strictly anatomical point of view, the basal ganglia are collections of neurons located within the hemispheres. In the upper illustration, the brain is viewed in the coronal plane partially sectioned midway between the frontal and occipital poles. The basal ganglia are seen in their anatomical location—on the proximal side and somewhat on the distal side.

The structures visualized include the **caudate** nucleus, the **putamen**, and the amygdala (in the temporal lobe; see Figure 2.9A and Figure 2.10A). The caudate and putamen are also called the **neostriatum**; histologically, these are the same neurons, but in the human brain they are partially separated from each other by projection fibers located within the internal capsule (see Figure 4.4). The development of the human brain includes the evolution of a temporal lobe, and many structures "migrate" into this lobe with the lateral ventricle. The caudate nucleus follows the curvature of the lateral ventricle into the temporal lobe (see Figure 2.5B and Figure 9.7).

LOWER ILLUSTRATION

If the basal ganglia of the proximal side are removed, another structure is seen on the distal side—the **globus pallidus**, which is also part of the basal ganglia. As seen in Figure 2.5B and in a horizontal cut through the hemisphere (see Figure 2.10A and Figure 2.10B), the putamen and globus pallidus are anatomically grouped together in

the human brain and form a lens-like configuration, hence their collective name, the **lentiform** or **lenticular nucleus**. The name is purely descriptive (and somewhat confusing) because the two nuclei are located together in the human nervous system, yet they are functionally quite distinct and do not constitute a true nucleus.

The amygdala, also called the amygdaloid nucleus, is classically one of the basal ganglia because it is a subcortical collection of neurons (in the temporal lobe). Most of the connections of the amygdala are with limbic structures (see Section 4), and so this nucleus is discussed in Section 4 (see Figure 9.6A and Figure 9.6B). Another functional area of the basal ganglia has now been recognized as highly important—the **nucleus accumbens**. Again, this nucleus has limbic connections and is discussed in Section 4 with the limbic system. Other subcortical nuclei located in the forebrain, particularly in the basal forebrain region, have not been grouped with the basal ganglia and will be described with the limbic system (in Section 4).

Functionally, the basal ganglia system acts as a sub-loop of the motor system by altering cortical activity (fully discussed in Section 2 with Figure 5.14, under the topic of Motor Modulation). In general terms, the basal ganglia receive much of their input from the cortex, including from the motor areas and from wide areas of association cortex, as well as from other nuclei of the basal ganglia system (described later). There are intricate connections among the various parts of the system, involving different neurotransmitters; the output is directed via the thalamus mainly to pre-motor, supplementary motor, and frontal cortical areas (see Figure 5.8 and Figure 5.18).

CLINICAL ASPECT

The functional role of this large collection of neurons is best illustrated by clinical conditions in which this system does not function properly—Parkinson's disease (discussed below) and Huntington's chorea (discussed with Figure 2.9A). These disease entities cause abnormal movements, such as chorea (jerky movements), athetosis (writhing movements), and tremors (rhythmic movements).

The most common condition, which affects this functional system of neurons is **Parkinson's** disease. The person with this disease has difficulty initiating movements, the face takes on a mask-like appearance with loss of facial expressiveness, there is muscular rigidity, a slowing of movements (bradykinesia), and a slow pill-rolling tremor of the hands *at rest* which goes away with purposeful movements (see also Figure 5.14). Some individuals with Parkinson's develop cognitive problems such as hallucinations and visuospatial problems and also emotional difficulties including anxiety and depression.

People with Parkinson's disease also develop **rigidity**. In rigidity, there is an increased resistance to passive movement of the limb, which involves both the flexors and extensors, and the response is not velocity dependent.

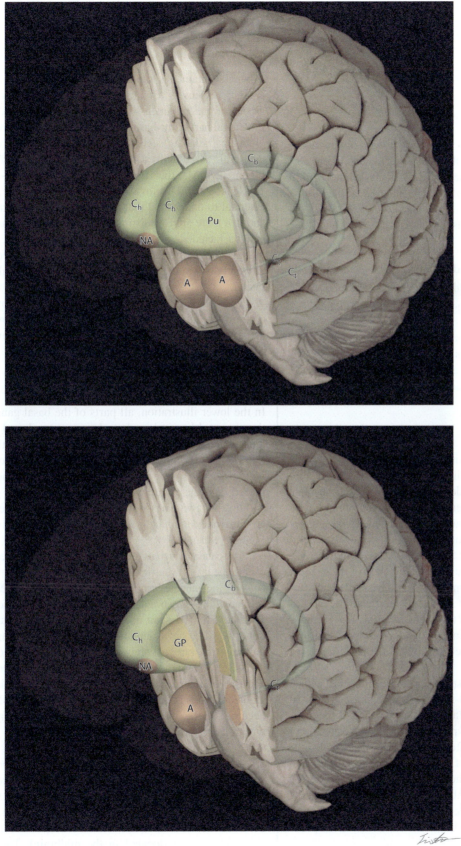

Caudate n.
C_h = Head
C_b = Body
C_t = Tail

NA = Nucleus
accumbens
Pu = Putamen
GP = Globus pallidus
A = Amygdala

FIGURE 2.5A: Basal Ganglia 1—Basal Ganglia: Orientation

FIGURE 2.5B—BASAL GANGLIA 2

BASAL GANGLIA: NUCLEI AND RELATIONSHIPS (WITH T1 MAGNETIC RESONANCE IMAGING SCANS)

The basal ganglia, from the point of view of functional neuroanatomy, consist of three major nuclei in each of the hemispheres (and these structures are located within the forebrain), the **caudate**, the **putamen**, and the **globus pallidus**—excluding the amygdala.

The caudate nucleus is described as having three portions:

- The head, located deep within the frontal lobe.
- The body, located deep in the parietal lobe.
- The tail, which goes in to the temporal lobe.

Each portion is associated anatomically with the lateral ventricle and follows its curvature (see Figure 9.7).

The basal ganglia are shown in this series from a lateral perspective, and adjacent to each is an MRI scan done at the level indicated.

UPPER ILLUSTRATION

Starting with the top illustration, the various parts of the caudate nucleus are easily recognized—head, body, and tail—on the proximal side. The relationship of the caudate nucleus with the lateral ventricle can readily be described—the large head of the caudate nucleus actually intrudes into the space of the anterior horn of the lateral ventricle (see Figure 2.2B, Figure 2.9A, Figure 2.9B, Figure 2.10A, and Figure 2.10B). The body of the caudate nucleus tapers and becomes considerably smaller, and it is found beside the body of the lateral ventricle. The tail follows the inferior horn of the lateral ventricle into the temporal lobe. This is a slender extended group of neurons, difficult to identify in sections of the temporal lobe (see Figure 9.5A).

From this lateral perspective, it is clear that the caudate nucleus is in continuity with another large nucleus situated laterally—this is the putamen. The caudate and the putamen contain the same types of neurons and have some of the same connections; often they are collectively called the **neostriatum**. Strands of neuronal tissue are often seen connecting the caudate nucleus with the putamen. A very distinct and important fiber bundle, the internal capsule, separates the head of the caudate nucleus from the putamen (see Figure 4.4). This fiber bundle fills the spaces between the cellular strands.

The adjacent axial (horizontal) T1 MRI image beside is done at the level of the foramen of Monro and is at the same level as the horizontal cut in Figure 2.10A and the accompanying (T2) MRI image shown in Figure 2.10B.

MIDDLE ILLUSTRATION

When the putamen is removed, as in the middle illustration, another nucleus is revealed that is attached to the inner aspect of the putamen—the **globus pallidus**, a distinct functional part of the basal ganglia. When the putamen and globus pallidus are seen in coronal and horizontal cuts of the brain (see lowest MRI; also Figure 2.9A and Figure 2.10A), the two nuclei form a lens-shaped "nucleus"—the **lentiform or lenticular nucleus**. It is essential to understand that this is a descriptive term only and that there are two distinct functional parts of the basal ganglia included—the putamen (laterally) and the globus pallidus (medially). The lentiform nucleus is situated laterally and deep in the hemispheres, within the central white matter.

The T1 MRI scan done at the level of the anterior commissure, in the horizontal (axial) plane, shows the actual connection between the caudate (head) and the putamen.

LOWER ILLUSTRATION

In the lower illustration, all parts of the basal ganglia of the proximal hemisphere have been removed, and one is looking at the distal side, from the medial perspective.

The lentiform nucleus is now seen to be composed of its two portions—the globus pallidus, which is medially placed, and the putamen lateral to it. In fact, the globus pallidus has two parts—an external (lateral) segment and an internal (medial) segment (see Figure 5.14 and Figure 5.18).

The caudate nucleus (of the distal side) is now seen adjacent to the lateral ventricle (of the distal side). (Note that the color of the caudate has changed somewhat because of the overlap of the "green" caudate with the "blue" ventricle.)

The T1 MRI scan accompanying this illustration is done in a coronal orientation also through the anterior commissure, but still shows the caudate nucleus ("bulging" slightly into the anterior horn of the lateral ventricle) and the putamen, separated by the internal capsule. The globus pallidus is difficult to distinguish radiographically because of the large numbers of fibers (white matter) within its midst.

Note to the Learner: From the functional point of view and based on the complex pattern of interconnections, two other nuclei that are not in the forebrain should be included with the description of the basal ganglia—the **subthalamic nucleus** (part of the diencephalon) and the **substantia nigra** (located in the midbrain). The functional connections of these nuclei are discussed as part of the motor system (see Figure 5.14 and Figure 5.18).

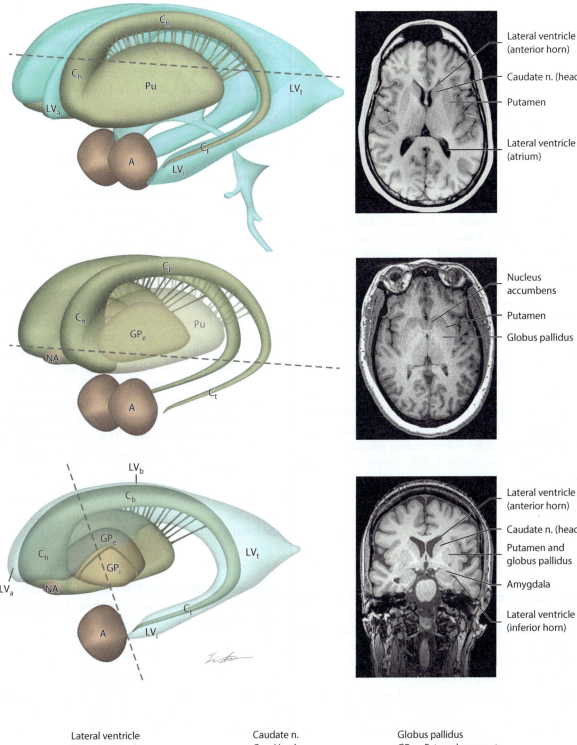

Lateral ventricle
LV_a = Anterior horn
LV_b = Body
LV_t = Atrium (trigone)
LV_i = Inferior horn

Caudate n.
C_h = Head
C_b = Body
C_t = Tail

NA = Nucleus
accumbens

Globus pallidus
GP_e = External segment
GP_i = Internal segment

A = Amygdala

FIGURE 2.5B: Basal Ganglia 2—Basal Ganglia: Nuclei and Relationships (with T1 magnetic resonance imaging scans)

FIGURE 2.6—THE DIENCEPHALON: THALAMUS

THALAMUS: ORIENTATION

The diencephalon, which translates as "between brain," is the next region of the brain to consider. The diencephalon, including both thalamus and hypothalamus and some other subparts, is situated between the brainstem and the cerebral hemispheres, deep within the brain.

As shown photographically (see Figure 1.7, Figure 1.8, and Figure 1.9) and diagrammatically (see Figure 3.1), the diencephalon sits on top of the brainstem. The enormous growth of the cerebral hemispheres in the human brain has virtually hidden or "buried" the diencephalon (somewhat like a weeping willow tree), so that it can no longer be visualized from the outside except from the inferior view (see the pituitary stalk and mammillary bodies, which are both part of the hypothalamus, in Figure 1.5 and Figure 1.6).

In this section of the atlas, we consider the **thalamus**, which makes up the bulk of the diencephalon. There are two thalami, one for each hemisphere of the brain (see Figure 2.9A and Figure 2.10A), and these are often connected across the midline by nervous tissue, the interthalamic adhesion (as seen in Figure 1.7 and Figure 3.2). As noted in Chapter 3, the 3rd ventricle is situated between the two thalami (see Figure 2.10A, Figure 2.10B, and Figure 3.1).

UPPER ILLUSTRATION

The thalamus of both hemispheres is shown from the same perspective as the basal ganglia shown previously (and still shown). The corpus callosum (cut) connecting the hemispheres is seen, below which are slit-sized (black) spaces, which are the lateral ventricles of each hemisphere. The thalamus of the proximal side is seen "behind"—that is medial to—the putamen (of the lentiform nucleus). On the distal side, the caudate nucleus is "ghosted." The globus pallidus on the distal side is seen, as has been explained previously.

The thalamus is usually described as the gateway to the cerebral cortex (see Figure 6.13). This description leaves out an important principle of thalamic function,

namely, that most thalamic nuclei that project to the cerebral cortex also receive input from that area—these are called reciprocal connections. This principle does not apply, however, to all the nuclei (see later). The various thalamic nuclei, and their functional component, are described in detail with Figure 4.3.

LOWER ILLUSTRATION

In this view, all structures of the proximal side have been removed, including the basal ganglia and the thalamus. The first structure seen adjacent to the midline is the thalamus (see Figure 1.7). It is said to be the size and shape of an almond. The next structure to be seen from this perspective is the globus pallidus (refer to the lower illustration in Figure 2.5A and the lowest illustration in Figure 2.5B). The amygdala is present in the temporal lobes, as previously shown.

Other parts of the diencephalon (not shown) include:

- The **hypothalamus**, one in each hemisphere (see Figure 1.7 and Figure 3.2). It is composed of a number of nuclei that regulate homeostatic functions of the body, including water balance. It is discussed with the limbic system in Section 4.
- The **pineal gland** (visible in Figure 1.9) is sometimes considered a part of the diencephalon (the epithalamus). This gland is thought to be involved with the regulation of our circadian rhythm.

The **subthalamic nucleus** is located below the thalamus. This nucleus is part of the circuitry associated with the basal ganglia (see Figure 5.14).

CLINICAL ASPECT

Many people are now taking melatonin, which is produced by the pineal, to regulate their sleep cycle and to overcome jet lag when travelling eastward.

ADDITIONAL DETAIL

As shown in the diagram, the diencephalon is situated within the brain below the level of the body of the lateral ventricles. In fact, the thalamus forms the "floor" of this part of the ventricle (see Figure 2.9A).

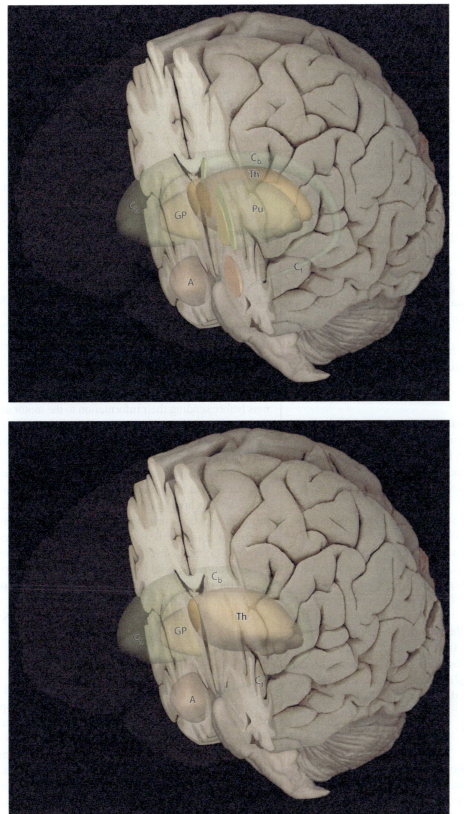

Th = Thalamus

Caudate n.
C$_h$ = Head
C$_b$ = Body
C$_t$ = Tail

Pu = Putamen
GP = Globus pallidus
A = Amygdala

FIGURE 2.6: The Diencephalon: Thalamus—Thalamus: Orientation

FIGURE 2.7—THE THALAMUS AND BASAL GANGLIA

ANATOMICAL RELATIONSHIPS (WITH FLAIR AND T1 MAGNETIC RESONANCE IMAGING SCANS)

UPPER ILLUSTRATION

This illustration has been created from the same perspectives as Figure 2.5B—from the lateral view. This illustration in fact looks almost identical to the top illustration in Figure 2.5B, showing the basal ganglia and ventricles of the proximal hemisphere. In this illustration, a structure is seen "behind" (i.e., on the medial aspect of the lentiform nucleus). This is the thalamus, which is located closest to the midline.

The magnetic resonance imaging (MRI) scan (a FLAIR image mode) has been done in the horizontal (axial) plane at the level of the interthalamic adhesion (see Figure 1.7) and with the thalamus indicated (compare with the similar cut in Figure 2.5B).

MIDDLE ILLUSTRATION

Removing the lentiform nucleus allows us to view the thalamus of the proximal hemisphere, as is shown in the middle illustration, and also a bit of the thalamus of the distal hemisphere. Looking back at Figure 2.5B, the basal ganglia nuclei are now in view—the (distal) globus pallidus and the putamen.

The (FLAIR) MRI scan beside this image that is done in the axial (horizontal) plane at the level just above the interthalamic adhesion (and includes the pineal gland) again shows the thalamus, one in each hemisphere, with the 3rd ventricle separating them.

Note to the Learner: In a horizontal section of the hemispheres, shown in the MRI scans beside the upper and middle illustrations (and in Figure 2.10B), the two thalami are located at the same level as the lentiform nucleus of the basal ganglia (see also Figure 2.10A). This important relationship is discussed with the internal capsule (see Figure 4.4).

LOWER ILLUSTRATION

The lower illustration shows the thalamus of the distal hemisphere, along with the caudate nucleus and the lateral ventricle of that side. The globus pallidus of that side is partially obscured by the thalamus. Note that the color of the thalamus is somewhat changed because this illustration includes the midline 3rd ventricle, situated "in front" of the thalamus in this view.

The T1 MRI scan beside this illustration is again done in a coronal plane at the level of the foramen of Monro and clearly shows how the thalamus forms the "floor" of the body of the lateral ventricles.

Recreating the location of these deep structures of the hemisphere, the thalamic nuclei occupy the most medial position adjacent to the midline and the 3rd ventricle, whereas the parts of the basal ganglia—the globus pallidus and the putamen—are located more laterally. Anteriorly, the head of the caudate nucleus is found in front of the thalamus (see Figure 2.10A and Figure 2.10B). The internal capsule separates these structures.

The major function of the thalamic nuclei is to process information before sending it on to the select area of the cerebral cortex (discussed with Figure 4.3; see also Figure 6.13). This is particularly so for all the sensory systems, except the olfactory sense. It is possible that crude forms of sensation including pain are "appreciated" in the thalamus, but localization of the sensation to a particular spot on the skin surface requires the involvement of the cortex. Similarly, two subsystems of the motor systems, the basal ganglia and the cerebellum, relay in the thalamus before sending their information to the motor areas of the cortex (see Figure 5.18). In addition, the limbic system has circuits that involve the thalamus (discussed in Section 4).

Other thalamic nuclei are related to areas of the cerebral cortex that are called association areas, vast areas of the cortex that are not specifically related either to sensory or motor functions (e.g., the dorsomedial nucleus and the prefrontal cortex, discussed with Figure 10.1B; see also Figure 6.13). Some nuclei of the thalamus, the intralaminar and reticular nuclei (see Figure 4.3), play an important role in the maintenance and regulation of the state of consciousness, and also possibly in attention, as part of the ascending reticular activating system (ARAS; see Figure 3.6A).

ADDITIONAL DETAIL

The nucleus accumbens, a functionally distinct part of the basal ganglia, is seen when looking at the medial aspect of the "distal" side, located where the head of the caudate nucleus becomes continuous with the putamen (in the middle illustration; see also Figure 2.5B). The caudate forms part of the dorsal striatum; by definition, the nucleus accumbens would then be part of the ventral striatum. The functional aspects of the nucleus accumbens are discussed with the limbic system (in Section 4).

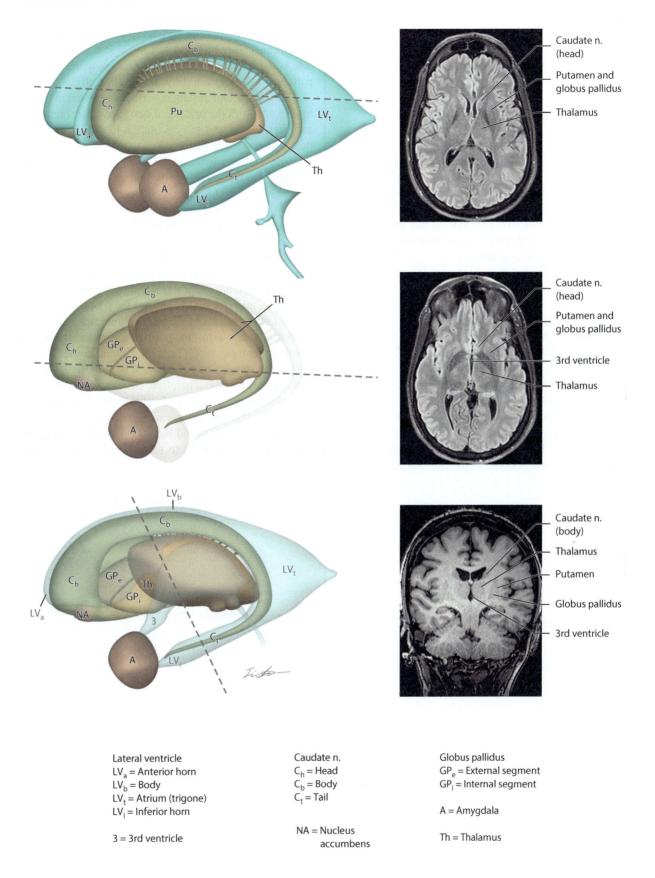

Lateral ventricle
LV$_a$ = Anterior horn
LV$_b$ = Body
LV$_t$ = Atrium (trigone)
LV$_i$ = Inferior horn

3 = 3rd ventricle

Caudate n.
C$_h$ = Head
C$_b$ = Body
C$_t$ = Tail

NA = Nucleus
    accumbens

Globus pallidus
GP$_e$ = External segment
GP$_i$ = Internal segment

A = Amygdala

Th = Thalamus

FIGURE 2.7: The Thalamus and Basal Ganglia—Anatomical Relationships (with FLAIR and T1 magnetic resonance imaging scans)

FIGURE 2.8—HEMISPHERES: INTERNAL STRUCTURES 1

ANTERIOR VIEW: VENTRICLES, WHITE MATTER, BASAL GANGLIA, THALAMUS

The composition of the hemispheres can now be better understood, with the introduction of the various components including the ventricular system, the white matter, the basal ganglia, and the thalamus. Note that the lateral ventricle appears twice on each side on this anterior view (see Figure 2.1B), first above within the hemispheres and again within the temporal lobe.

The only additional feature, already introduced previously, is the location of the next part of the ventricular system—the **third (3rd) ventricle**. This slit-like ventricle is located between the thalamic nuclei in the midline and is sometimes referred to as the ventricle of the diencephalic region of the brain (see Figure 1.9 and Figure 3.1).

The thalamus can now be visualized within the hemispheres adjacent to the midline, inferior to the lateral ventricle, and adjacent to the midline third ventricle.

Lateral to the thalamus (on both sides) is an area of white matter known as the **internal capsule**. Fibers from the spinal cord, the brainstem, and the thalamus *to* the cerebral cortex and fibers *from* the cerebral cortex to the thalamus, the brainstem, and the spinal cord pass through this "funnel" (see Figure 4.4). Lateral to the internal capsule is the lenticular (lentiform) nucleus, with the globus pallidus (medially) and the putamen (laterally).

The ventricular system is further discussed in the following part with the brainstem (see Figure 3.1, Figure 3.2, and Figure 3.3), and the circulation of the cerebrospinal fluid is described in Section 3.

The ventricles, white matter, basal ganglia and thalamus are shown in the following illustrations, anatomically with the brain sectioned in the coronal plane (see Figure 2.9A) and in the axial (horizontal) plane (see Figure 2.10A), and also radiographically (see Figure 2.9B and Figure 2.10B).

Note to the Learner: Many of the names of structures in the neuroanatomical literature are based upon earlier understandings of the brain, with terminology that is often descriptive and borrowed from other languages. As we learn more about the connections and functions of brain areas, this terminology often seems awkward if not obsolete, yet it persists. The term ganglia, in the strict use of the term, refers to a collection of neurons in the peripheral nervous system. Therefore, the anatomically correct name for the neurons in the forebrain should be the *basal nuclei*. Few texts use this term.

Most clinicians would be hard-pressed to change the name from basal ganglia to something else, so the traditional name remains.

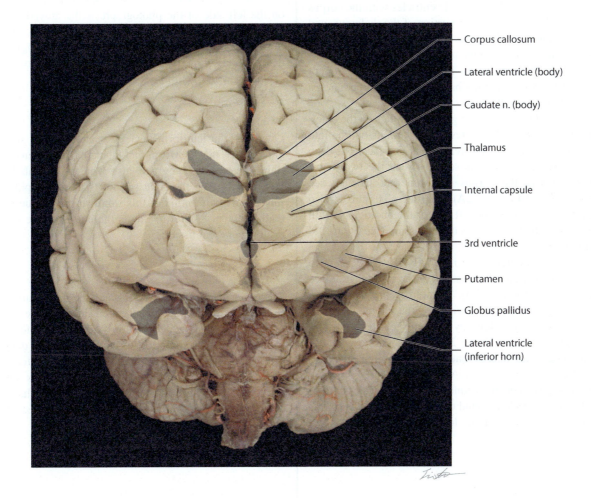

Corpus callosum

Lateral ventricle (body)

Caudate n. (body)

Thalamus

Internal capsule

3rd ventricle

Putamen

Globus pallidus

Lateral ventricle
(inferior horn)

FIGURE 2.8: Hemispheres: Internal Structures 1—Anterior View: Ventricles, White Matter, Basal Ganglia, Thalamus

FIGURE 2.9A—HEMISPHERES: INTERNAL STRUCTURES 2

CORONAL SECTION OF HEMISPHERES (PHOTOGRAPH)

This photographic view of the brain is sectioned in the coronal plane and shows the internal aspect of the hemispheres. On the dorsolateral view (the small figure on the upper left), the plane of section goes through both the frontal and the temporal lobes and would include the region of the basal ganglia. From the medial perspective (the small figure on the upper right), the section includes the body of the lateral ventricles with the corpus callosum above, the anterior portion of the thalamus, and the 3rd ventricle; the edge of the section also passes through the hypothalamus and the mammillary nucleus and includes the optic tracts. The section passes in front of the anterior part of the midbrain, the cerebral peduncles, and the front tip of the pons.

The cerebral cortex, the gray matter, lies on the external aspect of the hemispheres and follows its outline into the sulci in between, wherever there is a surface. The deep interhemispheric fissure is seen between the two hemispheres, above the corpus callosum (not labeled; see Figure 1.7 and Figure 2.2A). The lateral fissure is also present, well seen on the left side of the photograph with the insula within the depths of this fissure (see Figure 1.4 and Figure 6.3).

The white matter is seen internally; it is not possible to separate the various fiber systems of the white matter (see Figure 2.3 and Figure 2.4). Below the corpus callosum are the two spaces, the cavities of the lateral ventricle, represented at this plane by the body of the ventricles (see Figure 2.1A and Figure 2.1B). The small gray matter on the side of the lateral ventricle is the body of the caudate nucleus (see Figure 2.5A and Figure 2.5B). Because the section was not cut symmetrically, the inferior horn of the lateral ventricle is found only on the right side of this photograph, in the temporal lobe.

The brain is sectioned in the coronal plane through the diencephalic region. The gray matter on either side of the third ventricle is the thalamus (see Figure 1.9 and Figure 2.8). Lateral to this is a band of white matter, which by definition is part of the internal capsule, with the lentiform nucleus on its lateral side. The portion between the thalamus and lentiform nucleus is the posterior limb (refer to the section in the horizontal plane [Figure 2.10A, also Figure 4.4]).

The parts of the lentiform nucleus seen in this view include the putamen as well as the two portions of the globus pallidus, the external and internal segments. Because the brain has not been sectioned symmetrically, the two portions are more easily identified on the right side of the photograph.

The gray matter within the temporal lobe, best seen on the left side of the photograph, is the amygdala (see Figure 2.5A and Figure 2.6). It is easy to understand why this nucleus is considered one of the basal ganglia, by definition. Its function, as well as that of the fornix, is explained with the limbic system (see Section 4).

ADDITIONAL DETAIL

Lateral to the lentiform nucleus is another thin strip of gray matter, the claustrum (not labeled; better seen on the right side of the photograph; also seen and not labeled in Figure 2.10A, on the right side of the photograph). The functional contribution of this small strip of tissue is not really known. Lateral to this is the cortex of the insula, inside the lateral fissure.

CLINICAL ASPECT

The other major disease that affects the basal ganglia is **Huntington's chorea**, an inherited degenerative condition. This disease, which usually starts in mid-life, leads to severe motor dysfunction, as well as cognitive decline. The person whose name is most closely associated with this disease is Woody Guthrie, a legendary folk singer. There is now a genetic test for this disease that predicts whether the individual with a family history of Huntington's chorea will develop the disease.

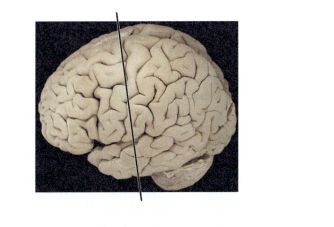
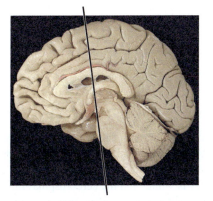

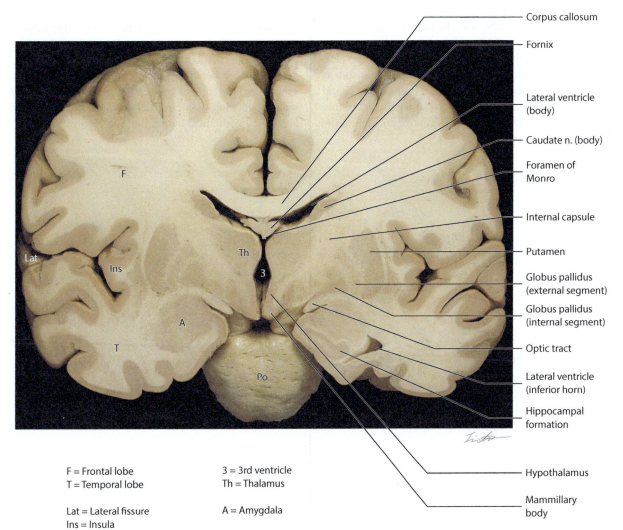

Corpus callosum

Fornix

Lateral ventricle
(body)

Caudate n. (body)

Foramen of
Monro

Internal capsule

Putamen

Globus pallidus
(external segment)

Globus pallidus
(internal segment)

Optic tract

Lateral ventricle
(inferior horn)

Hippocampal
formation

Hypothalamus

Mammillary
body

F = Frontal lobe
T = Temporal lobe

Lat = Lateral fissure
Ins = Insula

3 = 3rd ventricle
Th = Thalamus

A = Amygdala

Po = Pons

FIGURE 2.9A: Hemispheres: Internal Structures 2—Coronal Section of Hemispheres (photograph)

FIGURE 2.9B—HEMISPHERES: INTERNAL STRUCTURES 3

CORONAL VIEW (T1 MAGNETIC RESONANCE IMAGING SCAN)

This is a view of the brain similar to the brain section in Figure 2.9A, in the coronal plane but slightly more anteriorly. In this T1-weighted image, the cortex is gray, the white matter is white, and the cerebrospinal fluid is dark. Note that the tables of the skull are now dark, and the bone marrow is white. The dura of the meninges can also be visualized (see Figure 7.1 and Figure 7.2). The dural fold, the falx cerebri, is seen in the interhemispheric fissure (see Figure 7.4 and Figure 7.5), with the superior sagittal sinus seen in the midline, at the top of the falx cerebri.

The gray matter of the cerebral cortex and white matter can be easily differentiated. The corpus callosum is seen crossing the midline. The anterior horns of the lateral ventricles are seen, divided by the septum pellucidum into one for each hemisphere (see also Figure 8.6). Again, the plane of section has passed through the foramina of Monro, leading into the 3rd ventricle, which is situated between the thalamus on either side (see Figure 7.8). The head of the caudate nucleus is seen, as previously (see Figure 2.2B), protruding into the space of the anterior horn of the lateral ventricle. The lentiform nucleus is still present, and the thalamus (thalami) can be seen adjacent to the 3rd ventricle.

The section has passed through the posterior limb of the internal capsule (see Figure 4.4). Its fibers are seen as continuing to become the cerebral peduncle (see Figure 1.6, Figure 1.8, and Figure 3.1). The plane of section includes the lateral fissure and the insula (see lower illustration in Figure 2.5B, and Figure 2.7). The temporal lobe includes the hippocampal formation and the inferior horn of the lateral ventricle (see Figure 9.5A).

This view also includes the brainstem—the midbrain (the cerebral peduncles), the pontine region (the ventral portion), and the medulla. The tentorium cerebelli occupies the space between the inferior aspect of the hemispheres and the cerebellum (see Figure 7.5 and Figure 7.6), with its opening or incisura at the level of the midbrain (discussed with uncal herniation; see Figure 1.6).

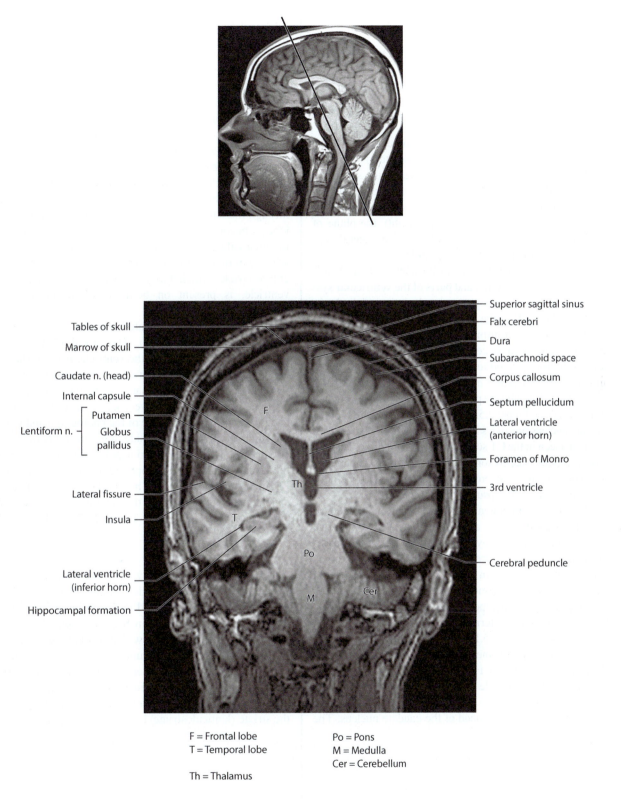

Tables of skull

Marrow of skull

Caudate n. (head)

Internal capsule

Putamen

Lentiform n.

Globus pallidus

Lateral fissure

Insula

Lateral ventricle (inferior horn)

Hippocampal formation

Superior sagittal sinus

Falx cerebri

Dura

Subarachnoid space

Corpus callosum

Septum pellucidum

Lateral ventricle (anterior horn)

Foramen of Monro

3rd ventricle

Cerebral peduncle

F = Frontal lobe
T = Temporal lobe

Th = Thalamus

Po = Pons
M = Medulla
Cer = Cerebellum

FIGURE 2.9B: Hemispheres: Internal Structures 3—Coronal View (T1 magnetic resonance imaging scan)

FIGURE 2.10A—HEMISPHERES: INTERNAL STRUCTURES 4

HORIZONTAL (AXIAL) SECTION OF HEMISPHERES (PHOTOGRAPH)

In this photograph, the brain has been sectioned in the horizontal (axial) plane. From the dorsolateral view (the small figure on the upper left), the level of the section is just above the lateral fissure, and at a slight angle downward from front to back. Using the medial view of the brain (the small figure on the upper right), the plane of section goes through the anterior horn of the lateral ventricle, the thalamus, and the occipital lobe.

This section exposes the white matter of the hemispheres, the basal ganglia, and parts of the ventricular system. Understanding this particular depiction of the brain is vital to the study of the forebrain. The structures seen in this view are also of immeasurable importance clinically, and this view is most commonly used in neuroimaging studies, both computed tomography (CT, see Figure 2.1A) and magnetic resonance imaging (MRI, see Figure 2.10B).

The basal ganglia are present when the brain is sectioned at this level (see Figure 2.5B and Figure 2.7). The head of the caudate nucleus protrudes into the anterior horn of the lateral ventricle (seen in the CT scan, Figure 2.1B, middle). The lentiform nucleus, shaped somewhat like a lens, is demarcated by white matter—anteromedially and posteromedially, which is the internal capsule (see Figure 4.4).

Because the putamen and caudate neurons are identical histologically (and also developmentally), the two nuclei have the same grayish coloration. The globus pallidus is functionally different; it contains many more fibers and therefore is lighter in color. Depending on the level of the section, it is sometimes possible (in this case on both sides) to see the two subdivisions of the globus pallidus, the internal and external segments (discussed with Figure 5.14 and Figure 5.18).

The white matter medial to the lentiform nucleus is the internal capsule (see Figure 2.9A and Figure 2.9B; also Figure 4.4). It is divisible into an anterior limb and a posterior limb and genu. The **anterior limb** separates the lentiform nucleus from the head of the caudate nucleus. The

posterior limb of the internal capsule separates the lentiform nucleus from the thalamus. Some strands of gray matter located within the internal capsule represent the strands of gray matter between the caudate and the putamen (as shown in Figure 2.5B). The base of the "V" which points medially is called the **genu**.

The anterior horn of the lateral ventricle is cut through its lowermost part and is seen in this photograph as a small cavity (see Figure 2.1A). The plane of the section has passed through the connections between each of the lateral ventricles and the 3rd ventricle, the foramina of Monro (see Figure 7.8). The section has also passed through the lateral ventricle as it curves into the temporal lobe to become the inferior horn of the lateral ventricle, the area called the atrium or trigone (better seen on the left side of this photograph). The choroid plexus of the lateral ventricle, which follows the inner curvature of the ventricle, is present on both sides (not labeled; see Figure 2.1B).

The 3rd ventricle is situated between the thalamus of both sides (see Figure 2.8). The pineal gland is seen attached to the back end of the ventricle. A bit of the cerebellar vermis is visible inferior to it (see also Figure 1.9).

The section is somewhat asymmetrical in that the occipital horn of the lateral ventricle is fully present in the occipital lobe on the left side of the photograph and not on the right side. On the right side, a group of fibers is seen streaming toward the posterior pole, and these represent the visual fibers, called the optic radiation (discussed with Figure 6.4 and Figure 6.5). The small size of the tail of the caudate nucleus alongside the lateral ventricle can be appreciated (see Figure 2.5A, Figure 2.5B, and Figure 9.7).

CLINICAL ASPECT

The major ascending sensory tracts from the thalamus to the cerebral cortex and the descending motor tracts from the cerebral cortex are found in the posterior limb of the internal capsule (reviewed in Section 2; see also Figure 4.4). This is the plane of view that would be used to look for small infarcts, called lacunes, in the posterior limb of the internal capsule (discussed with Figure 8.6 in Section 3). These infarcts are caused by occlusion of the small penetrating branches of the middle cerebral artery called the striate (lenticulostriate) branches (see Figure 8.6).

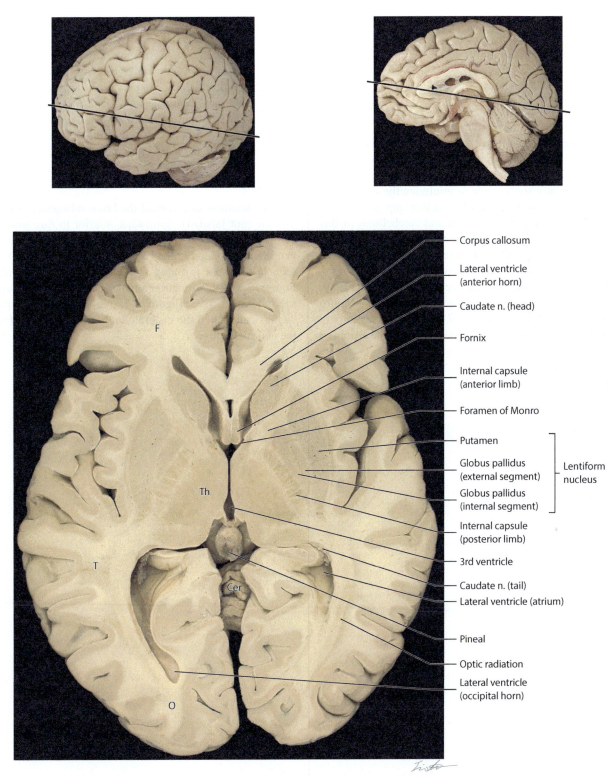

Corpus callosum

Lateral ventricle
(anterior horn)

Caudate n. (head)

Fornix

Internal capsule
(anterior limb)

Foramen of Monro

Putamen

Globus pallidus
(external segment)

Globus pallidus
(internal segment)

Lentiform
nucleus

Internal capsule
(posterior limb)

3rd ventricle

Caudate n. (tail)

Lateral ventricle (atrium)

Pineal

Optic radiation

Lateral ventricle
(occipital horn)

F = Frontal lobe   Th = Thalamus
T = Temporal lobe
O = Occipital lobe   Cer = Cerebellum

FIGURE 2.10A: Hemispheres: Internal Structures 4—Horizontal (Axial) Section of Hemispheres (photograph)

FIGURE 2.10B—HEMISPHERES: INTERNAL STRUCTURES 5

HORIZONTAL (AXIAL) VIEW (T2 MAGNETIC RESONANCE IMAGING SCAN)

This radiological view of the brain is not in exactly the same horizontal plane as the anatomical specimen shown in Figure 2.10A. The radiological images of the brain are often done at a slight angle to minimize the dense skull bones of the posterior cranial fossa that impair the viewing of the structures (brainstem and cerebellum) in this area. In this T2-weighted magnetic resonance image, the cerebrospinal fluid (CSF) is white, and the neuronal areas (cortex and basal ganglia) are gray.

The anterior horns of the lateral ventricle are present, and the section has passed through the foramina of Monro (see Figure 2.10A and Figure 7.8). The lateral ventricle posteriorly is cut at the level of its widening, the atrium or trigone, as it curves into the temporal lobe (see Figure 2.1A and Figure 2.10A).

The structures seen in the interior of the hemispheres are the ones already identified—caudate, putamen, and internal capsule. As discussed previously, the globus pallidus is sometimes idenfiable in CT and MRI scans. The 3rd ventricle is in the midline, between the thalami. The optic radiation can be visualized on the left side of this image (see Figure 6.4 and Figure 6.5).

ADDITIONAL DETAIL

A CSF cistern is seen behind the brain substance, in the midline (not labeled). Somewhat inferior to this would be the colliculi, also known as the tectum or the quadrigeminal plate (see Figure 1.9 and Figure 3.3). The CSF seen in this spot is found in a cistern (discussed with Figure 7.8)—the quadrigeminal cistern in the midsagittal views (see Figure 1.7, but not labeled); its "wings" are called the cisterna ambiens, a landmark for the neuroradiologist.

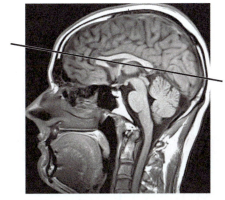

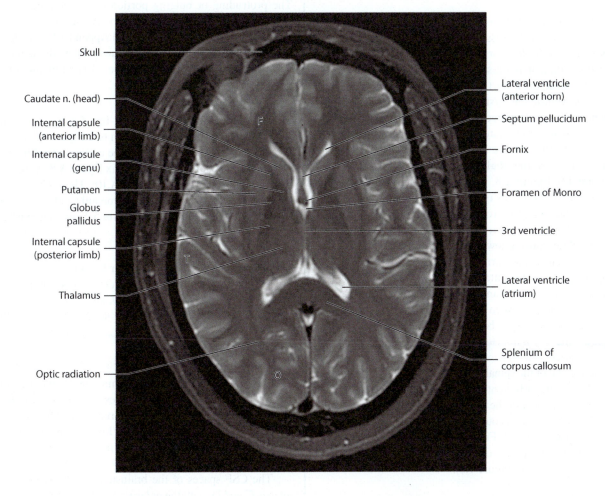

Skull

Caudate n. (head)

Internal capsule
(anterior limb)

Internal capsule
(genu)

Putamen

Globus
pallidus

Internal capsule
(posterior limb)

Thalamus

Optic radiation

Lateral ventricle
(anterior horn)

Septum pellucidum

Fornix

Foramen of Monro

3rd ventricle

Lateral ventricle
(atrium)

Splenium of
corpus callosum

F = Frontal lobe
T = Temporal lobe
O = Occipital lobe

FIGURE 2.10B: Hemispheres: Internal Structures 5—Horizontal (Axial) View (T2 magnetic resonance imaging scan)

Chapter 3

Internal Structures—Brainstem, Cerebellum, and Spinal Cord

FIGURE 3.1—BRAINSTEM 1

ANTERIOR VIEW (WITH T1 MAGNETIC RESONANCE IMAGING SCANS)

The illustration of the isolated brainstem and diencephalon is based on the dissection shown in Figure 1.8. It is used repeatedly for portraying aspects of the brainstem. At this point, it is presented with the focus now on its internal structures, starting with the ventricular system.

The ventricular system is shown "inside" the brainstem, where it is situated posteriorly.

The ventricular system continues from the diencephalon, where the slit-like **3rd ventricle** lies in the midline and separates the thalamus of each hemisphere (see Figure 1.9, Figure 2.8, Figure 2.9A, and Figure 2.10A). The system now enters into the brainstem. In the midbrain region, the system narrows considerably and is known as the **cerebral aqueduct**, otherwise called the aqueduct of the midbrain, also known as the aqueduct of Sylvius. (An appreciation of the narrow aqueduct can be obtained in Figure 3.2, a mid-sagittal view of the brainstem; see also Appendix Figure A.3 and Appendix Figure A.4.)

In the area between the pons and the cerebellum, the ventricular space expands considerably into a lozenge-shaped space—the **4th ventricle** (see Figure 3.3). At the bottom of this ventricle, the system narrows again and becomes the central canal of the spinal cord (discussed with Figure 7.8).

From the 4th ventricle, cerebrospinal fluid (CSF) leaves the interior of the brain and flows into the subarachnoid space—which is outside the brain—via three foramina, as discussed with Figure 3.2 and with Figure 7.8.

Other structures of the brainstem can now be explained (see also Figure 1.8).

THE CEREBRAL PEDUNCLE

The descending fibers present in the internal capsule (see Figure 4.4) continue in the brainstem (see the radiograph B below and also see Figure 2.9B, Figure 9.5A, and Figure 9.5B) as the cerebral peduncle (see Figure 4.2C; also Appendix Figure A.3 and Appendix Figure A.4). This includes the cortico-spinal tract (see Figure 5.9) and the cortico-pontine and cortico-bulbar fibers (see Figure 5.10 and Figure 5.15).

THE PONS

The protruding or bulging portion of the pontine region (see Figure 3.2) is created by a massive set of nuclei, the pontine nuclei (see Figure 4.2B and Appendix Figure A.6). These are relay nuclei in the pathway from the cerebral cortex to the cerebellum (see Figure 5.10 and Figure 5.15).

THE PYRAMID

The structure labeled the pyramid has this (approximate) shape when seen in an axial section (see Figure 4.2A). It is noteworthy because the fibers of the cortico-spinal tract in fact create the pyramid (see Appendix Figure A.8, Appendix Figure A.9, and Appendix Figure A.10). These fibers cross (decussate) at the lower level of the medulla and become the lateral cortico-spinal tract (see Figure 5.9 and Figure 6.12).

THE OLIVE

This nucleus of the medulla, better known as the inferior olivary nucleus, is large enough to create this "bump" (see Figure 4.2A, Appendix Figure A.9, and Appendix Figure A.10). It is one of the nuclei that sends afferents to the cerebellum (see Figure 5.15).

RADIOLOGY

The four magnetic resonance imaging scans are T1-weighted images of the hemispheres and the brainstem in the coronal plane, including the cerebellum. The shape of the brainstem can be seen in the "locator" image (and in Figure 3.2). The CSF in this instance is dark (as seen in the lateral ventricles of the hemispheres).

The CSF spaces of the brainstem can be seen in the images C and D, with the enlargement of the 4th ventricle.

Subsequent illustrations detail the various nuclei of the brainstem. In Section 2, all the pathways ascending and descending through the midbrain are described (see Figure 6.11 and Figure 6.12), as well as the cerebellar connections (see Figure 5.15 and Figure 5.17). Further details of the brainstem are found in the Appendix.

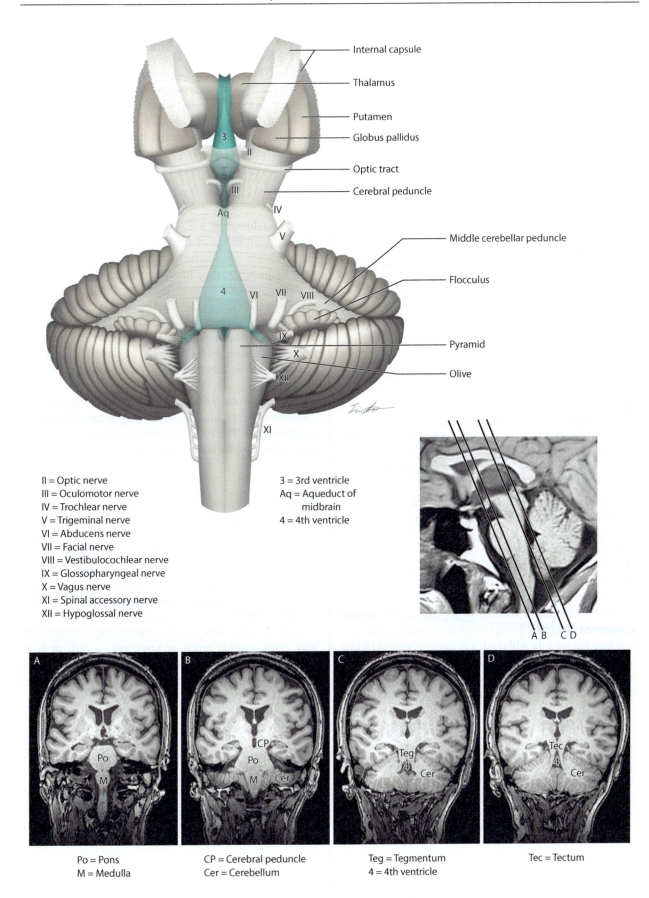

II = Optic nerve
III = Oculomotor nerve
IV = Trochlear nerve
V = Trigeminal nerve
VI = Abducens nerve
VII = Facial nerve
VIII = Vestibulocochlear nerve
IX = Glossopharyngeal nerve
X = Vagus nerve
XI = Spinal accessory nerve
XII = Hypoglossal nerve

3 = 3rd ventricle
Aq = Aqueduct of
       midbrain
4 = 4th ventricle

Po = Pons
M = Medulla

CP = Cerebral peduncle
Cer = Cerebellum

Teg = Tegmentum
4 = 4th ventricle

Tec = Tectum

FIGURE 3.1: Brainstem 1—Anterior View (with T1 magnetic resonance imaging scans)

FIGURE 3.2—BRAINSTEM 2

MID-SAGITTAL VIEW (PHOTOGRAPH WITH T1 MAGNETIC RESONANCE IMAGING SCAN)

This mid-sagittal view of the brain is a higher magnification of the view presented in Figure 1.7. The section of the brain goes through the interhemispheric fissure, the corpus callosum (see Figure 2.2A) and the midline 3rd ventricle, as well as sectioning the brainstem in the midline.

On this view, the thalamic portion of the diencephalon is separated from the hypothalamic part by a groove, the **hypothalamic sulcus**. This sulcus starts at the foramen of Monro (the interventricular foramen, discussed with the ventricles; see Figure 2.9B and Figure 7.8) and ends at the aqueduct of the midbrain. The optic chiasm is found at the anterior aspect of the hypothalamus, and behind it is the mammillary body (see Figure 1.5 and Figure 1.6).

The three parts of the brainstem can be distinguished on this view—the midbrain, the pons with its bulge anteriorly, and the medulla (refer to the ventral view shown in Figure 3.1 and in Figure 1.8). Through the midbrain is a narrow channel for cerebrospinal fluid (CSF), the aqueduct of the midbrain. The area of the midbrain behind the aqueduct includes the superior and inferior colliculi (see Figure 1.9), referred to as the tectum or tectal plate (see Figure 3.3).

The aqueduct connects the 3rd ventricle with the 4th ventricle, a space with CSF that separates the pons and medulla from the cerebellum (see later). CSF escapes from the ventricular system at the bottom of the 4th ventricle through the foramen of Magendie (see Figure 7.8), and the ventricular system continues as the narrow central canal of the spinal cord (see also Figure 7.8).

The cerebellum lies behind (or above) the 4th ventricle. It has been sectioned through its midline portion, the **vermis** (see Figure 1.9). Although it is not necessary to name all its various parts, it is useful to know two of them—the lingula and the nodulus. (The reason for this will become evident when describing the cerebellum; see Figure 5.17.)

The **tonsil** of the cerebellum can also be seen in this view (see Figure 1.7; also Figure 1.9 and Figure 5.16).

The cut edge of the tentorium cerebelli, one of the major main folds of dura, is seen separating the cerebellum from the occipital lobe (discussed with the meninges in Section 3; see Figure 7.4, Figure 7.5, and Figure 7.6). This view clarifies the separation of the supratentorial space, namely, the cerebral hemispheres, from the infratentorial space, the brainstem, and the cerebellum in the posterior cranial fossa.

RADIOLOGY

The lower image is a T1-weighted magnetic resonance imaging scan image of the brainstem in the mid-sagittal plane. This is an extremely important image, one that is most frequently seen in the clinical setting.

The corpus callosum is clearly seen, as well as the fornix, with the septum pellucidum attached between the two, thus separating one lateral ventricle from the other. The diencephalon is situated below the fornix outlining the "space" occupied by the 3rd ventricle. CSF (dark) can be seen in the aqueduct of the midbrain and filling the 4th ventricle.

The space outside the brainstem also contains CSF (explained with Figure 7.8), with enlargements of that space called cisterns. The large cistern below the cerebellum and behind the medulla is the important **cistern magna**.

CLINICAL ASPECT

Should there be an increase in the mass of tissue occupying the posterior cranial fossa (e.g., tumor, hemorrhage), the cerebellum would be pushed downward. This would force the cerebellar tonsils into the foramen magnum, thereby compressing the medulla. The compression, if severe, could lead to a compromising of function of the vital centers located in the medulla (discussed with Figure 3.6A).

The complete syndrome is known as **tonsillar herniation**, or coning. This is a life-threatening situation that may cause cardiac arrest or respiratory arrest, or both. (This will be further discussed in the context of Increased Intracranial Pressure, ICP, in the introduction to Section 3 and with Figure 7.1.)

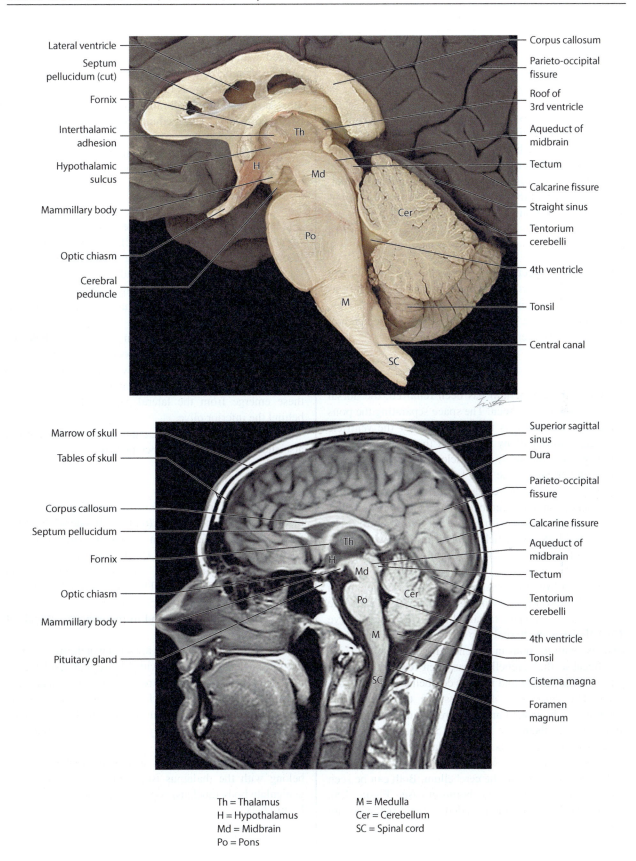

Lateral ventricle
Septum pellucidum (cut)
Fornix
Interthalamic adhesion
Hypothalamic sulcus
Mammillary body
Optic chiasm
Cerebral peduncle

Corpus callosum
Parieto-occipital fissure
Roof of 3rd ventricle
Aqueduct of midbrain
Tectum
Calcarine fissure
Straight sinus
Tentorium cerebelli
4th ventricle
Tonsil
Central canal

Th
H
Md
Po
Cer
M
SC

Marrow of skull
Tables of skull
Corpus callosum
Septum pellucidum
Fornix
Optic chiasm
Mammillary body
Pituitary gland

Superior sagittal sinus
Dura
Parieto-occipital fissure
Calcarine fissure
Aqueduct of midbrain
Tectum
Tentorium cerebelli
4th ventricle
Tonsil
Cisterna magna
Foramen magnum

Th
H
Md
Po
Cer
M
SC

Th = Thalamus
H = Hypothalamus
Md = Midbrain
Po = Pons

M = Medulla
Cer = Cerebellum
SC = Spinal cord

FIGURE 3.2: Brainstem 2—Mid-Sagittal View (photograph with T1 magnetic resonance imaging scan)

FIGURE 3.3—BRAINSTEM 3

DORSAL VIEW: CEREBELLUM REMOVED

This diagram shows the brainstem and 4th ventricle from the dorsal perspective, with the cerebellum removed. A similar view of the brainstem is used for some of the later diagrams (see Figure 6.11 and Figure 6.12). This dorsal perspective is useful for presenting the combined visualization of many of the cranial nerve nuclei and the various pathways of the brainstem.

MIDBRAIN LEVEL

The posterior aspect of the midbrain has the superior and inferior colliculi, as previously seen (see Figure 1.9), as well as the emerging fibers of cranial nerve (CN) IV, the trochlear nerve. The posterior aspect of the cerebral peduncle is also seen.

PONTINE LEVEL

Now that the cerebellum has been removed, the dorsal aspect of the pons is seen. The space separating the pons from the cerebellum is the 4th ventricle—the ventricle has been "unroofed" by removal of the cerebellum. The upper portion of the 4th ventricle is still covered by a sheet of nervous tissue that bears the name **superior medullary velum**; more relevant, it contains an important connection of the cerebellum, the superior cerebellar peduncles (discussed with Figure 5.17 and Figure 6.11). The choroid plexus, which is found in the lower half of the roof of the 4th ventricle (see Figure 7.8), has also been removed.

As seen from this perspective, the 4th ventricle has a "floor." Noteworthy are two large bumps, one on each side of the midline called the facial colliculi, where facial nerve CN VII makes an internal loop (discussed with Figure 6.12 and also with the pons in Appendix Figure A.7).

Because the cerebellum has been removed, the cut edges of the middle and inferior cerebellar peduncles are seen. The **cerebellar peduncles** are the connections between the brainstem and the cerebellum, and there are three pairs of them (see Figure 5.15). The **inferior** cerebellar peduncle connects the medulla and the cerebellum, and the prominent **middle** cerebellar peduncle brings fibers from the pons to the cerebellum. Both can be seen in the ventral view of the brainstem (see Figure 1.8). Details of the information carried in these pathways are

outlined when the functional aspects of the cerebellum are studied with the motor systems (see Figure 5.15). The **superior** cerebellar peduncles convey fibers from the cerebellum to the thalamus that pass through the roof of the 4th ventricle and the midbrain to synapse in the thalamus (see Figure 5.17). This peduncle can be visualized only from this perspective.

CN V emerges through the middle cerebellar peduncle (see also Figure 1.8 and Figure 3.4).

MEDULLARY LEVEL

The lower part of the 4th ventricle separates the medulla from the cerebellum (see Figure 3.2). The special structures below the 4th ventricle are two large protuberances on either side of the midline—the **gracilis** and **cuneatus nuclei**, relay nuclei that belong to the ascending somatosensory pathway (discussed with Figure 1.9, Figure 5.2, and Figure 6.11; see also Appendix Figure A.10).

The cranial nerves seen from this view include the entering nerve CN VIII (vestibulocochlear nerve). More anteriorly, from this oblique view, are the fibers of the glossopharyngeal (CN IX) and vagus (CN X) nerves as these emerge from the lateral aspect of the medulla, behind the inferior olive.

A representative cross-section of the spinal cord is also shown from this dorsal perspective.

ADDITIONAL DETAILS

There are fibers (not labeled) shown crossing the floor of the 4th ventricle, continuing (on the left side of the illustration) from CN VIII and an expansion (which is one of cochlear nuclei, see Figure 6.1). The fibers are part of the auditory projection, called the dorsal acoustic stria (described with Figure 6.1). These fibers of CN VIII, the auditory portion, take an alternative route to relay in the lower pons before they ascend to the inferior colliculi of the midbrain.

Two additional structures are shown in the midbrain—the red nucleus (described with Figure 4.2C and Figure 5.11; see also Appendix Figure A.3) and the brachium of the inferior colliculus, which is a connecting pathway between the inferior colliculus and the medial geniculate body, part of the auditory system (fully described with Figure 6.1 and Figure 6.2).

The medial and lateral geniculate bodies (nuclei) belong with the thalamus (see Figure 4.3). The lateral geniculate body (nucleus) is part of the visual system (see Figure 6.4).

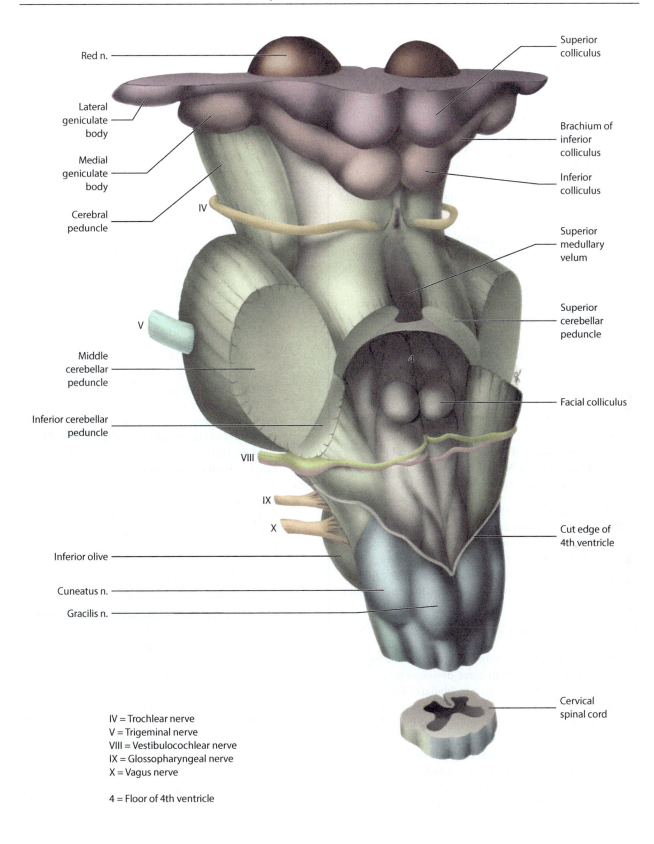

Red n.

Lateral
geniculate
body

Medial
geniculate
body

Cerebral
peduncle

IV

V

Middle
cerebellar
peduncle

Inferior cerebellar
peduncle

VIII

IX

X

Inferior olive

Cuneatus n.

Gracilis n.

Superior
colliculus

Brachium of
inferior
colliculus

Inferior
colliculus

Superior
medullary
velum

Superior
cerebellar
peduncle

Facial colliculus

Cut edge of
4th ventricle

Cervical
spinal cord

IV = Trochlear nerve
V = Trigeminal nerve
VIII = Vestibulocochlear nerve
IX = Glossopharyngeal nerve
X = Vagus nerve

4 = Floor of 4th ventricle

FIGURE 3.3: Brainstem 3—Dorsal View: Cerebellum Removed

FIGURE 3.4—BRAINSTEM 4

CRANIAL NERVE NUCLEI: SENSORY

The cranial nerve nuclei with sensory functions are discussed in this diagram (see also Appendix Figure A.1). The olfactory nerve (cranial nerve [CN] I) and the optic nerve (CN II) are not attached to the brainstem and are not considered at this stage.

Sensory information from the region of the head and neck includes the following:

- **Somatic afferents**: general sensations, consisting of touch (both discriminative and crude touch), pain, and temperature. These afferents come from the skin of the scalp and face and the mucous membranes of the head region via branches of the trigeminal nerve, CN V.

- **Visceral afferents**: sensory input from the pharynx and other homeostatic receptors of the neck (e.g., for blood pressure) and from the organs of the thorax and abdomen. This afferent input is carried mainly by the vagus, CN X, but also by the glossopharyngeal nerve, CN IX.

- **Special senses**: auditory (hearing) and vestibular (balance) afferents with the vestibulocochlear nerve, CN VIII, as well as the special sense of taste with CN VII and CN IX.

This diagram shows the location of the sensory nuclei of the cranial nerves superimposed on the ventral view of the brainstem (on one side only). These nuclei are also shown in Figure 6.11, in which the brainstem is presented from a dorsal perspective. The details of the location of the cranial nerve nuclei within the brainstem are described in the Appendix.

CN V: TRIGEMINAL NERVE

The major sensory nerve of the head region is the trigeminal nerve, CN V, through its three divisions peripherally (ophthalmic, maxillary, and mandibular). The sensory ganglion for this nerve, the trigeminal ganglion, is located inside the skull. The nerve supplies the skin of the scalp and face, the conjunctiva of the eye and the eyeball, the teeth, and the mucous membranes inside the head, including the surface of the tongue (but not taste—see later).

The sensory components of the trigeminal nerve are found at several levels of the brainstem (see trigeminal pathways in Figure 5.4, Figure 5.5, and Figure 6.11):

- The **principal nucleus**, which is responsible for the discriminative aspects of touch, is located at the mid-pontine level, adjacent to the motor nucleus of CN V.

- A long column of cells that relays pain and temperature information, known as the **spinal nucleus of V** or the **descending trigeminal nucleus**, descends through the medulla and reaches the upper cervical levels of the spinal cord.

- Another group of cells extends into the midbrain region, the **mesencephalic** nucleus of V. These cells appear to be similar to neurons of the dorsal root ganglia and are thought to be the sensory proprioceptive neurons for the muscles of mastication.

Note to the Learner: The location of the sensory nucleus of the CN V inside the brainstem does not correspond exactly to the level of attachment of the nerve to the brainstem as seen externally.

CN VIII: VESTIBULOCOCHLEAR NERVE

- **Cochlear nuclei**: The auditory fibers from the spiral ganglion in the cochlea are carried to the central nervous system in CN VIII, and these fibers form their first synapses in the cochlear nuclei as it enters the brainstem at the uppermost level of the medulla (see Figure 6.1). The auditory pathway is presented in Section 2 (see also Figure 6.2 and Figure 6.3).

- **Vestibular nuclei**: Vestibular afferents enter the central nervous system as part of CN VIII. There are four nuclei—the medial and inferior located in the medulla, the lateral located at the ponto-medullary junction, and the small superior nucleus located in the lower pontine region. The vestibular afferents terminate in these nuclei. The vestibular nuclei are further discussed in Section 2 with the motor systems (see Figure 5.13) and with the special senses (see Figure 6.8).

VISCERAL AFFERENTS AND TASTE: SOLITARY NUCLEUS

The special sense of taste from the surface of the tongue is carried in CN VII and CN IX, and these terminate in the solitary nucleus in the medulla (see Appendix Figure A.8).

CLINICAL ASPECT

Trigeminal neuralgia is discussed with the trigeminal pathways (see Figure 5.4).

ADDITIONAL DETAIL

The visceral afferents with CN IX and CN X from the pharynx, larynx, and internal organs are also received in the solitary nucleus (see Appendix Figure A.8).

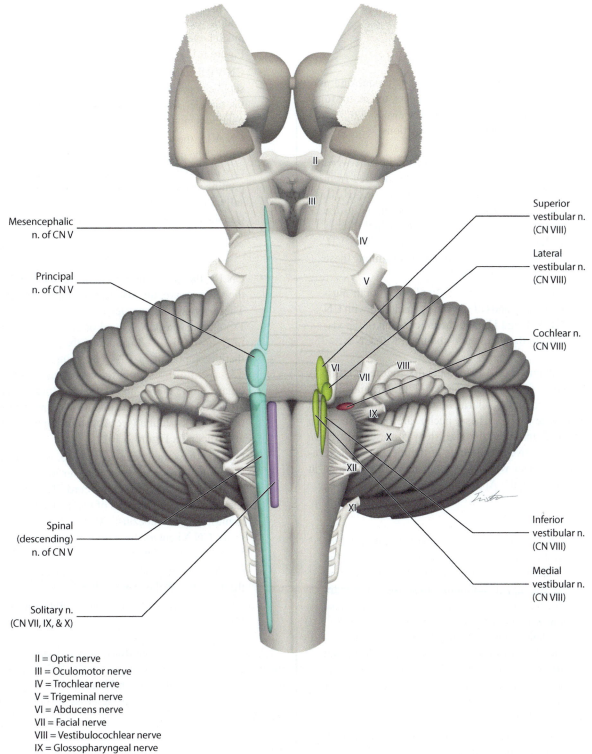

Mesencephalic
n. of CN V

Principal
n. of CN V

Spinal
(descending)
n. of CN V

Solitary n.
(CN VII, IX, & X)

Superior
vestibular n.
(CN VIII)

Lateral
vestibular n.
(CN VIII)

Cochlear n.
(CN VIII)

Inferior
vestibular n.
(CN VIII)

Medial
vestibular n.
(CN VIII)

II = Optic nerve
III = Oculomotor nerve
IV = Trochlear nerve
V = Trigeminal nerve
VI = Abducens nerve
VII = Facial nerve
VIII = Vestibulocochlear nerve
IX = Glossopharyngeal nerve
X = Vagus nerve
XI = Spinal accessory nerve
XII = Hypoglossal nerve

FIGURE 3.4: Brainstem 4—Cranial Nerve Nuclei: Sensory

FIGURE 3.5—BRAINSTEM 5

CRANIAL NERVE NUCLEI: MOTOR

Remembering that each cranial nerve is unique and may have one or more functional components, the motor functions of the cranial nerves are now reviewed.

There are two kinds of motor functions:

1. Most neuroanatomy texts distinguish between the **motor** supply to the muscles derived from **somites** (including cranial nerve [CN] III, IV, VI, and XII) and the motor supply to the muscles derived from the branchial arches (called **branchiomotor**, including CN V, VII, IX, and X). No distinction is made among these muscle types in this atlas.

2. The **parasympathetic** supply to smooth muscles and glands of the head is included with CN III, VII, and IX, and the innervation of the viscera in the thorax and abdomen with CN X.

This diagram (see also Appendix Figure A.1) shows the location of the motor nuclei of the cranial nerves, superimposed on the ventral view of the brainstem (again on one side only). These nuclei are also shown in Figure 6.12, in which the brainstem is presented from a dorsal perspective. The details of the location of the cranial nerve nuclei within the brainstem are described at various levels in the Appendix.

MIDBRAIN LEVEL

- CN III, the oculomotor nerve, has both motor and autonomic fibers. The motor nucleus that supplies most of the muscles of the eye is found at the upper midbrain level (see Appendix Figure A.3). The parasympathetic nucleus, known as the **Edinger-Westphal (E-W)** nucleus, supplies the pupillary constrictor muscle (the reflex response of the pupil to light) and the muscle that controls the curvature of the lens; both are part of the accommodation reflex (discussed with Figure 6.7).

- CN IV, the trochlear nerve, is a motor nerve to one eye muscle, the superior oblique muscle. The trochlear nucleus is found at the lower midbrain level (see Appendix Figure A.4).

PONTINE LEVEL

- CN V, the trigeminal nerve, has a motor component to the muscles of mastication (chewing). The nucleus is located at the mid-pontine level; the small motor nerve is attached to the brainstem at this level, along the route of the middle cerebellar peduncle, adjacent to the much larger sensory root (not shown here—see Figure 6.12).

- CN VI, the abducens nerve, is a motor nerve that supplies one extraocular muscle, the lateral rectus muscle. The nucleus is located in the lower pontine region.

- CN VII, the facial nerve, is a mixed cranial nerve. The motor nucleus that supplies the muscles of facial expression is found at the lower pontine level. The parasympathetic fibers, to salivary and lacrimal glands, are part of CN VII (see the additional details section later).

MEDULLARY LEVEL

- CN IX, the glossopharyngeal nerve, and CN X, the vagus nerve, are also mixed cranial nerves. These supply the muscles of the pharynx (CN IX) and larynx (CN X) and originate from the **nucleus ambiguus**. In addition, the parasympathetic component of CN X, coming from the **dorsal motor nucleus** of the vagus, supplies the organs of the thorax and abdomen. Both nuclei are found throughout the middle and lower portions of the medulla (see Appendix Figure A.9 and Appendix Figure A.10).

- CN XI, the spinal accessory nerve, originates from a cell group in the upper four to five segments of the cervical spinal cord. This nerve supplies the large muscles of the neck (the sternomastoid and trapezius). As mentioned previously, CN XI enters the skull and exits again, as if it were a true cranial nerve.

- CN XII, the hypoglossal nerve, innervates all the muscles of the tongue. It has an extended nucleus in the medulla situated alongside the midline.

Note to the Learner: In this diagram, it appears that the nucleus ambiguus is the origin for CN XII. This is not the case but is a visualization problem. A clearer view can be found in Figure 6.11 and in the cross-sectional views (see Appendix Figure A.9).

ADDITIONAL DETAIL

Two small parasympathetic nuclei are also shown but are rarely identified in brain sections—the superior and inferior salivatory nuclei. The superior nucleus supplies secretomotor fibers for CN VII (to the submandibular and sublingual salivary glands, as well as nasal and lacrimal glands). The inferior nucleus supplies the same fibers for CN IX (to the parotid salivary gland).

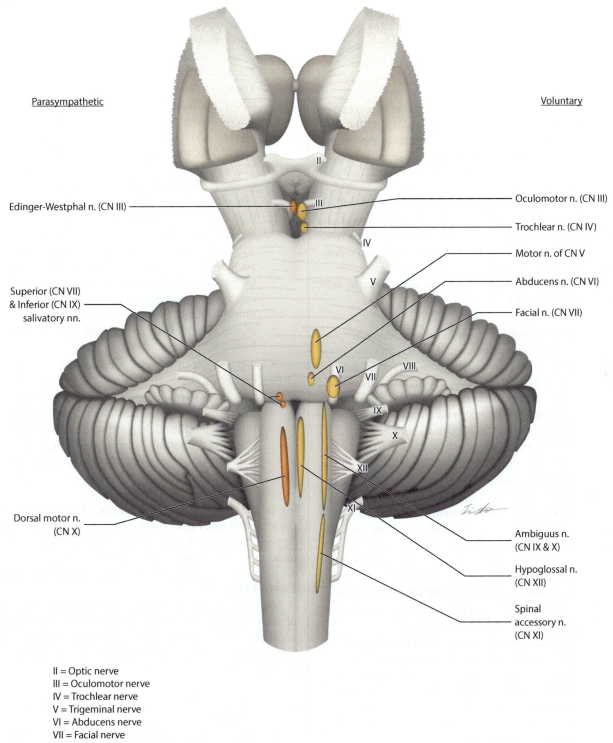

Parasympathetic

Voluntary

Edinger-Westphal n. (CN III)

Oculomotor n. (CN III)

Trochlear n. (CN IV)

Motor n. of CN V

Superior (CN VII)
& Inferior (CN IX)
salivatory nn.

Abducens n. (CN VI)

Facial n. (CN VII)

Dorsal motor n.
(CN X)

Ambiguus n.
(CN IX & X)

Hypoglossal n.
(CN XII)

Spinal
accessory n.
(CN XI)

II = Optic nerve
III = Oculomotor nerve
IV = Trochlear nerve
V = Trigeminal nerve
VI = Abducens nerve
VII = Facial nerve
VIII = Vestibulocochlear nerve
IX = Glossopharyngeal nerve
X = Vagus nerve
XI = Spinal accessory nerve
XII = Hypoglossal nerve

FIGURE 3.5: Brainstem 5—Cranial Nerve Nuclei: Motor

FIGURE 3.6A—RETICULAR FORMATION 1

RETICULAR FORMATION: ORGANIZATION

The **reticular formation (RF)** is the name of a group of neurons found throughout the brainstem. Using the ventral view of the brainstem, the RF occupies the central portion or core area of the brainstem, the tegmentum (see Appendix Figure A.2), from midbrain to medulla (see also brainstem cross-sections in the Appendix).

This collection of neurons is a phylogenetically old set of neurons that functions as a network or reticulum, from which it derives its name. The RF receives afferents from most of the sensory systems (see Figure 3.6B), and it projects to virtually all parts of the nervous system.

Functionally, it is possible to localize different subgroups within the RF:

- Cardiac and respiratory "centers": Subsets of neurons within the medullary RF and also in the pontine region are responsible for the control of the vital functions of heart rate and respiration. The importance of this knowledge was discussed in reference to the clinical emergency, tonsillar herniation (discussed with Figure 3.2 and Figure 7.1; also discussed in the context of increased intracranial pressure (ICP) in the Introduction to Section 3).

- Motor areas: Both the pontine and medullary nuclei of the reticular formation contribute to motor control via the cortico-reticulo-spinal system (discussed in Motor Systems—Introduction in Chapter 5 and also with Figure 5.12A and Figure 5.12B). In addition, these nuclei exert a very significant influence on muscle tone, which is very important clinically (discussed with Figure 5.12B).

- Ascending projection system: Fibers from the RF ascend to the thalamus and project to various non-specific thalamic nuclei (e.g., intralaminar; see Figure 4.3). From these nuclei, there is a diffuse distribution of connections to all parts of the cerebral cortex. This whole system is concerned with consciousness and is known as the **ascending reticular activating system (ARAS)**.

- Pre-cerebellar nuclei: There are numerous nuclei in the brainstem that are located within the boundaries of the RF that project to the cerebellum. These are not always included in discussions of the RF.

It is also possible to describe the RF topographically. The neurons appear to be arranged in three longitudinal sets; these are shown on the left side of this illustration:

- The **lateral group** consists of small neurons. These are the neurons that receive the various inputs to the RF, including those from the anterolateral system (pain and temperature; see Figure 5.3), the trigeminal pathway (see Figure 5.4), and auditory and visual input.

- The next group is the **medial group**. These neurons are larger and project their axons upward and downward. The ascending projection from the midbrain area is particularly involved with the consciousness system, the ARAS. Nuclei within this group, notably the nucleus gigantocellularis of the medulla and the pontine reticular nuclei, caudal (lower) and oral (upper) portions, give origin to the two reticulo-spinal tracts (discussed with Figure 3.6B; see also Figure 5.12A and Figure 5.12B).

- Another set of neurons occupies the midline region of the brainstem, the **raphe nuclei**, which use the monoamine serotonin for neurotransmission. The best-known nucleus of this group is the nucleus raphe magnus, which plays an important role in the descending pain modulation system (discussed with Figure 5.6).

In addition, both the periaqueductal gray located in the midbrain (see Figure 3.6B, Appendix Figure A.3, and Appendix Figure A.4) and the locus ceruleus in the upper pons (see Figure 3.6B and Appendix Figure A.5) are considered part of the RF (discussed with Figure 3.6B).

In summary, the RF is connected with almost all parts of the central nervous system (CNS). Although it has a generalized influence within the CNS, it also contains subsystems that are directly involved in specific functions. The most clinically significant aspects are:

- The cardiac and respiratory centers in the medulla.

- The descending systems in the pons and medulla that participate in motor control and influence muscle tone.

- The ascending pathways in the upper pons and midbrain that contribute to the consciousness system.

- The pain modulation pathway.

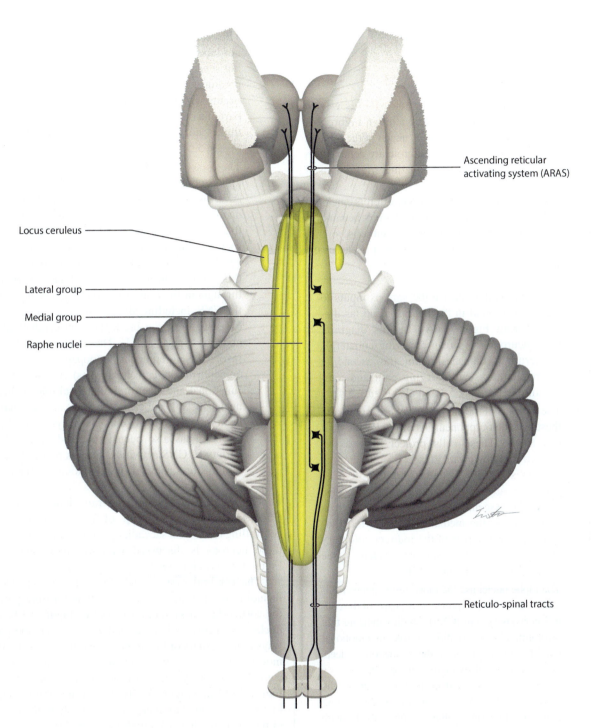

Ascending reticular activating system (ARAS)

Locus ceruleus

Lateral group

Medial group

Raphe nuclei

Reticulo-spinal tracts

FIGURE 3.6A: Reticular Formation 1—Reticular Formation: Organization

FIGURE 3.6B—RETICULAR FORMATION 2

RETICULAR FORMATION: NUCLEI

In this diagram, the **reticular formation (RF)** is viewed from the dorsal (posterior) perspective (as in Figure 3.3, Figure 6.11, and Figure 6.12). Various nuclei of the RF that have a significant (known) functional role are depicted, as well as the descending tracts emanating from some of these nuclei.

Functionally, there are afferent and efferent nuclei in the RF and groups of neurons that are distinct because of the catecholamine neurotransmitter used, either serotonin or norepinephrine. The afferent and efferent nuclei of the RF include the following:

- The neurons that receive the various inputs to the RF are found in the lateral group (as discussed with Figure 3.6A). In this diagram, these neurons are shown receiving collaterals (or terminal branches) from the ascending anterolateral system, carrying pain and temperature (see Figure 5.3 and also Figure 5.4).

- The neurons of the medial group are larger, and these are the output neurons of the RF, at various levels. These cells project their axons upward and downward. The **nucleus gigantocellularis** of the medulla and the **pontine reticular nuclei, caudal** and **oral** portions give rise to the descending tracts that emanate from these nuclei—the medial and lateral reticulospinal pathways, part of the indirect voluntary and non-voluntary motor system (see Figure 5.12A and Figure 5.12B).

- The raphe nuclei use the monoamine neurotransmitter serotonin and project to all parts of the central nervous system (CNS). Studies indicate that serotonin plays a significant role in emotional equilibrium, as well as in the regulation of sleep. One special nucleus of this group, the **nucleus raphe magnus** located in the upper part of the medulla, plays a special role in the descending pain modulation pathway (described with Figure 5.6).

Other nuclei in the brainstem appear functionally to belong to the RF yet are not located within the core region. These include the periaqueductal gray and the locus ceruleus.

- The **periaqueductal gray** of the midbrain (for its location, see Appendix Figure A.3 and Appendix Figure A.4) includes neurons that are found around the aqueduct of the midbrain. This area also receives input (illustrated but not labeled in this diagram) from the ascending sensory systems conveying pain and temperature, the anterolateral pathway; the same occurs with the trigeminal system. This area is part of a descending pathway to the spinal cord that is concerned with pain modulation (see Figure 5.6).

- The **locus ceruleus** is a small nucleus in the upper pontine region (see Appendix Figure A.5). In some species (including humans), the neurons of this nucleus accumulate a pigment that can be seen when the brain is sectioned (before histological processing; see the photograph of the pons in Figure 4.2B). Output from this small nucleus is distributed widely throughout the brain to virtually every part of the CNS, including all cortical areas, subcortical structures, the brainstem and cerebellum, and the spinal cord. The neurotransmitter involved is norepinephrine (dopamine and norepinephrine are catecholamines). Although the functional and electrophysiological role of this nucleus is still not clear, the locus ceruleus has been thought to act like an alarm system in the brain. It has also been implicated in a wide variety of CNS activities, such as mood, reaction to stress, and various autonomic activities.

One additional area—the paramedian pontine RF (PPRF)—comprises a group of neurons involved in controlling horizontal saccadic eye movements; the role of these neurons is discussed with eye movements (see Figure 6.8 and Figure 6.9).

The cerebral cortex sends fibers to the RF nuclei, including the periaqueductal gray, thus forming part of the cortico-bulbar system of fibers (see Figure 5.10). The nuclei that receive this input and then give off the pathways to the spinal cord form part of an indirect voluntary motor system—the cortico-reticulo-spinal pathways (discussed in Motor Systems—Introduction in Chapter 5; see Figure 5.12A and Figure 5.12B). In addition, this system is known to play an extremely important role in the control of muscle tone (discussed with Figure 5.12B).

CLINICAL ASPECT

Lesions of the cortical input to the RF in particular have a very significant impact on muscle tone. In humans, the end result is a state of increased muscle tone called spasticity, accompanied by hyper-reflexia, which is an increase in the responsiveness of the deep tendon reflexes (discussed with Figure 5.12B).

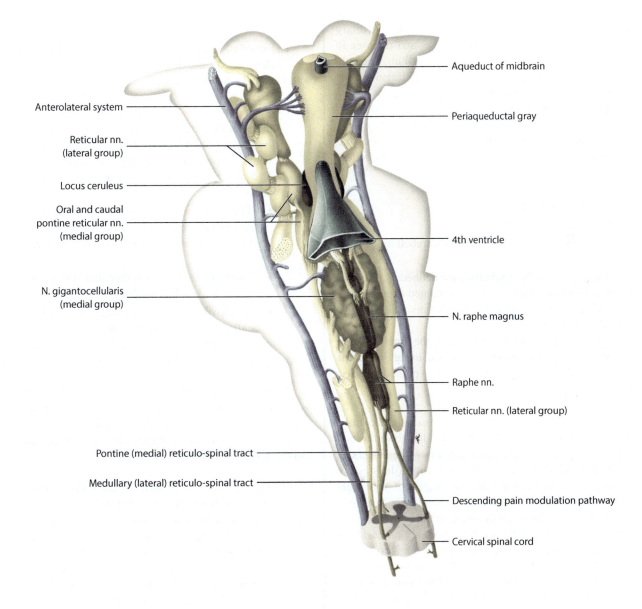

Aqueduct of midbrain

Anterolateral system

Reticular nn.
(lateral group)

Locus ceruleus

Oral and caudal
pontine reticular nn.
(medial group)

N. gigantocellularis
(medial group)

Pontine (medial) reticulo-spinal tract

Medullary (lateral) reticulo-spinal tract

Periaqueductal gray

4th ventricle

N. raphe magnus

Raphe nn.

Reticular nn. (lateral group)

Descending pain modulation pathway

Cervical spinal cord

FIGURE 3.6B: Reticular Formation 2—Reticular Formation: Nuclei

FIGURE 3.7—CEREBELLUM 1

ORGANIZATION: LOBES

The cerebellum has been subdivided anatomically according to some constant features and fissures (see Figure 1.9). In the midline, the worm-like portion is the **vermis**; the lateral portions are the **cerebellar hemispheres**. The horizontal fissure lies approximately at the division between the superior and the inferior surfaces. The deep primary fissure is found on the superior surface, and the area in front of it is the **anterior lobe** of the cerebellum. The only other parts to be noted are the nodulus and lingula of the vermis, as well as the tonsil.

To understand the functional anatomy of the cerebellum and its contribution to the regulation of motor control, it is necessary to subdivide the cerebellum into operational units. The three functional lobes of the cerebellum are:

- The vestibulocerebellum.
- The spinocerebellum.
- The neocerebellum or cerebrocerebellum.

These lobes of the cerebellum are defined by the areas of the cerebellar cortex involved, the related deep cerebellar nucleus, and the connections (afferents and efferents) with the rest of the brain.

There is a convention of portraying the functional cerebellum as if it is found in a single plane, using the lingula and the nodulus of the vermis as fixed points (see also Figure 1.7).

Note to the Learner: The best way to visualize this is to use the analogy of a book, with the binding toward you—representing the horizontal fissure. Place the fingers of your left hand on the edge of the front cover (the superior surface of the cerebellum) and the fingers of your right hand on the edges of the back cover (the inferior surface of the cerebellum), then (gently) open up the book to expose both the front and the back covers. Both are now laid out in a single plane; now, the lingula is at the "top" of the cerebellum and the nodulus is at the bottom of the diagram. This same "flattening" can be done with an isolated brainstem and attached cerebellum in the laboratory. (The Video on the web site titled Diencephalon, Brainstem and Cerebellum demonstrates how the exercise described is done with an actual specimen.)

Having done this, as is shown in the upper part of this figure, it is now possible to discuss the three functional lobes of the cerebellum.

- The **vestibulocerebellum** is the functional part of the cerebellum responsible for balance and gait. It is composed of two cortical components, the flocculus and the nodulus; hence it is also called the **flocculonodular lobe**. The flocculus is a small lobule of the cerebellum located on its inferior surface and oriented in a transverse direction below the middle cerebellar peduncle (see Figure 1.8 and Figure 3.1); the nodulus is part of the vermis.

The vestibulocerebellum sends its fibers to the **fastigial** nucleus, one of the deep cerebellar nuclei (discussed with Figure 3.8 and Figure 5.17).

- The **spinocerebellum** is concerned with coordinating the activities of the limb musculature. It receives information from muscles of the body via spino-cerebellar pathways (see Figure 5.15 and Figure 6.10). It is made up of three areas:
  - The **anterior lobe** of the cerebellum, the cerebellar area found on the superior surface, in front of the primary fissure (see Figure 1.9).
  - Most of the **vermis** (other than the parts mentioned earlier, see Figure 1.9).
  - A strip of tissue on either side of the vermis called the **paravermal** or **intermediate zone**—there is no anatomical fissure demarcating this functional area.

The output deep cerebellar nuclei for this functional part of the cerebellum are mostly the **intermediate** nuclei, the globose and emboliform nuclei (see Figure 3.8). Part of their role is to act as a *comparator* between the intended and the actual movements.

- The **neocerebellum** includes the remainder of the cerebellum, the areas behind the primary fissure and the inferior surface of the cerebellum (see Figure 1.9), with the exception of the vermis itself and the adjacent strip, the paravermal zone. This is the largest part of the cerebellum and the newest from an evolutionary point of view. It is also known as the **cerebrocerebellum** because most of its connections are with the cerebral cortex (see also Figure 5.15).

The output nucleus of this part of the cerebellum is the **dentate** nucleus (see Figure 3.8 and Figure 5.17). The neocerebellum is involved in the overall coordination of voluntary motor activities and also in motor planning.

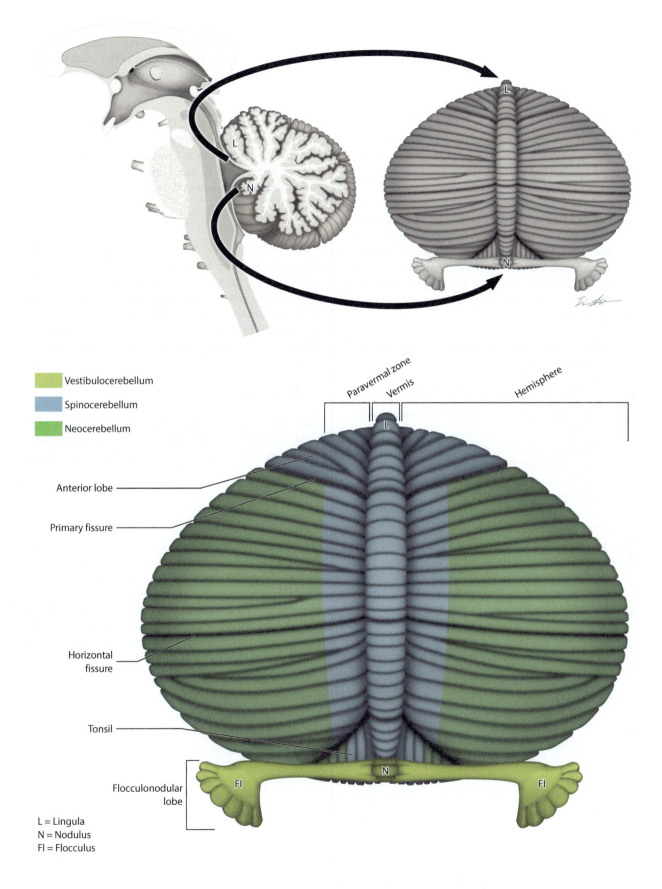

Vestibulocerebellum

Spinocerebellum

Neocerebellum

Paravermal zone

Vermis

Hemisphere

Anterior lobe

Primary fissure

Horizontal fissure

Tonsil

Flocculonodular lobe

L = Lingula
N = Nodulus
Fl = Flocculus

FIGURE 3.7: Cerebellum 1—Organization: Lobes

FIGURE 3.8—CEREBELLUM 2

INTRACEREBELLAR (DEEP CEREBELLAR) NUCLEI (WITH T2 MAGNETIC RESONANCE IMAGING SCAN)

The brainstem is presented from the anterior perspective, with the cerebellum attached (as in Figure 1.8, Figure 3.1, Figure 3.4, and Figure 3.5). This diagram shows the **intracerebellar** nuclei—also called the **deep cerebellar** nuclei—within the cerebellum.

There are four pair of deep cerebellar nuclei—the **fastigial** nucleus, the **intermediate** group (named the globose and emboliform), and the lateral or **dentate** nucleus. Each belongs to a different functional part of the cerebellum. These nuclei are the output nuclei of the cerebellum to other parts of the central nervous system (discussed with Figure 5.17).

- The fastigial (medial) nucleus is located next to the midline.
- The globose and emboliform nuclei are slightly more lateral; these are often grouped together and called the intermediate (or interposed) nucleus.
- The dentate nucleus, with its irregular margin, is the most lateral. This nucleus is sometimes called the lateral nucleus and is by far the largest (see also Figure 4.2A).

The nuclei are located within the cerebellum at the level of the junction of the medulla and the pons. Therefore, the cross-sections shown at this level (see Appendix Figure A.8) may include these deep cerebellar nuclei. Usually, only the dentate nucleus can be identified in sections of the gross brainstem and cerebellum done at this level.

Two of the afferent fiber systems are shown—representing cortico-ponto-cerebellar fibers and spino-cerebellar fibers (further described with Figure 5.15). It is important to note that the cortico-pontine/ponto-cerebellar afferents cross. All afferent fibers send collaterals to the deep cerebellar nuclei en route to the cerebellar cortex, and these are excitatory. Therefore, these neurons are maintained in a chronic state of activity.

Note to the Learner: The lateral vestibular nucleus functions as an additional deep cerebellar nucleus because its main input is from the vestibulocerebellum (shown in Figure 5.16 and Figure 5.17); its output is to the spinal cord non-voluntary motor system (see Figure 5.13).

RADIOLOGY

The accompanying magnetic resonance imaging scan (T2-weighted) in the axial plane shows the thin, leaf-like folia of the cerebellum hemispheres and the middle cerebellar peduncles, with the 4th ventricle (cerebrospinal fluid is white) separating the cerebellum from the pons anteriorly. The outline of the dentate nucleus can be discerned.

ADDITIONAL DETAIL

This "cut" shows the incoming vestibulocochlear nerve (both divisions are seen on the left side of the radiograph), as well as the horizontal semi-circular (fluid-filled) canal.

FUNCTIONAL ASPECTS

Although it is rather difficult to explain in words what the cerebellum does in motor control, damage to the cerebellum leads to quite dramatic alterations in ordinary movements (discussed with Figure 5.17). Lesions of the cerebellum result in the decomposition of the activity, or fractionation of movement, so that the action is no longer smooth and coordinated. Certain cerebellar lesions also produce a tremor that is seen when performing voluntary acts, better known as an intention tremor.

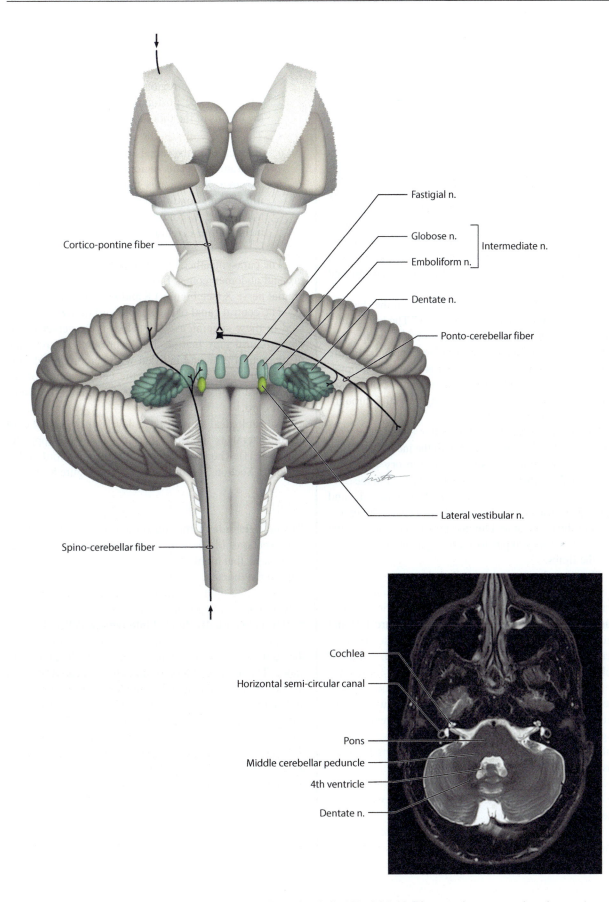

Fastigial n.

Globose n. ⎤
⎥ Intermediate n.
Emboliform n. ⎦

Dentate n.

Cortico-pontine fiber

Ponto-cerebellar fiber

Lateral vestibular n.

Spino-cerebellar fiber

Cochlea

Horizontal semi-circular canal

Pons

Middle cerebellar peduncle

4th ventricle

Dentate n.

FIGURE 3.8: Cerebellum 2—Intracerebellar (Deep Cerebellar) Nuclei (with T2 magnetic resonance imaging scan)

FIGURE 3.9—SPINAL CORD

SPINAL CORD: CROSS-SECTIONAL VIEWS (PHOTOGRAPHS AND DRAWINGS)

PHOTOGRAPHS

A human spinal cord was sectioned at four levels—cervical, thoracic, lumbar, and sacral—from the top of the page to the bottom, as shown on the left side. The gray matter of the spinal cord, which consists of the nuclei, is found on the inner aspect, surrounded by the white matter, which consists of tracts, with the pathways coursing up and down the spinal cord connecting the spinal cord with the higher centers of the brain, the brainstem and cerebral cortex.

The gray matter is said to be arranged in the shape of a butterfly or somewhat like the letter "H." The gray matter of the spinal cord contains a variety of cell groups (i.e., nuclei) that serve different functions (see Appendix Figure A.11). The division of the gray matter is shown in the upper drawing on the right side. Although rather difficult to visualize, these groups are continuous longitudinally throughout the length of the spinal cord.

The dorsal region of the gray matter, called the **dorsal** or **posterior horn**, is associated with the incoming (afferent) dorsal root and is thus related to sensory functions. The cell body of these sensory fibers is located in the **dorsal root ganglion** (see Figure 1.12, Figure 4.1, and Figure 5.1). The dorsal horn is quite prominent in the region of the cervical and lumbar plexuses because of the very large sensory input to these segments of the cord from the limbs.

The ventral gray matter, called the **ventral** or **anterior horn**, is the motor portion of the gray matter. The ventral horn has the large motor neurons, the anterior horn cells, that are *e*fferent to the muscles (see Figure 1.12 and Figure 5.7) via the ventral nerve roots. These neurons, because of their location in the spinal cord, which is "below" the brain, are also known as **lower motor neurons**. (The neurons in the cerebral cortex, at the "higher" level, are called upper motor neurons—discussed with Figure 5.8, Figure 5.9, and Figure 5.10.) The ventral horn is again prominent at the level of the limb plexuses because of the large number of motor neurons supplying the limb musculature.

The area in between is usually called the **intermediate gray** and has a variety of cell groups with some association-type functions.

The **autonomic** innervation to the organs of the chest, abdomen, and pelvis is controlled by neurons located in the spinal cord.

- The preganglionic **sympathetic** neurons form a distinctive protrusion of the gray matter called the **lateral horn**, which extends throughout the thoracic region, from spinal cord level T1 to L2. The post-ganglionic nerves supply the organs of the thorax, abdomen, and pelvis.

- The **parasympathetic** preganglionic neurons are located in the sacral area and do not form a separate horn. This region of the spinal cord in the area of the conus medullaris (the lowest illustration) controls bowel and bladder function, subject to commands from higher centers, including the cerebral cortex.

The parasympathetic control of the organs of the thorax and abdomen comes from the vagus nerve, which is cranial nerve X (see Figure 1.8 and Figure 3.5).

The central canal of the spinal cord (not well visualized in these specimens) is located in the center of the commissural gray matter. This represents the remnant of the neural tube and is filled with cerebrospinal fluid (see Figure 7.8). In the adult, the central canal of the spinal cord is probably not patent throughout the whole spinal cord.

A histological view of these levels of the spinal cord (except sacral) is shown in the Appendix (see Appendix Figure A.11). The blood supply of the spinal cord is discussed in Section 3 (with Figure 8.7 and Figure 8.8).

DRAWINGS

On the right side of the illustration are drawings of the cross-sections of the spinal cord, based on the actual specimens, that highlight certain features (e.g., the lateral horn in the thoracic region).

The white matter, which contains the ascending sensory and descending motor pathways, surrounds the gray matter and is usually divided into regions called **funiculi** (singular: funiculus) which are indicated in the drawing of the lumbar region: these have some functional significance. The posterior (dorsal) funiculus is located between the dorsal horns, the lateral funiculus between the dorsal and ventral horns, and the anterior funiculus between the anterior horn and the anterior median fissure. The distinct pathways will be described with the functional systems in Section 2; a summary diagram with all the tracts is shown after all the spinal cord pathways have been discussed in Section 2 (see Figure 6.10).

Note to the Learner: The cervical and lumbar levels illustrated on the left side are used, without the attached nerve roots, in the illustrations of all the pathways in Section 2.

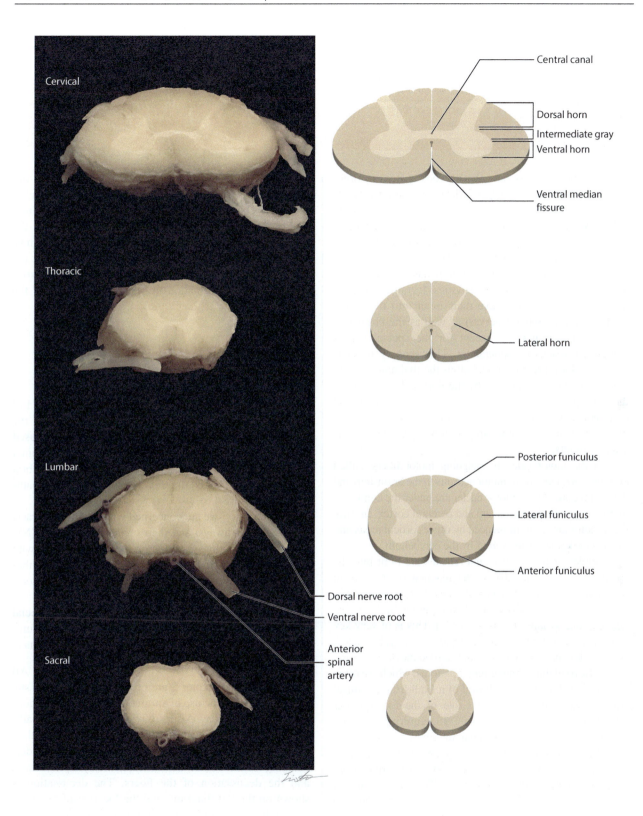

Cervical

Central canal

Dorsal horn

Intermediate gray

Ventral horn

Ventral median fissure

Thoracic

Lateral horn

Lumbar

Posterior funiculus

Lateral funiculus

Anterior funiculus

Dorsal nerve root

Ventral nerve root

Anterior spinal artery

Sacral

FIGURE 3.9: Spinal Cord—Spinal Cord: Cross-Sectional Views (photographs and drawings)

FUNCTIONAL SYSTEMS

This section explains how the nervous system is organized to assess sensory input and execute motor actions. Detailed knowledge of the various central nervous system (CNS) systems allows a clinician to deduce *whether* there is a problem involving the CNS. Usually (often), this type of problem is referred to a neurologist.

The functioning nervous system has a hierarchical organization to carry out its activities. Incoming sensory fibers, called afferents, have their input into the spinal cord (via the peripheral nerves) as well as the brainstem (via the cranial nerves), except for the special senses (which are discussed separately). This sensory input is processed by relay nuclei, including the thalamus, before the information is analyzed by the cortex. In the cortex, there are primary areas that receive the information, and other cortical association areas that elaborate the sensory information, and still other areas that integrate the various sensory inputs.

On the motor side, the outgoing motor fibers, called *e*fferents, originate from motor neurons in the brainstem and the spinal cord. These motor nuclei are under the control of motor centers in the brainstem and cerebral cortex. In turn, these motor areas are influenced by other cortical areas and by the basal ganglia, as well as by the cerebellum.

Simpler motor patterns and reflexes are built into the spinal cord. Most notable is the reaction of muscles in response to stretch—called the stretch reflex, also known as the **myotatic reflex** (discussed with Figure 5.7); this reflex has only one synapse (**monosynaptic**). This is of very key significance, both physiologically and clinically, because it is this reflex that is tested clinically, called the **deep tendon reflex**. Beyond that, simpler motor patterns, which are often reflexive, such as the withdrawal of a limb from a painful stimulus or stimulation of the sole of the foot (plantar reflex), involve processing in the CNS, requiring interneurons in the spinal cord, brainstem, thalamus, or cortex.

The processing of both sensory and motor activities, beyond simple reflexes, involves a series of neuronal connections, creating functional systems. These include nuclei of the CNS at the level of the spinal cord, brainstem, and thalamus. In almost all functional systems in humans, the cerebral cortex is also involved. The axonal connections among the nuclei in a functional system usually run together to form a distinct bundle of fibers, called a **tract** or **pathway**. These tracts are named according to the direction of the pathway (e.g., spino-thalamic means that the pathway is going from the spinal cord to the thalamus; cortico-spinal means that the pathway is going from the cortex to the spinal cord). Along their way, these axons may distribute information to several other parts of the CNS by means of **axon collaterals**.

ORGANIZATION OF THIS SECTION

Chapter 4 uses the information developed in Section 1 of the atlas (Chapters 1 to 3) to assemble the parts of the nervous system called system components, which are used to illustrate the pathways (tracts). These include the spinal cord, brainstem, thalamus, internal capsule and cerebral cortex. A standardized diagram is used to show the pathways (see Figure 4.6).

In **Chapter 5** we are concerned with the sensory tracts (also called pathways) and their connections in the CNS, and the motor pathways and brain regions concerned with movements, including the reticular formation and vestibular system. Included in this section is the issue of motor modulation by the basal ganglia and cerebellum.

Chapter 6 includes the presentation of the special senses (audition, vision, and vestibular), as well as a summary of all the pathways and their cortical connections.

Note on the Web site (www.atlasbrain.com): All the pathways in Chapter 5 are presented on the Web site supplemented by the use of animation demonstrating activation of the pathway. After studying the details of a pathway with the text and illustration, the learner should then view the same figure on the Web site for a better understanding of the course of the tract, the synaptic relays, and the decussation of the fibers. The decussation is shown on the left diagram, and the location of the pathway in the spinal cord and brainstem is shown in the axial cuts on the right side.

CLINICAL ASPECT

Destruction of the nuclei and pathways of the CNS as a result of disease or injury leads to a neurological loss of function. How does the physician (neurologist) diagnose what is wrong?

A thorough understanding of the structure and function of the nervous system is the foundation for clinical neurology. The neurologist's task is to analyze the history of the illness and the symptoms and signs of the patient, decide *whether* the problem is in fact neurological, configure the patient's complaints and the physical findings to establish *where* in the nervous system the problem is located (the localization), and then to ascertain a cause for the disease (*what*, the etiology). At some stage, laboratory investigations and imaging studies (computed tomography (CT), magnetic resonance imaging (MRI) and other specialized imaging modalities) are used to confirm the localization of the disease and to assist in establishing the diagnosis. An appropriate therapeutic plan would then be proposed, and the patient and family can be advised of the long-term outlook of his/her disease (the prognosis), and its impact on life, family and employment (psychosocial issues).

A simple mnemonic using the letter "w" helps to recall the basic steps necessary to establish a neurological diagnosis:

- **whether** the signs and symptoms are consistent with involvement of the nervous system, based upon a detailed history and a complete neurological examination.
- **where** the nervous system problem is located, i.e., **localization**.
- **what** is the etiology of the disease, its pathophysiological mechanism(s).

Skilled and knowledgeable expert clinicians can recognize more common neurological diseases based upon their presentation (for example, vascular lesions have a sudden onset versus a slower onset for tumors), the age of the patient, the part(s) of the nervous system involved, and the evolution of the disease process. These experts "know" when the disease presentation of a particular patient does not fit the expected pattern and will then change to a more analytic approach, such as the one proposed in the *Integrated Nervous System* text.

The *Integrated* text presents many case studies (both in the text and over 40 on its Web site) of neurological diseases of the various parts of the nervous system. These are analyzed using the recommended systematic approach presented in that text.

The task is more complex in children, depending on the age, because the nervous system continues to develop through infancy and childhood; diseases interfere with and interrupt this developmental pattern. Knowledge of normal growth and development is necessary to practice pediatric neurology.

Note to the Learner: There is an additional caveat—almost all the pathways cross the midline, each at a unique and different location; this is called a **decussation**. The important clinical correlate is that destruction of a pathway may affect the opposite side of the body, depending on the location of the lesion in relation to the level of the decussation.

Chapter 4

System Components

FIGURE 4.1—SPINAL CORD

LUMBAR AND CERVICAL LEVELS (WITH T2 MAGNETIC RESONANCE IMAGING SCANS)

The spinal cord was introduced in the orientation section of the atlas (Section 1; see Figure 1.10, Figure 1.11, and Figure 3.9). The organization of the nervous tissue in the spinal cord has the gray matter inside, in a typical "butterfly" or "H"-shaped configuration, with the white matter surrounding it.

The **dorsal horn** of the gray matter is associated with the incoming sensory information. Its nuclei are described with the sensory systems (see Figure 5.1).

The **ventral horn** of the gray matter is associated with the outgoing motor fibers. Its nuclei are described with the motor systems (see Figure 5.7).

The **white matter** surrounding the gray matter is divided into three funiculi with the ascending and descending tracts (see also Figure 3.9).

The two spinal cord levels that are used to delineate the pathways are lumbar (the lower illustration and radiology) and cervical (the upper illustration and radiology).

LUMBAR LEVEL (LOWER ILLUSTRATION AND MAGNETIC RESONANCE IMAGING SCAN)

This cross-sectional level of the spinal cord is used in the various illustrations of the pathways in Section 2. This cross-section is similar in appearance to the cervical section because both are innervating the limbs (see Figure 3.9).

There is, however, proportionately less white matter at the lumbar level than at the cervical level. The descending tracts are smaller because many of the fibers have terminated at higher levels. The ascending tracts are smaller because they are conveying information only from the lower regions of the body.

RADIOLOGY

This T2-weighted scan is taken through the lower vertebral thoracic region, which is at the level of spinal lumbar region (see Figure 1.10).

Note the orientation of the radiograph, with ventral above and dorsal below, which is the opposite of the usual way in which the spinal cord is depicted diagramatically.

The gray/white matter of the cord can be distinguished. The cord is surrounded by fibers descending to exit at a lower level; these fibers are seen as little black "dots" in the CSF surrounding the cord.

CERVICAL LEVEL (UPPER ILLUSTRATION AND MAGNETIC RESONANCE IMAGING SCAN)

This is a cross-section of the spinal cord through the cervical enlargement. This level is used in the illustrations of the various pathways. The incoming dorsal (afferent) root is associated with a group of nuclei in the dorsal horn (see Figure 5.3). The dorsal horn is large because of the amount of afferents coming from the skin, particularly of the fingers and hand. Given that the cervical enlargement contributes to the formation of the brachial plexus to the upper limb, the gray matter of the ventral horn is also very large because of the number of neurons involved in the innervation of the upper limb, particularly the muscles of the hand.

The white matter is comparatively larger at this level because all the ascending tracts are present and are carrying information from the lower parts of the body as well as the upper limb. In addition, all the descending tracts are fully represented as many of the fibers terminate in the cervical region of the spinal cord. In fact, some of them do not descend to lower levels.

RADIOLOGY

This T2-weighted scan is taken through the vertebral cervical region, which is at the level of spinal cervical region (see Figure 1.10).

Note the orientation of the radiograph, as with the lumbar region.

The gray and white matter can be distinguished (see also Figure 3.9). The roots are exiting at about the same level and are seen as black lines, particularly on the right side of the illustration.

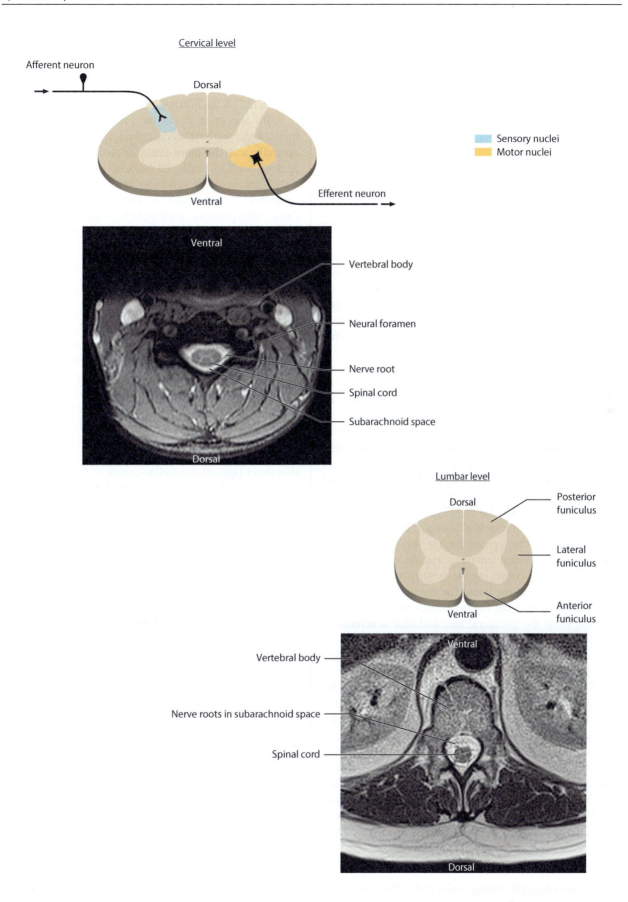

FIGURE 4.1: Spinal Cord—Lumbar and Cervical Levels (with T2 magnetic resonance imaging scans)

FIGURE 4.2A—BRAINSTEM A

MEDULLA (PHOTOGRAPH WITH T2 MAGNETIC RESONANCE IMAGING SCAN)

The description of the cross-sections (axial) through the brainstem follows a uniform "floor plan" (see Introduction to the Appendix and Appendix Figure A.2). This part of the brainstem has a distinct appearance because of the presence of two distinct structures: the pyramids and the inferior olivary nucleus (see Figure 1.8 and Figure 3.1).

The pyramids, located ventrally, are an elevated pair of structures located on either side of the midline. They contain the cortico-spinal fibers that have descended from the motor areas of the cortex (see Figure 5.9 and Figure 5.10). The olive (inferior olivary nucleus) is a prominent structure that has a distinct scalloped profile when seen in cross-section (see Appendix Figure A.8, Appendix Figure A.9, and Appendix Figure A.10). It is so large that it forms a prominent bulge on the lateral surface of the medulla. Its fibers relay to the cerebellum (see Figure 5.15).

The tegmentum is the area of the medulla that contains the cranial nerve nuclei, the nuclei of the reticular formation, the ascending tracts, and the inferior olivary nucleus.

Cranial nerves (CN) IX, X, and XII are attached to the medulla and have their nuclei here, as well as some nuclei of CN VIII (see Figure 3.4 and Figure 3.5). The reticular formation occupies the central region of the tegmentum.

Included in the tegmentum are the two ascending tracts, the large medial lemniscus and the small anterolateral system, both conveying the sensory modalities from the opposite side of the body. The spinal trigeminal tract and nucleus, conveying the modalities of pain and temperature from the ipsilateral face and oral structures, is also found throughout the medulla. The solitary nucleus and tract, which subserves both taste and visceral afferents, are similarly found in the medulla.

The nuclei gracilis and cuneatus, the relay nuclei for the dorsal column tracts, are found in the lower part of the medulla, on its dorsal aspect (see Figure 3.3 and Appendix Figure A.10). The 4th ventricle lies behind the tegmentum and separates the medulla from the cerebellum (see Figure 3.2). The roof of this (lower) part of the ventricle has choroid plexus (see Figure 7.8). Cerebrospinal fluid (CSF) escapes from the 4th ventricle via the various foramina located here and then flows into the subarachnoid space, the cisterna magna (see Figure 3.2 and Figure 7.8).

MEDULLA—MID (PHOTOGRAPH) AXIAL

This is a photographic image, enlarged, at the level of the middle of the medulla, with the cerebellum attached (see Figure 1.7 and Figure 1.8). The upper left image shows the level of the section through the medulla; the corresponding level is shown on a medial view of the brain. Many of the structures visible on this "gross" specimen are seen in more detail on the histological sections in the Appendix.

This specimen shows the principal identifying features of the medulla as described, the pyramids located ventrally and the more laterally placed inferior olivary nucleus, with its scalloped borders. Between the olivary nuclei, are two dense structures, the medial lemniscus, one on each side of the midline, which is the major sensory (ascending) pathway (see Figure 5.2). The medial longitudinal fasciculus (MLF) is still a distinct tract in its usual location (see Figure 6.8 and Figure 6.9). The other dense tract that is recognizable in this specimen is the inferior cerebellar peduncle located at the outer posterior edge of the medulla (see Figure 5.15).

The space behind is the 4th ventricle, narrowing in its lower portion (see Figure 3.1 and Figure 3.3). There is no "roof" to the ventricle in this section, and it is most likely that the plane of the section has passed through the median aperture, the foramen of Magendie (see Figure 7.8).

The cerebellum remains attached to the medulla, with the prominent vermis and the large cerebellar hemispheres. The cerebellar lobe adjacent to the medulla is the tonsil (see Figure 1.9). The extensive white matter of the cerebellum is seen, as well as the thin outer layer of cerebellar cortex.

The medulla is represented by three sections in the Appendix:

- Upper medulla: The upper level typically includes CN VIII (both parts) and its nuclei (see Appendix Figure A.8).
- Mid-medulla: This section through the mid-medulla includes the nuclei of CN IX, X, and XII (see Appendix Figure A.9).
- Lower medulla: The lowermost section is at the level of the dorsal column nuclei, the nuclei gracilis, and cuneatus (see Appendix Figure A.10).

RADIOLOGY (AXIAL)

The T2-weighted magnetic resonance image of the medulla shown in the lower part of the illustration is a mirror image of the specimen, so that the ventral (anterior) aspect is at the top and the cerebellum (with its narrow folia) is below; the CSF (white) of the 4th ventricle is in between. The medulla is surrounded by CSF (in the subarachnoid space).

The small size of the medulla is notable. Other than the size and relationships, the medullary outline is recognizable by the "bump" of the pyramids and by the "bump" of the inferior olivary nucleus immediately behind.

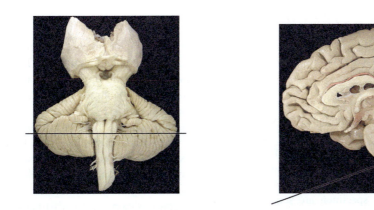

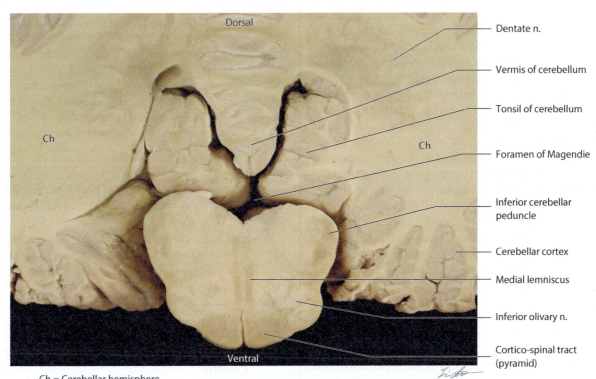

Dorsal

Dentate n.

Vermis of cerebellum

Tonsil of cerebellum

Foramen of Magendie

Inferior cerebellar
peduncle

Cerebellar cortex

Medial lemniscus

Inferior olivary n.

Cortico-spinal tract
(pyramid)

Ch

Ch

Ventral

Ch = Cerebellar hemisphere

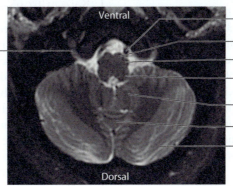

Ventral

Glossopharyngeal (IX) and
vagus (X) nerves

Vertebral artery

Pyramid

Olive

Inferior cerebellar peduncle

Tonsil of cerebellum

Vermis of cerebellum

Cerebellar hemisphere

Dorsal

FIGURE 4.2A: Brainstem A—Medulla (photograph with T2 magnetic resonance imaging scan)

FIGURE 4.2B—BRAINSTEM B

PONS (PHOTOGRAPH WITH T2 MAGNETIC RESONANCE IMAGING SCAN)

The pons is characterized by its protruding anterior (ventral) portion, the pons proper (see Figure 1.8 and Figure 3.1). The upper left image shows the level of the section through the mid- to upper pons; the corresponding level is shown on a medial view of the brain. Many of the structures visible on this "gross" specimen are seen in more detail on the histological sections in the Appendix.

The basilar portion of the pons (or pons proper) contains the pontine nuclei, the site of relay of the cortico-pontine fibers (see Figure 5.10 and Figure 5.15); the ponto-cerebellar fibers then cross and enter the cerebellum via the large middle cerebellar peduncle. Intermingled with the pontine nuclei are the dispersed fibers that belong to the cortico-spinal system (see Figure 5.9).

Behind the pons proper is the tegmentum, which seems quite compressed. This region of the brainstem contains the cranial nerve nuclei, most of the ascending and descending tracts, and the nuclei of the reticular formation. The cranial nerves attached to the pons include the trigeminal (cranial nerve [CN] V) at the mid-pontine level and the abducens (CN VI), the facial (CN VII), and part of CN VIII (the vestibulocochlear) at the lowermost pons; the fibers of CN VII form an internal loop over the abducens nucleus in the pons (see Figure 6.12 and Appendix Figure A.7). The fibers of CN VII and CN VIII are located adjacent to each other at the cerebello-pontine angle (see Figure 1.8 and Figure 3.4).

The ascending tracts present in the tegmentum are those conveying sensory information from the body and face. These include the medial lemniscus and the antero-lateral fibers (see Figure 5.2 and Figure 5.3).

One of the distinctive nuclei of the upper pons is the locus ceruleus, a pigment-containing nucleus (discussed with Figure 3.6B and Appendix Figure A.5). The nuclei of the reticular formation of the pons have their typical location in the tegmentum (see Figure 3.6A and Figure 3.6B).

The 4th ventricle in the pontine region begins as a widening of the aqueduct and then continues to enlarge so that it is widest at about the level of the junction between the pons and medulla (see Figure 3.1 and Figure 3.3). This ventricle separates the pons and medulla anteriorly from the cerebellum posteriorly (see Figure 3.2).

There is no pontine nucleus dorsal to the 4 ventricle; the cerebellum is located above (posterior to) the roof of the ventricle.

The thin folia of the cerebellum are easily recognized, with an inner strip of white matter bounded on either side by the thin gray matter of the cerebellar cortex.

PONS—MID/UPPER (PHOTOGRAPH) AXIAL

This is a photographic image, enlarged, of the pontine region, with the cerebellum attached. The section is done at the level of the mid-upper pons, as indicated in the upper images of the ventral view of the brainstem and in the mid-sagittal view.

The unique nucleus present at this level is the locus ceruleus, a small nucleus whose cells have pigment (see Appendix Figure A.5), much like those of the substantia nigra, pars compacta. As with that nucleus, the pigment is lost during histological processing.

The ventral region has the distinctive appearance of the pontine nuclei, with the cortico-spinal and cortico-pontine fibers dispersed among them. The pontine tegmentum seems quite compressed. The space in the middle of the tissue section is the 4th ventricle, as it begins to widen. Behind the ventricle is a small area of white matter called the superior medullary velum and the superior cerebellar peduncles (see Figure 3.2 and Figure 3.3). The thin folia of the cerebellum are easily recognized, with an inner strip of white matter bounded on either side by the thin gray matter of the cerebellar cortex.

The pons is represented by three sections in the Appendix:

* Upper pons: The uppermost pons is at the level of the locus ceruleus (see Appendix Figure A.5).
* Mid-pons: The mid-pons (middle pons) is at the level of the attachment of the trigeminal nerve. It includes the massive middle cerebellar peduncles (see Appendix Figure A.6).
* Lower pons: The lowermost pons is just above the junction with the medulla (see Appendix Figure A.7). This lowermost level has the nuclei of CN VI and VII and parts of both divisions of CN VIII.

RADIOLOGY (AXIAL)

The magnetic resonance imaging (T2-weighted) scan of the pons shown in the lower part of the figure is a mirror image of the specimen, so that the ventral (anterior) aspect is at the top and the cerebellum (with its narrow folia) below; the cerebrospinal fluid (CSF) (white) of the 4th ventricle is in between. The distinctive feature of the ventral pons, the pons proper, can be recognized in the radiograph. The large trigeminal nerve (CN V) is seen. (The locus ceruleus is not seen.) The pons is surrounded by CSF (in the subarachnoid space). The black dot in the midline ventrally (at the top, within the CSF) is the basilar artery (see Figure 8.1). The temporal lobe and internal carotid arteries are also visualized.

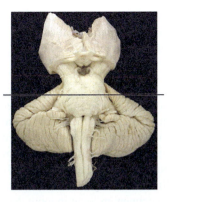

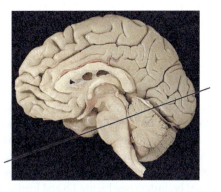

Dorsal

Ch

Ch

Ventral

Ch = Cerebellar hemisphere

Superior
medullary velum

4th ventricle

Superior cerebellar
peduncle

Locus ceruleus

Medial lemniscus

Pontine nuclei

Middle cerebellar
peduncle

Ponto-cerebellar
fibers

Vestibulocochlear
nerve (CN VIII)

Cortico-spinal
fibers

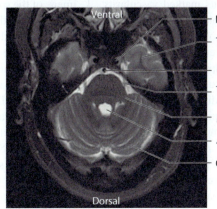

Ventral

Internal carotid artery

Temporal lobe

Basilar artery

Trigeminal nerve (CN V)

Middle cerebellar peduncle

4th ventricle

Cerebellar hemisphere

Dorsal

FIGURE 4.2B: Brainstem B—Pons (photograph with T2 magnetic resonance imaging scan)

FIGURE 4.2C—BRAINSTEM C

MIDBRAIN (PHOTOGRAPH WITH T2 MAGNETIC RESONANCE IMAGING SCAN)

The midbrain is the smallest of the three parts of the brainstem. The temporal lobes of the hemispheres tend to obscure its presence on an anterior and inferior view of the whole brain (see Figure 1.1 and Figure 2.8). The midbrain area is easily recognizable from the anterior view in a dissected specimen of the isolated brainstem (see Figure 1.8 and Figure 3.1).

The ventral portion of the midbrain contains the massive cerebral peduncles located anteriorly, with a fossa in between. These peduncles contain axons that are a direct continuation of the fiber systems of the internal capsule (see Figure 2.4 and Figure 4.4). Within them are found the pathways descending from the cerebral cortex to the brainstem (cortico-bulbar), cortico-pontine to the cerebellum via the pons (see Figure 5.10), and to the spinal cord (cortico-spinal; see Figure 5.9).

The tegmentum contains the nucleus of cranial nerve (CN) III to the eye muscles and the associated Edinger-Westphal (EW, parasympathetic) nucleus and the nucleus of CN IV (to one eye muscle), as well as two special nuclei in the midbrain region—the substantia nigra and the red nucleus, both involved in motor control.

The **substantia nigra** is found throughout the midbrain and is located behind the cerebral peduncles. It derives its name from the dark melanin-like pigment found (not in all species) within its neurons in a freshly dissected specimen, as seen in the present illustration (see also Figure 1.5). (The nucleus has thus been color-coded in black.) The pigment is not retained when the tissue is processed for sectioning (discussed with Appendix Figure A.3). The function of the substantia nigra is related to the basal ganglia (see Figure 5.14 and Figure 5.18).

The **red nucleus** derives its name from the observation that this nucleus has a reddish color in a freshly dissected specimen, presumably because of its marked vascularity. (This nucleus has therefore been color-coded in red.) The red nucleus is found at the superior collicular level. Its function is discussed with the motor systems (see Figure 5.11).

The aqueduct of the midbrain helps to identify this cross-section as the midbrain area (see Figure 3.2 and Figure 7.8). The **periaqueductal gray**, surrounding the aqueduct, has been included as part of the reticular formation (see Figure 3.6B) and is thought to be important in the maintenance of consciousness as part of the ARAS (discussed with Figure 3.6A); this area participates as part of the descending control system for pain modulation (see Figure 5.6). The reticular formation is found in the core area of the tegmentum.

Posterior to the aqueduct are the two pair of colliculi, which can also be seen on the dorsal view of the isolated brainstem (see Figure 1.9 and Figure 3.3). The four nuclei together form the tectal plate, or tectum, also called the quadrigeminal plate.

The **pretectal region**, located in front of and somewhat above the superior colliculus, is the nuclear area for the pupillary light reflex (see Figure 6.7).

MIDBRAIN—UPPER (PHOTOGRAPH) AXIAL

This is a photographic image, enlarged, of the sectioned midbrain. As shown in the upper left image, the brainstem was sectioned at the level of the cerebral peduncles; the corresponding level is shown on a medial view of the brain, thus indicating that the section is through the superior colliculus. Many of the structures visible on this "gross" specimen are seen in more detail on the histological sections in the Appendix.

The distinctive features identifying this section as midbrain are anteriorly, the outline of the cerebral peduncles with the fossa in between. Immediately behind is a dark band, the substantia nigra, pars compacta, with pigment present in the cell bodies. A faint outline of the red nucleus can be seen in the tegmentum, which identifies this section as the superior collicular level. In the middle toward the back of the specimen is a narrow channel, which is the aqueduct of the midbrain, surrounded by the periaqueductal gray. The gray matter behind the "ventricle" is the superior colliculus at this level.

There are two levels presented for a study of the midbrain in the Appendix:

- Upper midbrain: This includes the CN III nucleus and the superior colliculus (see Appendix Figure A.3).
- Lower midbrain: This is at the level of the CN IV nucleus and the inferior colliculus and the decussation of the superior cerebellar peduncles (see Appendix Figure A.4).

RADIOLOGY (AXIAL)

The magnetic resonance imaging (T2-weighted) scan of the midbrain shown at the bottom of the figure is a mirror image of the specimen, so that the ventral (anterior) aspect is at the top; the cerebrospinal fluid (CSF) (white) of the cerebral aqueduct is not really visible.

The shape of the midbrain is quite distinct, with the cerebral peduncles prominent. The space between the peduncles, the interpeduncular fossa, is where the oculomotor nerve (CN III) emerges (see Figure 1.8 and Figure 3.5).

The midbrain is surrounded by CSF (in the subarachnoid space); the area behind it (below in the image) is an enlargement of the CSF space, a cistern, called the quadrigeminal cistern. It is of some importance as a radiological landmark.

The level of the image includes the temporal lobe, with the amygdala and inferior horn of the lateral ventricle. The optic chiasm and mammillary bodies are also seen (see Figure 1.5 and Figure 1.6). The middle cerebral artery is also indicated (see Section 3).

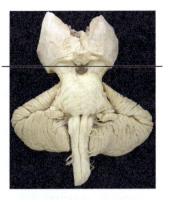
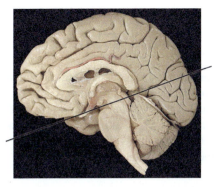

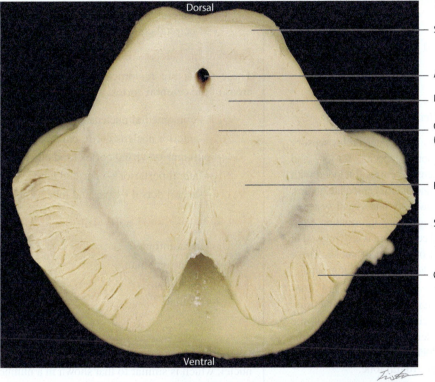

Dorsal

Superior colliculus

Aqueduct of midbrain

Periaqueductal gray

Oculomotor nucleus
(CN III)

Red nucleus

Substantia nigra

Cerebral peduncle

Ventral

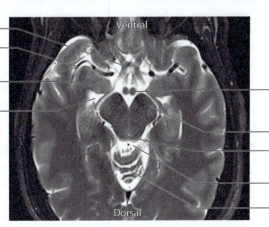

Optic chiasm

Middle cerebral artery

Amygdala

Lateral ventricle (inferior horn)

Ventral

Mammillary body

Cerebral peduncle

Tectum

Quadrigeminal cistern

Vermis of cerebellum

Dorsal

FIGURE 4.2C: Brainstem C—Midbrain (photograph with T2 magnetic resonance imaging scan)

FIGURE 4.3—THALAMUS

ORGANIZATION AND NUCLEI

To lay the groundwork for understanding the functional organization of the sensory and motor pathways (see Chapter 5), it is necessary to have a familiarity with the nuclei of the thalamus, their organization, and their names.

There are two ways of dividing up the nuclei of the thalamus, namely, topographically and functionally.

TOPOGRAPHICALLY

The thalamus is subdivided by bands of white matter into a number of component parts. The main white matter band that runs within the thalamus is called the **internal medullary lamina**, and it is shaped like the letter "Y." It divides the thalamus into a lateral group with its lower ventral set of nuclei, a medial group, and an anterior group of nuclei.

FUNCTIONALLY

The thalamus has three different types of nuclei:

- **Specific relay nuclei:** These nuclei relay sensory and motor information to specific sensory and motor areas of the cerebral cortex. Included with these are the medial and lateral geniculate bodies, relay nuclei for the auditory and visual systems. In addition, motor regulatory information from the basal ganglia and cerebellum is also relayed in the thalamus as part of this set of nuclei. These nuclei are located in the ventral nuclear group.

- **Association nuclei:** These nuclei are connected to broad areas of the cerebral cortex known as the association areas. One of the most important nuclei of this group is the dorsomedial nucleus, located in the medial group, which projects to the frontal lobe.

- **Non-specific nuclei:** These scattered nuclei have other or multiple connections including the cerebral cortex, basal ganglia, other thalamic nuclei, the brainstem and the spinal cord. Some of these nuclei are located within the internal medullary lamina and are often referred to as the **intralaminar** nuclei. Some of these nuclei form part of the ascending reticular activating system, which is involved in the regulation of our state of consciousness and arousal (discussed with Figure 3.6A). The **reticular nucleus**, which lies on the outside of the thalamus, is also part of this functional system.

The following detailed classification system is given at this point but will be understood only as the functional systems of the central nervous system are described (see the Note to the Learner at the end of this section).

Specific Relay Nuclei (and Function)

The cortical connections (reciprocal) of these nuclei are given at this point (to be used with the description of the pathways—see note below).

VA—ventral anterior (motor) ↔ premotor area and supplementary motor area.

VL—ventral lateral (motor) ↔ precentral gyrus and premotor area.

VPL—ventral posterolateral (somatosensory) ↔ postcentral gyrus.

VPM—ventral posteromedial (trigeminal) ↔ postcentral gyrus.

MGB—medial geniculate (body) nucleus (auditory) ↔ temporal cortex.

LGB—lateral geniculate (body) nucleus (vision) ↔ occipital cortex.

Association Nuclei (and Association Cortex)

These nuclei are reciprocally connected to association areas of the cerebral cortex.

DM—dorsomedial nucleus ↔ prefrontal cortex.

AN—anterior nucleus ↔ limbic lobe.

P—pulvinar ↔ visual cortex (and other areas).

LP—lateral posterior ↔ parietal lobe.

LD—lateral dorsal ↔ parietal lobe.

Non-Specific Nuclei (diffuse projections)

IL—intralaminar.

CM—centromedian.

Mid—midline.

R—reticular.

ADDITIONAL DETAIL

For schematic purposes, this presentation of the thalamic nuclei, which is similar to that shown in a number of textbooks, is quite usable. Histological sections through the thalamus are challenging and beyond the scope of an Introductory course.

Note to the Learner: The thalamus is introduced at this point because it is involved throughout the study of the brain. The learner should become familiar with the names and understand the general organization of the various nuclei at this point. It is advised to consult this diagram as the cerebral cortex is described in the following illustrations. Each of the specific relay nuclei involved in one of the pathways will be introduced again with the functional systems, and at that point the student should return to this illustration. A summary diagram showing the thalamus and the cortex with the detailed connections is presented in Chapter 6 (see Figure 6.13). Various nuclei are also involved with the limbic system (see Section 4).

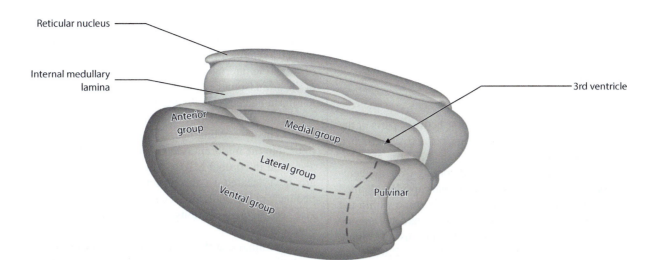

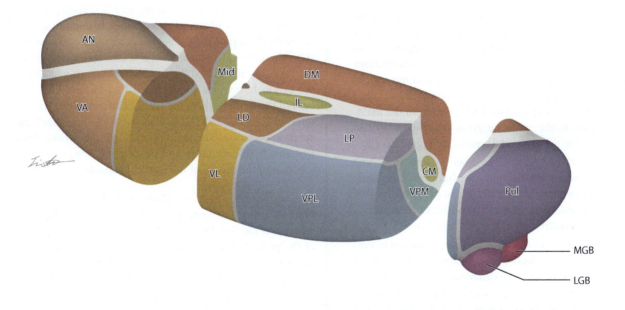

AN = Anterior nuclei

DM = Dorsomedial

Mid = Midline
IL = Intralaminar
CM = Centromedian

LD = Lateral dorsal
LP = Lateral posterior

VA = Ventral anterior
VL = Ventral lateral
VPL = Ventral posterolateral
VPM = Ventral posteromedial

Pul = Pulvinar

MGB = Medial geniculate body
LGB = Lateral geniculate body

FIGURE 4.3: Thalamus—Organization and Nuclei

FIGURE 4.4—INTERNAL CAPSULE

PROJECTION FIBERS

The white matter bundles that course between parts of the basal ganglia and the thalamus are collectively grouped together and called the internal capsule. These are projection fibers, axons going to and coming from the cerebral cortex (see Figure 2.4). The internal capsule is defined as a group of fibers located at a specific plane within the cerebral hemispheres in a region that is situated among the head of the caudate, the lentiform nucleus (see Figure 2.10A), and the thalamus (see Figure 2.10A and Figure 2.10B; also the dissection, Figure 9.4).

The internal capsule has three parts:

- **Anterior limb**: A group of fibers separates the two parts of the neostriatum from each other—the head of the caudate from the putamen. This fiber system carries axons that are coming down from the cortex, mostly to the pontine region, which are then relayed to the cerebellum. Other fibers in the anterior limb relay from the thalamus to the cingulate gyrus (see Figure 10.1A) and to the prefrontal cortex (see Figure 10.1B).
- **Posterior limb**: The fiber system that runs between the thalamus (medially) and the lentiform nucleus (laterally) is the posterior limb of the internal capsule. The posterior limb carries three extremely important sets of fibers.
  - Sensory information from thalamus to cortex, as well as the reciprocal connections from cortex to thalamus.
  - Most of the descending fibers to the brainstem (cortico-bulbar) and spinal cord (cortico-spinal). In addition, there are fibers from other parts of the cortex that are destined for the cerebellum, after synapsing in the pontine nuclei.
- **Genu**: In a horizontal section, the internal capsule (of each side) is seen to be "V"-shaped (see Figure 2.10A). Both the anterior limb and the posterior limb have been described—the bend of the "V" is called the genu and it points medially (also seen with neuroradiological imaging, both computed tomography [CT, see Figure 2.1A] and magnetic resonance imaging [MRI, see Figure 2.10B]).

The internal capsule fibers are shown from the medial perspective in a dissection in which the thalamus has been partially removed (see Figure 2.7 and Figure 9.1B). The fibers of the internal capsule are also seen in a dissection of the brain from the lateral perspective, just medial to the lentiform nucleus (see Figure 9.4).

Below the level of the internal capsule is the midbrain. The descending fibers of the internal capsule continue into the midbrain (see Figure 5.15 and Figure 9.5A, also seen in coronal MRIs in Figure 2.9B and 9.5B) and are next located in the structure called the cerebral peduncle of the midbrain (see Figure 1.8 and Figure 4.2C; also seen in cross-sections of the brainstem in Appendix Figure A.3).

In summary, at the level of the internal capsule, there are both the ascending fibers from thalamus to cortex and descending fibers from widespread areas of the cerebral cortex to the thalamus, the brainstem and cerebellum, and the spinal cord. These ascending and descending fibers, called projection fibers, are sometimes likened to a funnel, with the top of the funnel the cerebral cortex and the stem the cerebral peduncle. The wider upper portion of the funnel is also called the *corona radiata,* as part of the projection fibers (see Figure 2.2B and Figure 2.4). The base of the funnel, where the funnel narrows, would be the internal capsule, and the bottom stem of the funnel the cerebral peduncle. The main point is that the various fiber systems, both ascending and descending, are condensed together in the region of the internal capsule.

Note to the Learner: Many students have difficulty understanding the concept of the internal capsule and where it is located. One way of thinking about it is to look at the projection fibers as a busy two-lane highway. The internal capsule represents one section of this pathway, analogous to highways where many lanes of traffic are reduced to a narrowed roadway.

CLINICAL ASPECT

The posterior limb of the internal capsule is a region that is apparently particularly vulnerable to small vascular events (discussed with Figure 8.6). These small occlusions lead to the destruction of the fibers supplied. Because of the high packing density of the axons in this region, a small lesion can cause extensive disruption of descending motor and/or ascending sensory pathways. This is one of the most frequent types of cerebrovascular accidents (often shortened to CVA), commonly called a "lacunar stroke." (The details of the vascular supply to this region are discussed with Figure 8.6.)

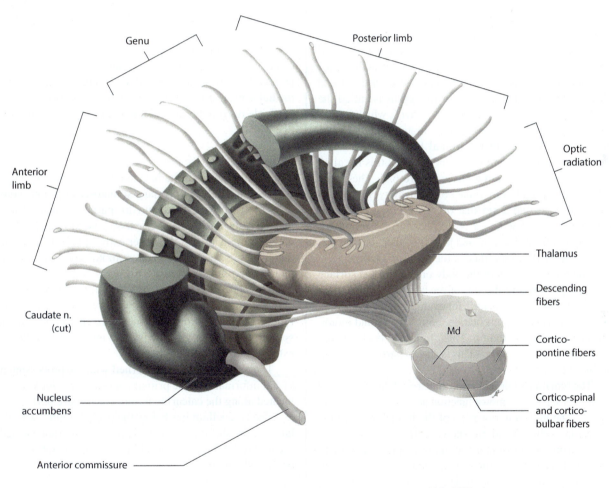

FIGURE 4.4: Internal Capsule—Projection Fibers

FIGURE 4.5—CEREBRAL CORTEX

FUNCTIONAL AREAS

DORSOLATERAL VIEW (PHOTOGRAPH)

This is a photographic image of the same brain as shown previously (see Figure 1.3). The **central fissure** (often called the fissure of Rolando) divides the frontal lobe anteriorly from the parietal lobe posteriorly. The deep **lateral fissure** is clearly visible (see later).

Some cortical areas are functionally directly connected with either a sensory or motor system; these are known as the **primary areas.** The gyrus in front of the central fissure is called the **precentral** gyrus, also called *area 4,* and it is the primary **motor** area, specialized for the control of voluntary movements (see Figure 5.9 and Figure 6.13). The area in front of this gyrus is called the **(lateral) premotor area,** also called *area 6,* which is similarly involved with voluntary motor actions (see Figure 5.8). An area in the frontal lobe (colored) has a motor function in regard to eye movements; this is called the **frontal eye field** (area 8; see Figure 6.9). The gyrus behind the central fissure is the **postcentral gyrus**, including areas 3, 1, and 2 (in that order; see Figure 5.2 and Figure 6.13), and it has a **somatosensory** function for information from the skin (and joints). (Other sensory primary areas will be identified at the appropriate time.)

The representation of the body on the cortical surface is *not* proportional to the size of the body part but is more representative of the "usage"; for example, the hand has an extensive cortical representation, both sensory and motor, and particularly the thumb. This "figurine" is classically known as a **homunculus.** (This is shown particularly well in *The Integrated Nervous System.*)

The remaining cortical areas that are not directly linked to either a sensory or motor function are called **association cortex**. The most anterior parts of the frontal lobe are the newest in evolution and are known as the **prefrontal cortex** (in front of the frontal eye fields, as previously mentioned). This broad cortical area seems to be the chief "executive" part of the brain. The **parietal areas** are connected to sensory inputs and have a major role in integrating sensory information from the various modalities. In the parietal lobe are two special gyri, the **supramarginal** and **angular gyri**; these areas, particularly on the non-dominant side, seem to be involved in visuo-spatial activities.

Some cortical functions are not equally divided between the two hemispheres. One hemisphere is therefore said to be dominant for that function. This is the case for **language** ability, which, in most people, is located in the left hemisphere. This photograph of the left hemisphere shows the two language areas: **Broca's area** for the motor aspects of speech and **Wernicke's area** for the comprehension of written and spoken language (near the auditory area).

The lateral fissure (also known as the fissure of Sylvius) divides the temporal lobe below from the frontal and parietal lobes above. Extending the line of the lateral fissure posteriorly continues the demarcation between the temporal and parietal lobes. The **temporal lobe** seen on this view is a large area of association cortex whose function is still being defined, other than the portions involved with the auditory system (see Figure 6.2 and Figure 6.3) and language (on the dominant side). Other portions of the temporal lobe include the inferior parts (see Figure 1.5) and the medial portion, which is part of the limbic system (see Section 4).

The location of the parieto-occipital fissure is indicated on this photograph. This fissure, which separates the parietal lobe from the occipital lobe, is best seen when the medial aspect of the brain is visualized after dividing the hemispheres (see Figure 1.7 and the next paragraph).

MEDIAL VIEW (PHOTOGRAPH)

The medial aspect of the hemisphere is also shown (as in Figure 1.7). Both the primary somatosensory and motor areas of the cortex extend onto this part of the hemisphere, located in the interhemispheric fissure (see Figure 2.2A)—with representation of the lower extremity. In front of the motor area is the supplementary motor area, involved in motor planning (see Figure 5.8).

The prefrontal cortex extends onto the medial aspect of the brain, as well as onto its inferior (orbital) aspect (see Figure 1.5 and Figure 1.6; also discussed with the limbic system in Section 4). Likewise, the temporal lobe has an extensive inferior aspect.

The occipital lobe is concerned with the processing of visual information. The primary sensory area for vision is located along the calcarine fissure (see Figure 6.5).

The cerebellum lies below the occipital lobe, with the large dural sheath, the tentorium cerebelli (not labeled; see Figure 7.4, Figure 7.5, and Figure 7.6) separating these parts of the brain.

CLINICAL ASPECTS

It is most important to delineate anatomically the functional areas of the cortex. This forms the basis for understanding the clinical implications of damage (called lesions) to the various parts of the brain, including loss of the blood supply (clinically, an ischemic stroke). Clinicians are now being assisted in their tasks by modern imaging techniques, including CT and MRI.

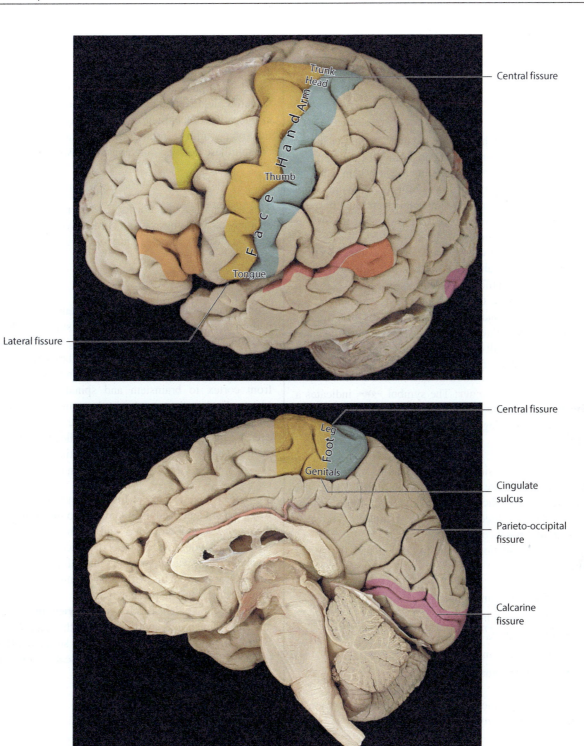

FIGURE 4.5: Cerebral Cortex—Functional Areas: Dorsolateral and Medial Views (photographs)

FIGURE 4.6—PATHWAYS AND CROSS-SECTIONS

ORIENTATION TO DIAGRAMS

The illustrations of the sensory and motor pathways in this section of the atlas are all done in a standard manner.

On the left side: The central nervous system (CNS) is depicted, including spinal cord, brainstem, thalamus, and a coronal section through the hemispheres, with small diagrams of the hemisphere at the top showing the area of the cerebral cortex involved.

The diagram of the hemispheres is a coronal section, similar to the one already described in Chapter 3, at the plane of the lentiform nucleus (see Figure 2.8 and Figure 2.9A). Note the basal ganglia, the thalamus, the internal capsule, and the ventricles; these labels are not repeated in the following diagrams. This diagram is used to convey the overall course of the tract, and particularly at what level the fibers cross (i.e., decussate).

Note to the Learner: The symbol ✳ indicates a synaptic relay in the pathway. The symbol ↩ indicates a decussation (crossing) of the pathway (from one side to the other).

On the right side: Cross-sections of the brainstem and spinal cord are shown, at standardized levels; the exact levels are indicated on the left. In all, there are five cross-sections—three through the brainstem and two through the spinal cord.

The cross-sections of the brainstem and the spinal cord include:

- The midbrain—upper.
- The pons—mid-upper.
- The medulla—mid.
- Cervical spinal cord.
- Lumbar spinal cord.

The exact position of the tract under consideration is indicated in these cross-sections.

These brainstem and spinal cord cross-sections are the same as those shown in the Appendix. In the Appendix, details of the histological anatomy of the brainstem and spinal cord are given. We have titled the Appendix "Neurological Neuroanatomy" because it allows the precise location of the tracts that is necessary for the localization of an injury or disease. The learner may wish to consult these detailed diagrams at this stage.

CLINICAL ASPECT

This section is a foundation for the student in correlating the anatomy of the pathways with clinical symptoms following lesions caused by disease, injury or vascular problems.

LEARNING PLAN

Studying pathways in the CNS necessitates visualizing the pathways, a challenging task for many students. The pathways that are under study extend longitudinally through the CNS, going from spinal cord and brainstem to thalamus and cortex for sensory (ascending) pathways, and from cortex to brainstem and spinal cord for motor (descending) pathways. As is done in other texts and atlases, diagrams are used to facilitate this visualization exercise for the student; color adds to the ability to visualize these pathways, as does the illustration—**with animation**—on the **Web site** (www.atlasbrain.com).

Note to the Learner: In this overview of the pathways, the student is advised to return to the description of the **cortex** (see Figure 1.3 and Figure 4.5). This will inform the student which areas of the cerebral cortex are involved in the various sensory modalities and the relevant motor areas. This will assist in integrating the anatomical information presented in the previous section. Similarly, the student is advised to review the **thalamus** and its connections with each of the pathways (see Figure 4.3 and Figure 6.13). It is also suggested that the learner refer to Figure 6.10 which presents a summary of all the pathways in the spinal cord.

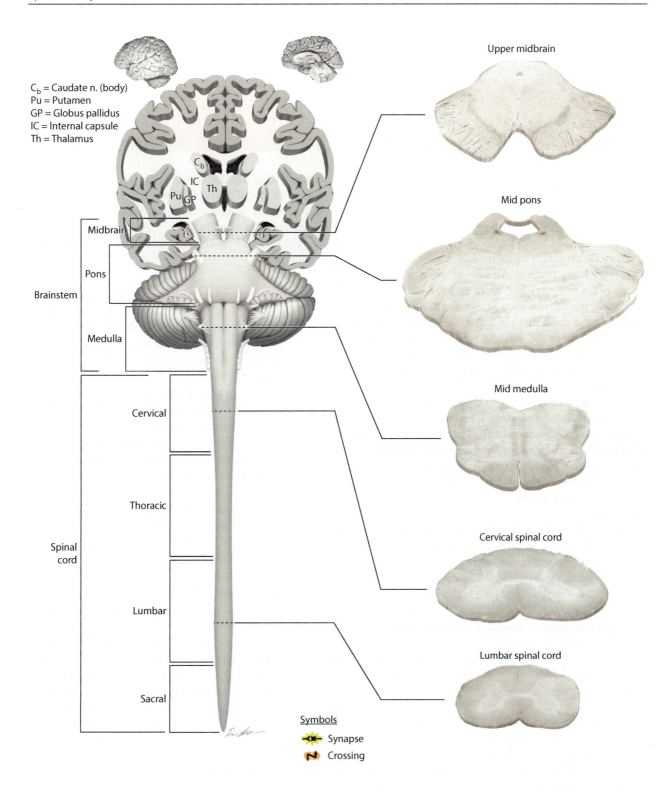

C_b = Caudate n. (body)
Pu = Putamen
GP = Globus pallidus
IC = Internal capsule
Th = Thalamus

Upper midbrain

Mid pons

Mid medulla

Cervical spinal cord

Lumbar spinal cord

Symbols
Synapse
Crossing

FIGURE 4.6: Pathways and Cross-Sections—Orientation to Diagrams

Chapter 5

Sensory and Motor Pathways

SENSORY SYSTEMS

INTRODUCTION

Sensory systems, also called modalities (singular, **modality**), share many features. All sensory systems begin with **receptors**, sometimes free nerve endings and others that are highly specialized, such as those in the skin for touch and vibration sense and the hair cells in the cochlea for hearing, as well as the rods and cones in the retina. These receptors activate the peripheral sensory fibers appropriate for that sensory system. The peripheral nerves have their cell bodies in **sensory ganglia** that belong to the **peripheral nervous system (PNS)**. For the body (neck down), these are the dorsal root ganglia, located in the intervertebral spaces (see Figure 1.12). The trigeminal system, which is the sensory nerve for the face and head, has its ganglion inside the skull. The central process of these peripheral neurons enters the CNS and synapses in the nucleus appropriate for that sensory system (this is hard-wired).

Generally speaking, the older systems both peripherally and centrally involve axons of small diameter that are thinly myelinated or unmyelinated, with a slow rate of conduction. In general, these pathways consist of fibers-synapses-fibers, with collaterals, creating a multisynaptic chain with many opportunities for spreading the information, but thereby making transmission slow and quite insecure.

The various forms of sensation in this category include **pain** and **temperature**, as well as related sensations of "itch" and perhaps also sensations of a "pleasurable" or sexual nature.

- A clinician examines the pain pathways by using the tip of a (clean, sterile) pin—gently, sometimes using the dull side to check whether the patient is reporting accurately.
- Temperature is more difficult to check clinically and involves the use of objects that are either cold (e.g., metal) or warm (e.g., water).

The newer pathways that have evolved have larger axons that are more thickly myelinated and therefore conduct more rapidly. These form rather direct connections with few, if any, collaterals. The latter type of pathway transfers information more securely and is more specialized functionally.

The various forms of sensation in this category include the following.

- **Discriminative Touch** is defined as the ability to sense fine touch such as a cloth or tissue (with the eyes closed). This should not be confused with cortical analysis of sensation such as stereognosis, two-point discrimination or graphesthesia.
- **Stereognosis**: the recognition of an object using only tactile information (e.g., a coin such as a 10 cent denomination in the U.S. or Canada).
- **Two-point discrimination**: the perception of being touched by 2 points simultaneously (finer in the hand where there are more receptors than on the back).
- **Graphesthesia**: the recognition of letters or numbers "written" on the skin (with a blunt object) which would again be better where there are more receptors.
- **Joint position** is tested by moving the bones about a joint and asking the patient to report the direction of the movement (again with the eyes closed).
- **Vibration** is tested by placing a 256 Hz tuning fork that has been set into motion onto a joint space (e.g., the interphalangeal joint of hands or feet, wrist, the ankle). These sensory receptors in the joint capsule are quite specialized; the fibers carrying the afferents to the CNS are rapidly conducting large diameter and thickly myelinated axons, meaning that the information is carried quickly and with a high degree of fidelity.

Because of the upright posture of humans, the sensory systems go upward or ascend to the cortex—the **ascending systems**. The sensory information is "processed" by various nuclei along the pathway. Three pathways are concerned with carrying sensory information from the skin (and the joints), two from the body region, and one (with subparts) from the head:

- The **dorsal column–medial lemniscus** pathway, a newer pathway for the somatosensory sensory modalities of discriminative touch, joint position, and "vibration."
- The **anterolateral system**, an older system that carries pain and temperature and some less discriminative forms of touch sensations; formerly called the lateral spino-thalamic and ventral (anterior) spino-thalamic tracts, respectively.
- The **trigeminal pathway**, carrying sensations from the face and head area (including discriminative touch, pain, and temperature) and involving both newer and older types of sensation.

All the sensory pathways including the special senses, except for olfaction, relay in the thalamus before going on to the cerebral cortex (see Figure 5.5 and Figure 6.13); the olfactory system (smell) is considered with the limbic system in Section 4.

CLINICAL ASPECT

The aim of the sensory examination is to establish whether the deficit, if there is any, involves the peripheral nerves supplying a region *or* whether there is a level (spinal or brainstem) below which some or all or some sensory sensations are not perceived. Both sides need to be tested.

ADDITIONAL DETAIL

The sensory systems from the periphery (including the head region) have traditionally been described as a system of neurons:

- **First order neurons**—these are the neuronal cell bodies of the peripheral nerve fibers themselves, located in the dorsal root ganglia (DRG, part of the peripheral nervous system, the PNS) for all the somatosensory systems from the body. For the trigeminal afferents (CN V from the face and head region), the sensory ganglion is located inside the skull. The central processes of these neurons enter the central nervous system (CNS), the spinal cord and brainstem respectively. Once within the CNS, the oligodendrocyte becomes the glial cell responsible for myelin, whereas the Schwann cell is responsible for myelination in the PNS.
- **Second order neurons**—these neurons are located in the central nervous system and are the ones that project their axons upward. In the dorsal column—medial lemniscal pathway— they are located in the lowermost brainstem (see Figure 5.2). In the anterolateral pathway, they are found in the spinal cord (see Figure 5.3). In the trigeminal system, they are located either in the pons or in the lower brainstem (see Figure 5.4). It is important to note that there may be synaptic relays preceding the activation of the second order neuron. In all of these systems, the axons of the second order neurons cross the midline—**decussate**—and ascend to the thalamus. As will be shown, the crossing (decussation) occurs at different levels for each of the pathways.
- **Third order neurons**—these are located in the thalamic relay nuclei and project to the cortex (see Figure 4.3 and Figure 6.13).

This analysis of the sensory systems is more difficult to apply to the special senses and it is probably best to consider each of these—vision, audition, olfaction (smell)—separately.

FIGURE 5.1—SPINAL CORD CROSS-SECTION

SENSORY: NUCLEI AND AFFERENTS

This is a representation of a spinal cord cross-section, at the cervical level (see Figure 3.9 and Figure 4.1), with a focus on the sensory afferent side. All levels of the spinal cord have the same sensory organization, although the size of the nuclei varies with the number of afferents.

The dorsal horn of the spinal cord has a number of nuclei related to sensory afferents, particularly pain and temperature, as well as crude touch (see upper illustration). The first nucleus encountered is the posteromarginal, where some sensory afferents terminate. The next and most prominent nucleus is the **substantia gelatinosa**, composed of small cells, where many of the pain afferents terminate. Medial to this is the **proper sensory nucleus**, which is a relay site for these fibers; neurons in this nucleus project across the midline and give rise to a tract—the anterolateral tract (see later and Figure 5.3).

A small local tract that carries pain and temperature afferents up and down the spinal cord for a few segments is called the **dorsolateral fasciculus** (of Lissauer).

The dorsal nucleus (of Clarke) is a relay nucleus for cerebellar afferents (see Figure 5.15 and Figure 6.10).

The illustrations show the difference at the entry level between the two sensory pathways—the dorsal column tracts and the anterolateral system. The cell bodies for both these sensory nerves are located in the **dorsal root ganglion (DRG)** (see Figure 1.12).

UPPER ILLUSTRATION

On the left side, the afferent fibers carrying discriminative touch, position sense, and vibration enter the dorsal horn and immediately turn upward. The fibers may give off local collaterals (e.g., to the intermediate gray), but the information from these rapidly conducting heavily myelinated fibers is carried upward in the two tracts that lie between the dorsal horns, called collectively the **dorsal columns** (see Figure 6.10). The first synapse in this pathway occurs at the level of the lower medulla (see Figure 5.2).

LOWER ILLUSTRATION

On the same side, the afferents carrying the pathways for pain, temperature, and crude touch enter and synapse in the nuclei of the dorsal horn. The nerves conveying this sensory input into the spinal cord are thinly myelinated or unmyelinated and conduct slowly. After several synapses, these fibers *cross* the midline *(decussate)* in the white matter in front of the commissural gray matter (the gray matter joining the two sides), called the **ventral (anterior) white commissure** (see upper illustration). The fibers then ascend as the spino-thalamic tracts, called collectively the **anterolateral system** (see Figure 5.3 and Figure 6.10).

CLINICAL ASPECT

A lesion of one side of the spinal cord therefore affects the two sensory systems differently because of this arrangement. The sensory modalities of the dorsal column system are disrupted on the same side (ipsilateral). The pain and temperature pathway, having crossed, leads to a loss of these modalities on the opposite side. This is known as a **dissociated sensory deficit**.

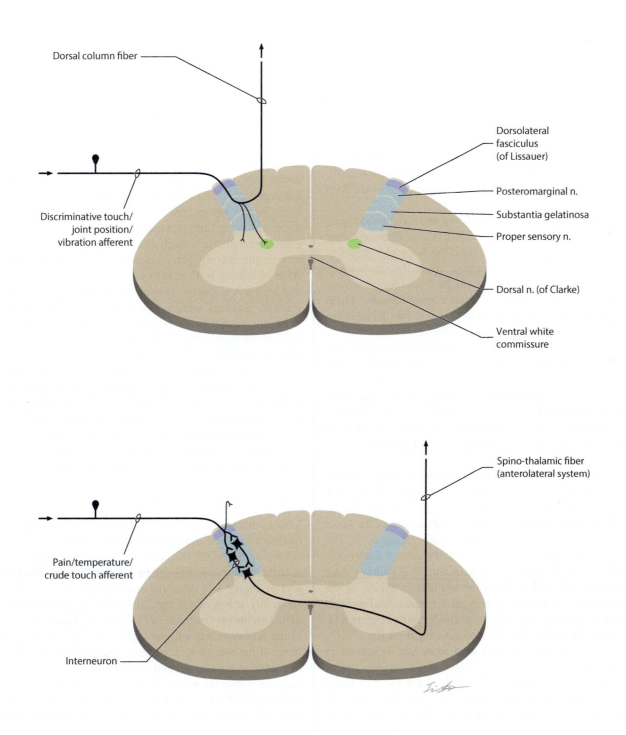

Dorsal column fiber

Discriminative touch/
joint position/
vibration afferent

Dorsolateral
fasciculus
(of Lissauer)

Posteromarginal n.

Substantia gelatinosa

Proper sensory n.

Dorsal n. (of Clarke)

Ventral white
commissure

Spino-thalamic fiber
(anterolateral system)

Pain/temperature/
crude touch afferent

Interneuron

FIGURE 5.1: Spinal Cord Cross-Section—Sensory: Nuclei and Afferents

FIGURE 5.2—DORSAL COLUMN–MEDIAL LEMNISCUS PATHWAY

DISCRIMINATIVE TOUCH, JOINT POSITION, VIBRATION

This pathway carries the modalities **discriminative touch, joint position**, and the somewhat artificial "sense" of **vibration** from the body. Receptors for these modalities are generally specialized endings in the skin and joint capsule.

The axons enter the spinal cord and turn upward, with no synapse (see also Figure 5.1). Those fibers entering below spinal cord level T6 form the **fasciculus gracilis**, the gracile tract; those entering above T6, particularly those from the upper limb, form the **fasciculus cuneatus**, the cuneate tract, which is situated more laterally. These tracts ascend the spinal cord between the two dorsal horns and form the **dorsal column** (see Figure 6.10).

The first synapse in this pathway is found in two nuclei located in the lowermost part of the medulla, in the **nuclei gracilis and cuneatus** (see Appendix Figure A.10; also Figure 1.9 and Figure 3.3). Topographical representation, also called **somatotopic** organization, is maintained in these nuclei, meaning that there are distinct populations of neurons that are activated by areas of the periphery that were stimulated.

After neurophysiological processing, axons emanate from these two nuclei that cross the midline (decussate). This stream of fibers, called the **internal arcuate fibers**, can be recognized in suitably stained sections of the lower medulla (see Figure 6.11 and Appendix Figure A.10). The fibers then group together to form the **medial lemniscus**, which ascends through the brainstem. This pathway does not give off collaterals to the reticular formation in the brainstem. This pathway changes orientation and position as it ascends through the pons and midbrain (see Figure 6.11, Appendix Figure A.6, and Appendix Figure A.10).

The medial lemniscus terminates (i.e., synapses) in the **ventral posterolateral nucleus** of the thalamus **(VPL)** (see Figure 4.3, Figure 5.5, and Figure 6.13). The fibers then enter the internal capsule, its posterior limb (see Figure 4.4), and travel to the somatosensory cortex, to terminate along the **post-central gyrus areas 1, 2, and 3**

(see Figure 1.3, Figure 4.5, and Figure 6.13). The representation of the body on this gyrus is not proportional to the size of the area being represented; for example, the fingers, particularly the thumb, are given a much larger area of cortical representation than the trunk; this is called the **sensory "homunculus"** (see Figure 4.5). The lower limb, represented on the medial aspect of the hemisphere, has little cortical representation.

NEUROLOGICAL NEUROANATOMY

In the spinal cord, the pathways are found between the two dorsal horns as a well-myelinated bundle of fibers called the dorsal column (see Figure 4.1 and Figure 6.10). The tracts have a topographical organization, with the lower body and lower limb represented in the medially placed gracile tract (thoracic level) and the upper body and upper limb in the laterally placed cuneate tract (cervical level). After synapsing in their respective nuclei (see Figure 1.9 and Figure 3.3) and the crossing of the fibers in the lower medulla (internal arcuate fibers, see Appendix Figure A.10), the medial lemniscus tract is formed. This heavily myelinated tract, which is easily seen in myelin-stained sections of the brainstem, is located initially between the inferior olivary nuclei and is oriented in the dorsoventral position. The tract moves more posteriorly, shifts laterally, and also changes orientation as it ascends. The fibers are topographically organized, with the leg represented laterally and the upper limb medially. The medial lemniscus is joined by the anterolateral system and trigeminal pathway in the upper pons (see Figure 5.5 and Figure 6.11).

CLINICAL ASPECT

Lesions involving this tract result in the loss of the sensory modalities carried in this pathway. A lesion of the dorsal column in the spinal cord causes a loss on the same side; note that this area is supplied by the posterior spinal arteries (see Figure 8.7). After the crossing in the lower brainstem, any lesion of the medial lemniscus results in the deficit on the opposite side of the body. Lesions occurring in the midbrain and internal capsule usually involve the fibers of the anterolateral pathway, as well as the modalities carried in the medial lemniscus and trigeminothalamic pathways (to be discussed with Figure 5.3 and Figure 5.4). With cortical lesions, the area of the body affected is determined by the area of the post-central gyrus involved (see Figure 4.5).

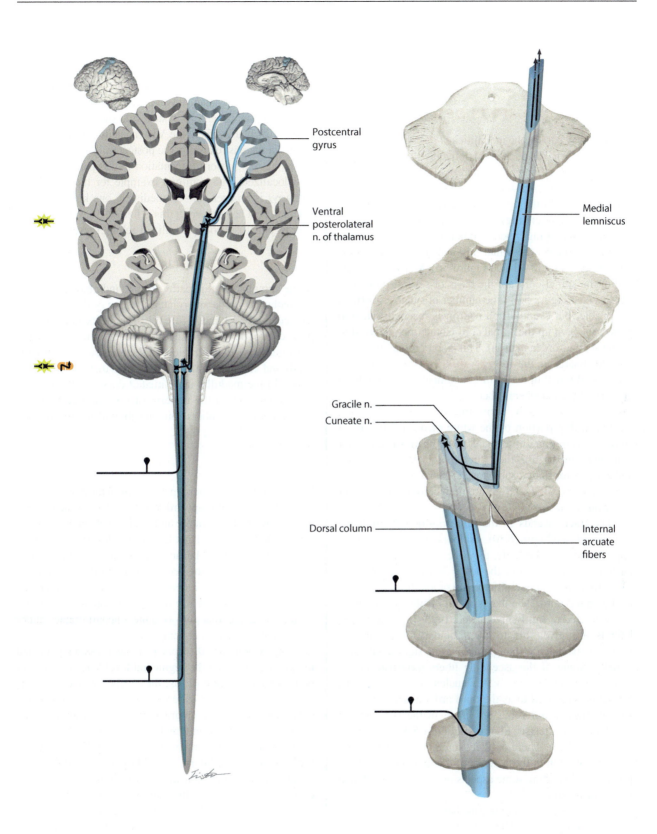

Postcentral gyrus

Ventral posterolateral n. of thalamus

Medial lemniscus

Gracile n.

Cuneate n.

Dorsal column

Internal arcuate fibers

FIGURE 5.2: Dorsal Column–Medial Lemniscus Pathway—Discriminative Touch, Joint Position, Vibration

FIGURE 5.3—ANTEROLATERAL SYSTEM

PAIN, TEMPERATURE, CRUDE TOUCH

This pathway carries the modalities of **pain** and **temperature** and a form of touch sensation called **crude** or **light touch**. The sensations of itch and tickle, and other forms of sensation (e.g., pleasurable/"sexual") are likely carried in this system. In the periphery the receptors are usually simply free nerve endings, without any specialization.

These incoming fibers (sometimes called the first order neuron) enter the spinal cord and synapse in the dorsal horn (see Figure 4.1 and Figure 5.1). Many collaterals within the spinal cord are the basis of several protective reflexes (see Figure 5.7). The number of synapses formed is variable, but eventually a neuron is reached that will project its axon up the spinal cord (sometimes referred to as the second order neuron). This axon crosses the midline, decussates, in the **ventral (anterior) white commissure**, usually within two to three segments above the level of entry of the peripheral fibers.

These axons now form the **anterolateral tract**, located in that portion of the white matter of the spinal cord. It was traditional to speak of two pathways, one for pain and temperature, the **lateral spino-thalamic tract**, and another for light (crude) touch, the **anterior (ventral) spino-thalamic tract**. Both are now considered together under one name.

The tract ascends in the same position through the spinal cord (see Figure 6.10). As fibers are added from the upper regions of the body, they are positioned medially, pushing the fibers from the lower body more laterally. Thus, there is a topographic organization to this pathway in the spinal cord. The axons of this pathway are either unmyelinated or thinly myelinated. In the brainstem, collaterals are given off to the reticular formation (see Figure 3.6B) that are thought to be quite significant functionally. Some of the ascending fibers terminate in the **ventral posterolateral (VPL)** nucleus of the thalamus (sometimes referred to as the third order neuron in a sensory pathway), and some terminate in the non-specific intralaminar nuclei (see Figure 4.3, Figure 5.5, Figure 6.11, and Figure 6.13). There is likely some processing of sensory information in these nuclei of the thalamus, not simply a relay, including some aspects of a "crude" touch and particularly pain.

There is a general consensus that acute pain sensation has two components. The older (also called the **paleo-spino-thalamic**) pathway involves the reported sensation of an ache, or diffuse pain that is poorly localized. The fibers underlying this pain system are likely unmyelinated

both peripherally and centrally, and the central connections are probably very diffuse; most likely these fibers terminate in the non-specific thalamic nuclei and influence the cortex widely. The newer pathway, sometimes called the **neo-spino-thalamic** system, involves thinly myelinated fibers in the peripheral nervous system and the central nervous system, likely ascends to the VPL nucleus of the thalamus, and from there is relayed to the postcentral (sensory) gyrus. Therefore, the sensory information in this pathway can be well localized. The common example for these different pathways is a paper cut—immediately one knows exactly where the cut has occurred; this is followed several seconds later by a diffuse, poorly localized aching sensation.

NEUROLOGICAL NEUROANATOMY

In the spinal cord, this pathway is found among the various pathways in the anterolateral region of the white matter (see Figure 4.1 and Figure 6.10), hence its name. Its two parts cannot be distinguished from each other or from the other pathways in that region. In the brainstem, the tract is small and cannot usually be seen as a distinct bundle of fibers. In the medulla, it is situated dorsal to the inferior olivary nucleus; in the uppermost pons and certainly in the midbrain, the fibers join the medial lemniscus (see Figure 5.5 and Figure 6.11).

CLINICAL ASPECT

Lesions of the anterolateral pathway from the point of crossing in the spinal cord upward result in a loss of the modalities of pain and temperature and crude touch on the opposite side of the body. The exact level of the lesion can be quite accurately ascertained because the sensation of pain can be quite simply tested at the bedside by using the end of a pin. The best instrument is a new unused safety pin, which can then be discarded after use on a single patient. (The tester should be aware that this is quite uncomfortable and/or unpleasant for the patient being tested.)

Any lesion that disrupts just the crossing pain and temperature fibers at the segmental level leads to a loss of pain and temperature of just the levels affected. There is an uncommon disease, called **syringomyelia**, that involves a pathological cystic enlargement of the central canal, called a **syrinx** (pipe in Greek). The cause of this disease is largely unknown, but sometimes it may be related to a previous traumatic injury. The enlargement of the central canal interrupts the pain and temperature fibers in their crossing anteriorly in the anterior white commissure. Usually, this occurs in the cervical region, and patients complain of the loss of these modalities in the upper limbs and hand, in what is called a cape-like distribution. The enlargement can be visualized with magnetic resonance imaging (MRI).

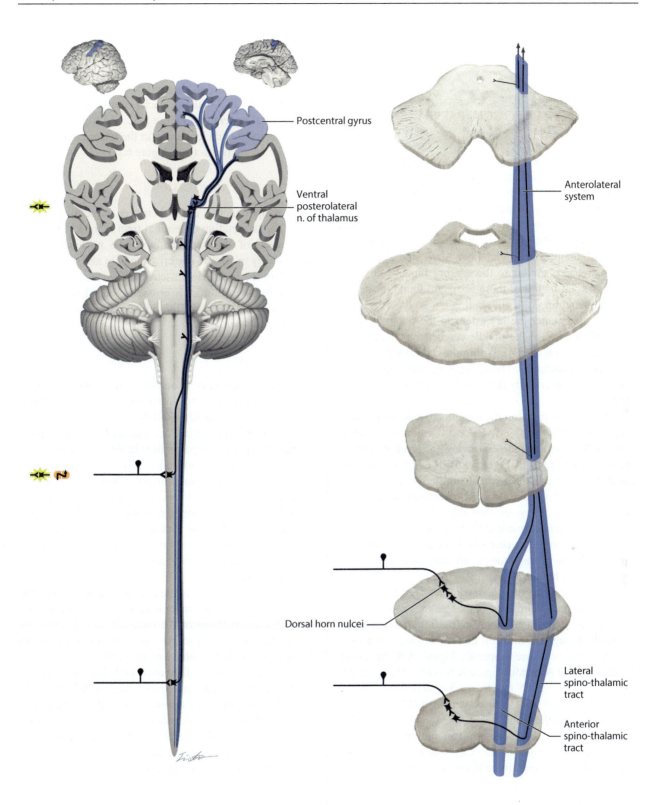

Postcentral gyrus

Ventral posterolateral n. of thalamus

Anterolateral system

Dorsal horn nulcei

Lateral spino-thalamic tract

Anterior spino-thalamic tract

FIGURE 5.3: Anterolateral System—Pain, Temperature, Crude Touch

FIGURE 5.4—TRIGEMINAL PATHWAYS

DISCRIMINATIVE TOUCH, PAIN, TEMPERATURE

The sensory input carried in the trigeminal nerve comes from the face, particularly from the lips, all the mucous membranes inside the mouth including the tongue, the teeth, the mucous membranes of the paranasal (air) sinuses, and, the conjunctiva of the eye. The sensory fibers include the modalities of discriminative touch, as well as pain and temperature. The fiber sizes and degree of myelination are similar to the sensory inputs below the neck. The cell bodies of these fibers are found in the trigeminal ganglion inside the skull.

The fibers enter the brainstem in the pontine region (along the course of the middle cerebellar peduncle; see Figure 1.8 and Figure 3.4). Within the central nervous system there is a differential handling of the modalities, comparable to the previously described pathways in the spinal cord.

Those fibers carrying the sensations of discriminative touch synapse in the **principal (main) nucleus** of cranial nerve (CN) V, in the mid-pons, below the level of entry of the nerve (see Appendix Figure A.6). The fibers then cross the midline and join the medial lemniscus, to terminate in the **ventral posteromedial (VPM)** nucleus of the thalamus (see Figure 4.3). They are then relayed via the posterior limb of the internal capsule to the postcentral gyrus, where the face area is represented on the dorsolateral surface (see Figure 4.5); the lips and tongue are very well represented on the sensory homunculus.

Those fibers carrying the modalities of pain and temperature descend within the brainstem on the same side. They form a tract that starts at the mid-pontine level, descends through the medulla, and reaches the upper level of the spinal cord (see Appendix Figure A.8, Appendix Figure A.9, and Appendix Figure A.10) called the **descending or spinal tract of V** (see Figure 3.4), and also called the **spinal trigeminal tract**. Immediately medial to this tract is a nucleus with the same name, the spinal (descending) nucleus of CN V. The fibers terminate in this nucleus and, after synapsing, cross to the other side and ascend (see Figure 6.11). Therefore, these fibers decussate over a wide region and do not form a compact bundle of crossing fibers; they also send collaterals to the reticular formation. These trigeminal fibers join with those carrying touch to form the **trigeminal pathway** in the mid-pons. They terminate in the VPM and other thalamic nuclei, similar to those of the anterolateral system (see Figure 5.5). The trigeminal pathway joins the medial lemniscus in the upper pons, as does the anterolateral pathway (se Figure 6.11).

NEUROLOGICAL NEUROANATOMY

The principal nucleus of CN V is seen at the mid-pontine level. The descending trigeminal tract is found in the lateral aspect of the medulla, with the nucleus situated immediately medially. The crossing pain and temperature fibers join the medial lemniscus over a wide area and are thought to have completely crossed by the lower pontine region. The collaterals of these fibers to the reticular formation are shown.

CLINICAL ASPECT

Trigeminal neuralgia is an affliction of the trigeminal nerve of uncertain origin that causes severe "lightning" pain in one of the branches of CN V; often there is a trigger such as moving the jaw or an area of skin. The shooting pains may occur in paroxysms lasting several minutes. An older name for this affliction is *tic douloureux*. There is often a history of trauma to the CN V such as dental work or facial trauma; in some cases a vascular loop from the basilar artery can impinge on the CN V. Treatment of these cases, which cause enormous pain and suffering, is difficult. There are surgical options for treatment based on the underlying cause. Most patients can be managed with medical therapy.

An ischemic infarct of the lateral medulla disrupts the descending pain and temperature fibers and results in a loss of these sensations on the same side of the face while leaving the fibers for discriminative touch sensation from the face intact. This lesion, known as the **lateral medullary syndrome** (of Wallenberg), includes other deficits (see Figure 6.11 and discussed with Appendix Figure A.9). A lesion of the medial lemniscus above the pontine level involves all trigeminal sensations on the opposite side. Internal capsule and cortical lesions cause a loss of trigeminal sensations from the opposite side of the face, as well as involving other pathways.

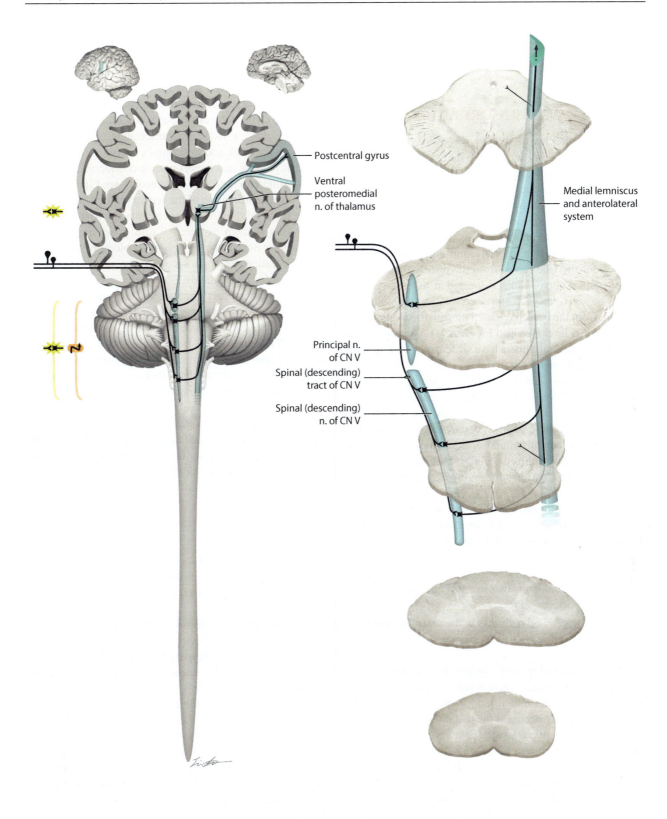

Postcentral gyrus

Ventral posteromedial n. of thalamus

Medial lemniscus and anterolateral system

Principal n. of CN V

Spinal (descending) tract of CN V

Spinal (descending) n. of CN V

FIGURE 5.4: Trigeminal Pathways—Discriminative Touch, Pain, Temperature

FIGURE 5.5—SENSORY SYSTEMS AND THALAMUS

THALAMUS WITH SENSORY PATHWAYS

This diagram presents all the somatosensory pathways, the dorsal column-medial lemniscus, the anterolateral pathway, and the trigeminal pathway as they pass through the midbrain region into the thalamus and onto the cortex. The view is a dorsal perspective (as in Figure 1.9 and Figure 3.3).

The pathway that carries discriminative touch sensation and information about joint position (as well as vibration) from the body is the medial lemniscus. The equivalent pathway for the face comes from the principal nucleus of the trigeminal nerve, which is located at the mid-pontine level. The anterolateral pathway conveying pain and temperature from the body has joined up with the medial lemniscus by this level (see also Figure 6.11). The trigeminal pain and temperature fibers have similarly joined up with the other trigeminal fibers.

The various sensory pathways are all grouped together at the level of the midbrain (see the inset cross-section; note the dorsal perspective). At the level of the lower midbrain, these pathways are located near the surface, dorsal to the substantia nigra; as they ascend they are found deeper within the midbrain, dorsal to the red nucleus (shown in Appendix Figure A.3 and Appendix Figure A.4).

The two pathways carrying the modalities of fine touch and position sense (and vibration) terminate in different specific relay nuclei of the thalamus (see Figure 4.3 and Figure 6.13):

- The medial lemniscus in the ventral posterolateral nucleus (VPL).
- The trigeminal pathways in the ventral posteromedial nucleus (VPM).

Sensory modality and topographic information is retained in these nuclei. There is physiological processing of the sensory information, and some type of sensory "perception" likely occurs at the thalamic level.

After the synaptic relay, the pathways continue as the (superior) thalamo-cortical radiation through the **posterior limb** of the **internal capsule**, between the thalamus and lentiform nucleus (see Figure 2.10A, Figure 2.10B, and Figure 4.4). The fibers are then found within the white matter of the hemispheres. The somatosensory information is distributed to the cortex along the **postcentral gyrus** (see the small diagrams of the brain), also called **S1**. Precise localization and two-point discrimination are cortical functions.

The information from the face and hand is topographically located on the dorsolateral aspect of the hemispheres (see Figure 1.3 and Figure 4.5). The information from the lower limb is localized along the continuation of this gyrus on the medial aspect of the hemispheres (see Figure 1.7 and Figure 4.5). This cortical representation is called the sensory "**homunculus**," a distorted representation of the body and face with the trunk and lower limbs having very little area, whereas the face and fingers receive considerable representation (similar to the motor homunculus, see Figure 4.5).

Further elaboration of the sensory information occurs in the **parietal association** areas adjacent to the postcentral gyrus (known as **areas 5 and 7**; see Figure 6.13). This allows us to learn to recognize objects by tactile sensations, called stereognosis (e.g., the denomination of various coins in the hand).

The pathways carrying pain and temperature from the body (the anterolateral system) and the face (spinal trigeminal system) terminate in part in the specific relay nuclei, VPL and VPM, but mainly in the intralaminar nuclei. These latter terminations may well be involved with the emotional correlates that accompany many sensory experiences (e.g., pleasant or unpleasant).

The fibers that have relayed pain information project from these nuclei to several cortical areas, including the post-central gyrus, SI, and area **SII** (a secondary sensory area), which is located in the lower portion of the parietal lobe, as well as other cortical regions. The output from the intralaminar nuclei of the thalamus goes to widespread cortical areas.

CLINICAL ASPECT

Knowledge of the exact location of the various pathways in the brainstem assists with the localization of various lesions (including vascular) affecting that region (see the Appendix).

Lesions of the thalamus may sometimes give rise to pain syndromes (also discussed with Figure 5.3).

There are three principal thalamic syndromes. The lateral thalamic syndrome causes deficits to the contralateral sensation (refer to Figure 4.3). The anterior thalamic syndrome causes clinical effects similar to Korsakoff's syndrome (psychosis) due to the disconnection of the thalamus to the hippocampus via the fornices (see Figure 10.1A and Figure 10.2). The medial thalamic syndrome results in a anhedonic syndrome (see definition below) similar to a frontal lobe syndrome (discussed with Figure 10.1B).

On occasion, language can be affected with thalamic lesions.

Note to the Learner: The dictionary definition of *anhedonia* is an "insensitiveness to pleasure," and also an "incapacity for experiencing happiness."

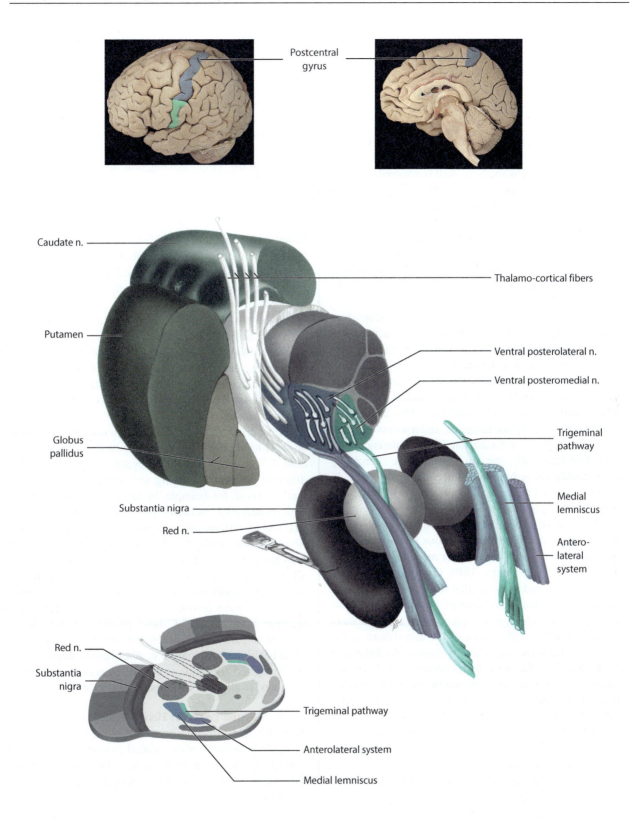

FIGURE 5.5: Sensory Systems and Thalamus—Thalamus with Sensory Pathways

FIGURE 5.6—PAIN MODULATION SYSTEM

RETICULAR FORMATION PATHWAY

Pain from physical sources is recognized by the nervous system at multiple levels. Localization of pain, knowing which part of the limbs and body wall is involved, requires the cortex of the postcentral gyrus (SI); SII is likely also involved in the perception of pain (discussed with Figure 5.5). There is good evidence that some "conscious" perception of pain occurs at the thalamic level.

We have a built-in system for dampening the influences of pain from the spinal cord level—the descending pain modulation pathway. This system apparently functions in the following way:

The neurons of the periaqueductal gray (see Figure 3.6B and Appendix Figure A.3) can be activated in a number of ways. It is known that many ascending fibers from the anterolateral system and trigeminal system activate neurons in this area (only the anterolateral fibers are shown in this illustration) either as collaterals or as direct endings of these fibers in the midbrain. This area is also known to be rich in opiate receptors, and it seems that neurons of this region can be activated by circulating endorphins. Experimentally, one can activate these neurons by direct stimulation or by a local injection of morphine. In addition, descending cortical fibers (cortico-bulbar; see Figure 5.10) may activate these neurons.

The axons of some of the neurons of the periaqueductal gray descend and terminate in one of the serotonin-containing raphe nuclei in the upper medulla, the **nucleus raphe magnus** (see Figure 3.6B). From here, there is a descending, crossed, pathway that is located in the dorsolateral white matter (funiculus) of the spinal cord. The serotonergic fibers terminate in the substantia gelatinosa of the spinal cord, a nuclear area of the dorsal horn of the spinal cord where the pain afferents synapse (see Figure 3.9 and Figure 5.1). The descending serotonergic fibers are thought to terminate on small interneurons that contain enkephalin. There is evidence that these enkephalin-containing spinal neurons *inhibit* the transmission of the pain afferents entering the spinal cord from peripheral pain receptors. Thus, descending influences are thought to modulate a local circuit.

There is a proposed mechanism that these same interneurons in the spinal cord can be activated by stimulation of other sensory afferents, particularly those from the mechanoreceptors in the skin and joints; these are anatomically large, well-myelinated peripheral nerve fibers. This is the physiological basis for the **gate control theory of pain**. In this model, the same circuit is activated at a segmental level.

It is useful to think about multiple gates for pain transmission. We know that mental states and cognitive processes can affect, positively and negatively, the experience of pain and our reaction to pain. The role of the limbic system and the "emotional reaction" to pain are discussed in Section 4.

CLINICAL ASPECT

In our daily experience with local pain, a bump or small cut, the common response is to rub the limb or the affected region vigorously. What we may be doing is activating the local spinal segmental circuits via the mechanoreceptors to decrease the pain sensation.

Some of the current treatments for pain are based on the structures and neurotransmitters discussed here. The gate control theory underlies the use of transcutaneous stimulation, one of the current therapies offered for the relief of pain. More controversial and certainly less certain is the postulated mechanism for the use of acupuncture in the treatment of pain.

Most discussions concerning pain refer to **acute** pain, or short-term pain caused by an injury or dental procedure. **Chronic** pain should be regarded from a somewhat different perspective. Living with pain on a daily basis that is caused, for example, by arthritis or cancer or diabetic neuropathy, including post-herpetic neuralgia (see later), is an unfortunately tragic state of being for many people. Those involved with pain therapy and research on pain have proposed that the central nervous system actually re-wires itself in reaction to chronic pain and may in fact become more sensitized to pain the longer the pain pathways remain active; some of this may occur at the receptor level. Many of these people are now being referred to pain clinics, where a team of physicians and other health professionals (e.g., anesthetists, neurologists, psychologists) try to assist people, by using a variety of therapies, to alleviate their disabling condition.

Note to the Learner: Herpes zoster is a not uncommon affliction affecting primarily older individuals and particularly persons with reduced immunity (e.g., those undergoing chemotherapy for cancer). Often these individuals are afflicted with recurrent (persistent) pain following the acute phase, called **post-herpetic neuralgia**. There is now a specific vaccine available—as a preventive—to boost the immunity for this vulnerable population.

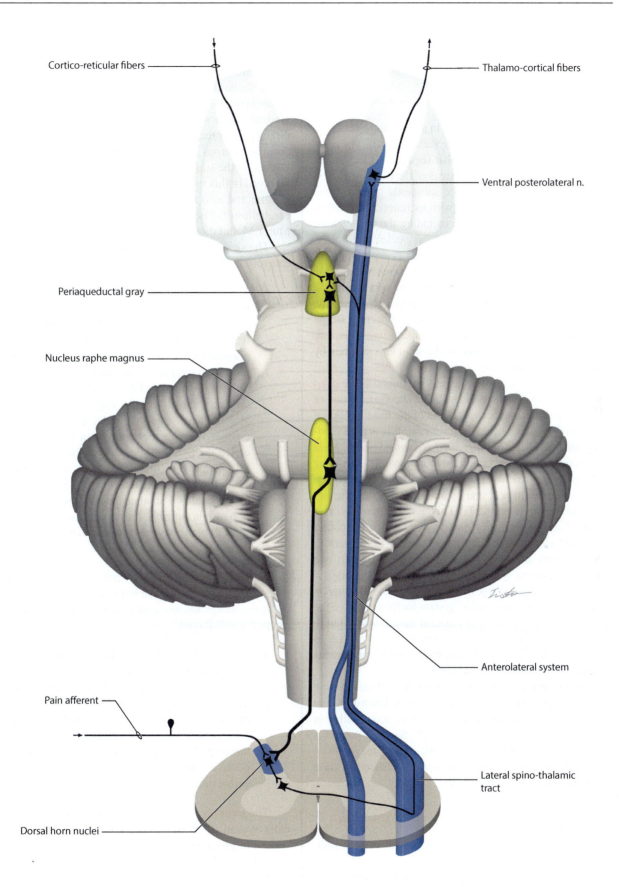

Cortico-reticular fibers

Thalamo-cortical fibers

Ventral posterolateral n.

Periaqueductal gray

Nucleus raphe magnus

Anterolateral system

Pain afferent

Lateral spino-thalamic tract

Dorsal horn nuclei

FIGURE 5.6: Pain Modulation System—Reticular Formation Pathway

MOTOR SYSTEMS

INTRODUCTION

There are multiple areas involved in motor control, and this is the reason for the title motor systems (plural). The parts of the central nervous system that regulate the movement of our muscles include: motor areas of the cerebral cortex; the basal ganglia (including the substantia nigra and the subthalamus); the cerebellum (with its functional subdivisions); nuclei of the brainstem including the red nucleus, the reticular formation, and vestibular nuclei; and finally, the output motor neurons of the cranial nerve motor nuclei and the motor neurons of the spinal cord, anatomically the anterior horn cell in the ventral horn of the spinal cord.

The anterior horn cell is the neuron that completes the pathway from command to action by connecting to the muscle; clinically, it is called the **lower motor neuron**. This neuron reacts to stretch of the muscle it supplies by contracting, known simply as the **stretch reflex**, also called the **myotatic reflex**, and is clinically examined as the **deep tendon reflex** (**DTR**, see Figure 5.7). The examination of the degree of reflex reactivity is one of the most important aspects of the neurological examination. It is graded clinically (discussed in *The Integrated Nervous System*). Most important, this reflex circuit is monosynaptic, involving only *one synapse* between the afferent fiber from the stretch receptors in the muscle and its anterior horn cell (further discussed with Figure 5.7). This reflex circuit is also involved in the control of muscle tone, which is another important aspect of the clinical neurological assessment.

Voluntary motor activity can be separated into two systems—a newer system (from the evolutionary perspective) for the control of fine motor movements of the fingers and hand and another, "older" system for the control of the proximal (large) joints and postural (axial) musculature that is partly under voluntary control. It is important to recognize that almost all voluntary motor movements involve some postural adjustments.

The motor pathways (tracts) are called descending because they commence in the cortex or brainstem and influence motor cells lower down in the neuraxis, either in the brainstem or the spinal cord. Several pathways "descend" through the spinal cord; only some of them cross (decussate), whereas others are uncrossed or have fibers that descend on both sides.

VOLUNTARY MOTOR SYSTEM

Voluntary motor control involves both direct and indirect pathways:

The **direct voluntary pathway**, for the control of fine motor movements, is controlled from the cerebral cortex. Areas of the frontal lobe are involved in motor planning, which precedes the activation of the motor commands that occurs in the motor strip. The descending pathways include the cortico-bulbar fibers to cranial nerve nuclei and the cortico-spinal tract to spinal cord motor neurons. In this atlas, this pathway may be considered the neo-motor system because it is controlled directly from the (neo)cortex. In the clinical milieu, the cortical neurons giving rise to this direct pathway are known as the **upper motor neuron**. The cortical areas involved in this motor control are modulated by the basal ganglia and also by the evolutionary newer areas of the cerebellum (to be discussed later in this chapter under Motor Modulation).

The descending tracts or pathways in the neo/direct category include:

- **Cortico-spinal tract**—This pathway originates in motor areas of the cerebral cortex. The cortico-spinal tract, from cortex to spinal cord, is a relatively new (from an evolutionary perspective) tract and the most important for voluntary movements in humans, particularly of the hand and digits—the direct voluntary motor pathway. As explained later, it is also known as the pyramidal tract because it courses through the medulla in the structure called the pyramid. Because of the location of this pathway in the spinal cord, it continues under the name the lateral cortical-spinal tract. There is also a smaller anterior (ventral) portion of this pathway, the anterior cortico-spinal tract, which may have a different functional contribution.

Note to the Learner: This pathway synapses sometimes directly on the anterior horn cells of the spinal cord, the *lower motor neuron*, and sometimes via an interneuron in the spinal cord.

- **Cortico-bulbar fibers**—This is a descriptive term that is poorly defined and includes all fibers that go to the brainstem, both cranial nerve nuclei and other brainstem nuclei. The fibers that go to the reticular formation include those that form part of the indirect voluntary motor pathway (discussed next). The cortico-pontine fibers are described with the cerebellum.

The **indirect voluntary pathways** are functionally an older system (from the evolutionary perspective) for the control of proximal joint movements and axial (postural) musculature. The motor cortex is controlling movements via the reticular formation and other nuclei of the brainstem. This functional part of motor control could be considered the paleo-motor system. It is most important to reiterate that almost all voluntary motor movements involve some postural adjustments. The various nuclei of the brainstem (the red nucleus, the vestibular nuclei, and the reticular formation) are also regulated by the functionally older parts of the cerebellum.

The descending tracts or pathways involved in the paleo/indirect category include:

- **Rubro-spinal tract**—The red nucleus of the midbrain gives rise to the rubro-spinal tract. Its connections are such that it may play a role in voluntary and non-voluntary motor activity; this may be the case in higher primates, but its precise role in humans is not clear. Nevertheless, the red nucleus region appears to be of clinical importance in the localization of lesions causing comatose states and abnormal reflex posturing of the limbs—named decorticate and decerebrate rigidity or posturing (see Figure 5.13).

- **Reticulo-spinal tracts**—These tracts are involved both in the indirect voluntary pathways and in non-voluntary motor regulation (see later), as well as in the underlying regulation of muscle tone and reflex responsiveness. Two tracts descend from the reticular formation—one from the pontine region, the medial reticulo-spinal tract, and one from the medulla, the lateral reticulo-spinal tract.

Note to the Learner: This older paleo/indirect system controls the reactivity of the lower motor neuron and hence influences the reactivity of the stretch reflex (the tendon reflex) and muscle tone, most important both functionally and clinically.

NON-VOLUNTARY MOTOR SYSTEM

This very old system (from the evolutionary perspective) serves for adjustment of the body to vestibular rotational and gravitational changes, via the vestibular apparatus in the inner ear. It also responds to visual input, motor responses that do not involve volitional actions but are essential for our daily activities. These would constitute the archi-motor system.

The descending tracts or pathways involved in the archi-motor category include:

- **Lateral and medial vestibulo-spinal tracts**— The lateral vestibular nucleus of the pons gives

rise to the lateral vestibulo-spinal tract. It is under the control of the oldest parts of the cerebellum, not the cerebral cortex. The medial vestibular nucleus sends some of its fibers downward in the medial vestibulo-spinal tract.

- **Reticulo-spinal tracts**—These tracts are involved both in the indirect voluntary pathways (see earlier) and in non-voluntary motor regulation.

LEARNING PLAN

The next part of this chapter considers the motor areas of the cerebral cortex, and the nuclei of the brainstem and reticular formation involved in motor regulation. The same standardized diagram of the nervous system is used as with the sensory systems, as well as the select cross-sections of the spinal cord and brainstem.

The last part of this chapter will discuss **Motor Modulation** by the basal ganglia and cerebellum. Both subsystems of motor control, acting via the thalamus, provide feedback to the motor cortex and modify motor movements. There are additional areas of the cerebral cortex involved in other aspects of motor activity. Broca's area for the motor control of speech is situated on the dominant side for language (usually the left hemisphere), on the dorsolateral surface, a little anterior to the lower portions of the motor areas (see Figure 4.5). The frontal eye field, in front of the premotor area, controls voluntary eye movements (see Figure 4.5 and discussed with Figure 6.8).

CLINICAL ASPECT

The conceptual approach to the motor system as comprising an upper motor neuron and a lower motor neuron is most important for clinical neurology. Injuries and diseases (all called lesions) in humans rarely if ever affect a single component of these pathways. A typical human lesion of the brain (e.g., vascular, trauma, tumor) usually affects cortical and/or subcortical areas, involving both the direct and indirect motor systems. The end results are alterations in several of the descending systems that lead to a mixture of deficits of movement, changes of the stretch reflexes (hyperreflexia, hyporeflexia, or absent reflexes), and a change in muscle tone (flaccidity or spasticity).

The plantar reflex and its abnormal response, now called the **extensor plantar response**—no longer using the terminology of a positive or negative Babinski sign—are extremely important clinically and are discussed with Figure 5.9.

FIGURE 5.7—SPINAL CORD CROSS-SECTIONS

MOTOR-ASSOCIATED NUCLEI

UPPER ILLUSTRATION

The motor regions of the spinal cord in the ventral horn are shown in this diagram. The lateral motor nuclei supply the distal musculature (e.g., the hand), and, as would be expected, this area is largest in the region of the limb plexuses (brachial and lumbosacral; see Figure 3.9 and Figure 4.1). The medial group of neurons supplies the axial musculature.

LOWER ILLUSTRATION

In the spinal cord, the neurons that are located in the ventral or anterior horn and are (histologically) the anterior horn cells, are usually called the **lower motor neuron**. Physiologists call these neurons the **alpha motor neurons**. In the brainstem, these neurons include the motor neurons of the cranial nerves (see Figure 3.5). Because all the descending influences converge on the lower motor neurons, these neurons have also been called, in a functional sense, the **final common pathway**.

The axons of these neurons leave the spinal cord though a series of rootlets (see Figure 1.11, Figure 1.12, and Figure 8.7). The lower motor neuron and its axon and the muscle fibers that it activates are collectively called the **motor unit**. The intactness of the motor unit determines the reflex reactivity, muscle strength, and muscle function. The ventral and dorsal roots unite outside the vertebral canal to form the mixed spinal nerve (see Figure 1.12 and 7.3).

MOTOR REFLEXES

The **myotatic reflex** is elicited by stretching a muscle (e.g., by tapping on its tendon), and this causes a contraction of the same muscle that was stretched; the receptor for this stretch is the muscle spindle. Thus, the reflex is also known as the **stretch reflex**, the **deep tendon reflex**, often simply called the **DTR**. In this reflex arc (shown on the left side), the information from the muscle spindle (afferent) ends directly on the anterior horn cell (efferent); there is only one synapse (i.e., a **monosynaptic** reflex).

Note to the Learner: This reflex is discussed in *The Integrated Nervous System,* and an animation of the reflex is shown on the Web site of that text.

All other reflexes, even a simple withdrawal reflex (e.g., touching a hot surface), involves some central processing (more than one synapse, multisynaptic) in the spinal cord, before the response (shown on the right side). All these reflexes involve hard-wired circuits of the spinal cord, but they can be and are influenced by information descending from higher levels of the nervous system.

Studies indicate that complex motor patterns are present in the spinal cord, such as stepping movements with alternating movements of the limbs, and that influences from higher centers provide the organization for these built-in patterns of activity.

CLINICAL ASPECT

The deep tendon reflex is a monosynaptic reflex and perhaps the most important for a neurological examination. The degree of reactivity of the lower motor neuron is influenced by higher centers, also called descending influences, particularly by the reticular formation (discussed with Figure 5.12B). An increase in this reflex responsiveness is called **hyperreflexia**, and a decrease is **hyporeflexia**. The state of activity of the lower motor neuron also influences **muscle tone**—the "feel" of a muscle at rest and the way in which the muscles react to passive stretch (by the examiner); again, there may be hypertonia or hypotonia.

Disease or destruction of the anterior horn cells results in weakness or paralysis of the muscles supplied by those neurons. The extent of the weakness depends on the extent of the neuronal loss and is rated on a clinical scale, called the **MRC** (Medical Research Council). There is also a decrease in muscle tone, as well as a decrease in reflex responsiveness (hyporeflexia) of the affected segments; the plantar response is normal.

The clinical usefulness of this information is discussed in *The Integrated Nervous System.*

The specific disease that affects the lower motor neurons is **poliomyelitis** (usually called simply **polio**), a infectious disease usually affecting children carried in fecally contaminated water. This disease entity has almost been totally eradicated in the industrialized world by immunization of all children.

In adults, the disease that affects both the upper and lower motor neurons specifically (including cranial nerve motor neurons) is **amyotrophic lateral sclerosis (ALS)**, also known as Lou Gehrig disease. In this progressive degenerative disease, there is a loss of the motor neurons in the cerebral cortex (the upper motor neuron) *and* the anterior horn cells (the lower motor neuron). The clinical picture depends on the degree of loss of the neurons at both levels. People afflicted with this devastating disease suffer a continuous march of loss of function, including swallowing and respiratory function, that leads to their death. Their intellectual functions are not affected. This exacts a heavy emotional toll on the person afflicted and on family and friends. Researchers are actively seeking ways to arrest the degenerative loss of these neurons.

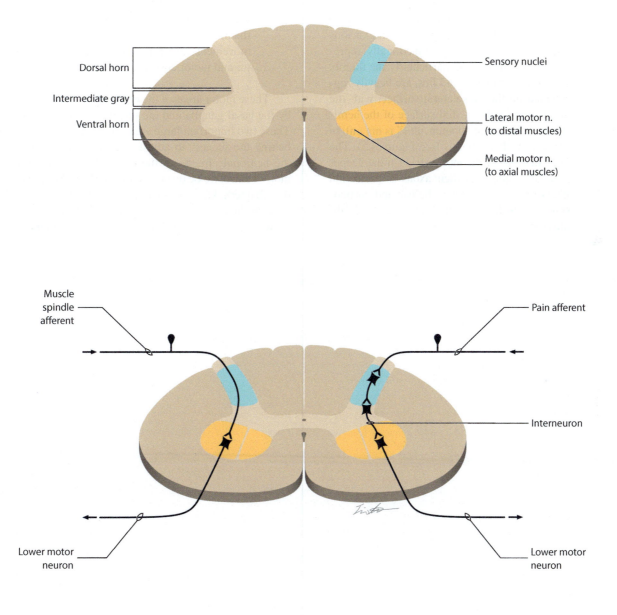

Dorsal horn

Intermediate gray

Ventral horn

Sensory nuclei

Lateral motor n.
(to distal muscles)

Medial motor n.
(to axial muscles)

Muscle
spindle
afferent

Pain afferent

Interneuron

Lower motor
neuron

Lower motor
neuron

FIGURE 5.7: Spinal Cord Cross-Sections—Motor-Associated Nuclei

FIGURE 5.8—MOTOR CORTEX

MOTOR, PREMOTOR, SUPPLEMENTARY

There are three areas of the cerebral cortex directly involved in motor control (see Figure 1.3 and Figure 4.5):

- The **motor cortex** is otherwise known as the precentral gyrus, anatomically **area 4** (also called the motor strip), in which the various portions of the body are functionally represented; the fingers and particularly the thumb, as well as the tongue and lips, are heavily represented on the dorsolateral surface, with the lower limb on the medial surface of the hemisphere. This motor "homunculus" is not unlike the sensory homunculus (see Figure 4.5). The large neurons of the motor strip (in the deeper cortical layers) send their axons as projection fibers to form the cortico-bulbar and cortico-spinal tracts. It is this cortical strip that contributes most to voluntary movements.

- Anterior to this is another wedge-shaped cortical area, the **premotor cortex**, **area 6**, with a less definite body representation. This cortical area sends its axons to the motor cortex, as well as to the cortico-spinal tract, and likely its function has more to do with proximal joint control and postural adjustments needed for movements.

- The **supplementary motor cortex** is located on the dorsolateral surface and mostly on the medial surface of the hemisphere, anterior to the motor areas. This is an organizing area for movements and its axons are sent to the premotor and motor cortex.

These motor areas of the cerebral cortex are regulated by the basal ganglia and (newer) parts of the cerebellum. These two important large areas of the brain are "working behind the scenes" to adjust and calibrate, to modulate, the neuronal circuits of the cerebral cortex involved in motor control (to be discussed under Motor Modulation in this chapter). All these areas also receive input from other parts of the cerebral cortex, particularly from the sensory postcentral gyrus, as well as from the parietal lobe.

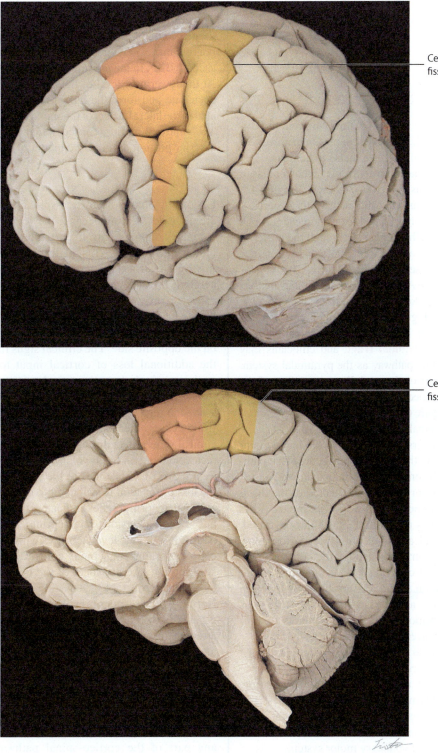

Central fissure

Central fissure

■ Primary motor cortex
■ Premotor cortex
■ Supplementary motor area

FIGURE 5.8: Motor Cortex—Motor, Premotor, Supplementary

FIGURE 5.9—CORTICO-SPINAL TRACT: THE PYRAMIDAL SYSTEM

DIRECT VOLUNTARY PATHWAY

The cortico-spinal tract, a direct pathway linking the cortex with the spinal cord, is the most important pathway for precise fine voluntary motor movements in humans.

This pathway originates mostly from the motor areas of the cerebral cortex, areas 4 and 6 (see Figure 1.3, Figure 4.5, and Figure 5.8; also discussed earlier with Motor Systems—Introduction). The well-myelinated axons descend through the white matter of the hemispheres, through the posterior limb of the internal capsule (see Figure 4.4), continue through the midbrain (see Appendix Figure A.3) and pons (as disbursed fibers; see Appendix Figure A.6), and are then found within the medullary pyramids (see Figure 1.8, Figure 3.1, Appendix Figure A.8, Appendix Figure A.9, and Appendix Figure A.10). Hence, the cortico-spinal pathway is often called the **pyramidal tract**, and clinicians may sometimes refer to this pathway as the pyramidal system. At the lowermost part of the medulla, most (90%) of the cortico-spinal fibers decussate (cross) in the **pyramidal decussation** (see Figure 6.12) and form the **lateral cortico-spinal tract** in the spinal cord (see Figure 6.10).

Many of these fibers end directly on the lower motor neuron, particularly in the cervical spinal cord. This pathway is involved with controlling the individualized movements, especially of our fingers and hands (i.e., the distal limb musculature). Experimental work with monkeys has shown, after a lesion is placed in the medullary pyramid, muscle weakness and a loss of ability to perform fine movements of the fingers and hand (on the opposite side); the animals were still capable of voluntary gross motor movements of the limb. There was no change in the deep tendon reflexes, and a decrease in muscle tone was reported. The innervation for the lower extremity is similar, but it clearly involves less voluntary activity.

Those fibers that do not cross in the pyramidal decussation form the **anterior (or ventral) cortico-spinal tract**. Many of these axons cross before terminating, whereas others supply motor neurons on both sides. The ventral pathway is concerned with movements of the proximal limb joints and axial movements, similar to other pathways of the indirect voluntary motor system.

Other areas of cortex contribute to the cortico-spinal pathway—these include the sensory cortical areas, the postcentral gyrus (also discussed with Figure 5.10).

NEUROLOGICAL NEUROANATOMY

After emerging from the internal capsule, the cortico-spinal tract is found in the mid-portion of the cerebral peduncles in the midbrain (see Figure 1.8 and Figure 5.10). The cortico-spinal fibers are then dispersed in the pontine region and are seen as bundles of axons among the pontine nuclei. The fibers collect again in the medulla as a single tract, in the pyramids on each side of the midline. At the lowermost level of the medulla, 90% of the fibers decussate and form the lateral cortico-spinal tract, situated in the lateral aspect of the spinal cord (see Figure 6.10). The ventral (anterior) cortico-spinal tract is found in the anterior portion of the white matter of the spinal cord.

CLINICAL ASPECT

Lesions involving the cortico-spinal pathway in humans are quite devastating because they rob the individual of voluntary motor control, particularly the fine skilled motor movements. This pathway is quite commonly involved in strokes, as a result of vascular lesions of the cerebral arteries or of the deep arteries to the internal capsule (reviewed with Figure 8.4, Figure 8.5, and Figure 8.6). This lesion results in an upper motor neuron pattern of weakness (paresis) or paralysis of the muscles on the opposite side. The clinical signs in humans reflect the additional loss of cortical input to the brainstem nuclei, particularly to the reticular formation.

Damage to the tract in the spinal cord is seen after traumatic injuries (e.g., automobile and diving accidents). In this case, other pathways would be involved, and the clinical signs reflect this damage, with the loss of the indirect voluntary and non-voluntary tracts (discussed with Figure 5.12B). If one-half of the spinal cord is damaged, the loss of function is ipsilateral to the lesion.

One abnormal reflex indicates, in humans, that there has been a lesion interrupting the cortico-spinal pathway—at any level (cortex, white matter, internal capsule, brainstem, spinal cord). The reflex involves stroking the lateral aspect of the bottom of the foot (a most uncomfortable sensation for most people). Normally, the response involves flexion of the toes (note especially the big toe)—the **plantar reflex**, and often an attempt to withdraw the limb. Testing this same reflex after a lesion interrupting the cortico-spinal pathway results in an upward movement of the big toe (extension) and a fanning apart of the other toes. The abnormal response is called an **extensor plantar response** (this terminology will be used in the atlas), and it can be elicited almost immediately after any lesion that interrupts any part of the cortico-spinal pathway, from cortex through to spinal cord (except in the period immediately after an acute transaction of the spinal cord, called spinal shock; discussed with Figure 6.10).

Note to the Learner: The term *positive Babinski sign*—not reflex—is no longer being used because generations of students often misused the terminology of positive and negative Babinski (see also Clinical Cases).

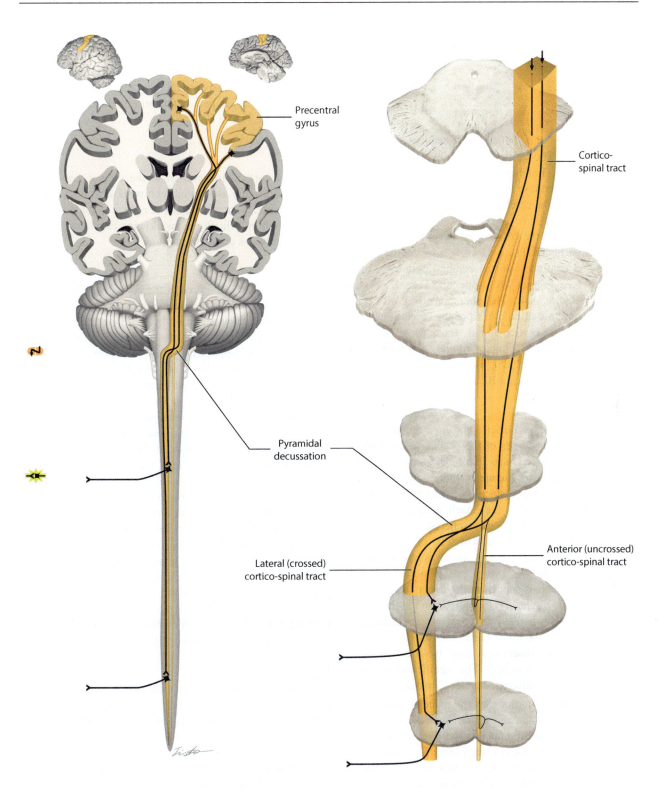

Precentral
gyrus

Cortico-
spinal tract

Pyramidal
decussation

Lateral (crossed)
cortico-spinal tract

Anterior (uncrossed)
cortico-spinal tract

FIGURE 5.9: Cortico-Spinal Tract: The Pyramidal System—Direct Voluntary Pathway

FIGURE 5.10—CORTICO-BULBAR FIBERS

NUCLEI OF THE BRAINSTEM

The word "bulb" (i.e., bulbar) is descriptive and refers to the brainstem. The cortico-bulbar fibers do not form a single pathway. The fibers end in a wide variety of nuclei of the brainstem; those fibers ending in the pontine nuclei to be relayed to the cerebellum are considered separately (see Figure 5.15).

Wide areas of the cortex send fibers to the brainstem as projection fibers. These axons course via the internal capsule and continue into the cerebral peduncles of the midbrain (see Figure 1.8, Appendix Figure A.3, and Appendix Figure A.4). The fibers involved with motor control occupy the middle third of the cerebral peduncle along with the cortico-spinal tract (described with Figure 5.9) supplying the motor cranial nerve nuclei of the brainstem (see Figure 3.5), the reticular formation (see Figure 3.6A and Figure 3.6B), and other motor-associated nuclei of the brainstem.

- Cranial nerve nuclei—The motor neurons of the cranial nerves of the brainstem are lower motor neurons (see Figure 3.5); the cortical motor cells are the upper motor neuron. These motor nuclei are generally innervated by fibers from both sides (i.e., each nucleus receives input from both hemispheres). The major exception to this rule, which is very important in the clinical setting, involves the cortical input to the **facial nucleus**. The portion of the facial nucleus supplying the upper facial muscles (the forehead) is supplied from both hemispheres, whereas the part of the nucleus supplying the lower facial muscles (around the mouth) is innervated only by the opposite hemisphere (crossed). The significance of this clinically is discussed under Clinical Aspect.
- Brainstem motor control nuclei—Cortical fibers influence all the brainstem motor nuclei, including the red nucleus (see Figure 5.11) and the substantia nigra (see Figure 5.14), and particularly the reticular formation (see Figure 5.12A and Figure 5.12B), but not the lateral vestibular nucleus (see Figure 5.13, Figure 5.16, and Figure 5.17). The cortico-reticular fibers are extremely important for voluntary movements of the proximal joints (indirect voluntary pathway) and for the regulation of muscle tone.
- Other brainstem nuclei—The cortical input to the sensory nuclei of the brainstem is consistent with cortical input to all relay nuclei; this includes the somatosensory nuclei, the nuclei cuneatus and gracilis (see Figure 5.2). There is also cortical input to the periaqueductal gray, as part of the pain modulation system (see Figure 5.6).

CLINICAL ASPECT

Loss of cortical innervation to the cranial nerve motor nuclei is usually associated with weakness, not paralysis, of the muscles supplied. For example, a lesion on one side may result in difficulty in swallowing or phonation; often these problems dissipate in time.

Facial Movements

A lesion of the facial area of the cortex or of the cortico-bulbar fibers affects the muscles of the face differentially. Due to bilateral innervation by the upper motor neurons supplying the upper facial muscles (including the frontalis muscles of the forehead), a patient with such a lesion is able to wrinkle his or her forehead normally on both sides when asked to look up. She/he will not be able to show the teeth or smile symmetrically on the side opposite the lesion because the innervation to the part of the motor nucleus of CN VII which supplies the lower facial muscles is only from the opposite side. Because of the marked weakness of the muscles of the lower face, there is a drooping of the lower face on the side opposite the lesion. This also affects the muscle of the cheek (the buccinator muscle) and may cause some difficulties with drinking, eating, and chewing (the food becomes stuck in the cheek and often has to be manually removed); sometimes there is also drooling.

This clinical situation must be distinguished from a lesion of the **facial nerve** itself, a lower motor neuron lesion, most often seen with **Bell's palsy** (a lesion of the facial nerve as it emerges from the skull). In this case, the movements of the muscles of the upper face, midface, and lower face are lost on one (ipsilateral) side.

Tongue Movements

In the clinical setting, movement of the tongue is assessed by asking the patient to "stick out the tongue" and move it from side to side. A defect in movement of the tongue rarely occurs in isolation. A lesion affecting the hypoglossal nucleus or nerve is a lower motor lesion of one half of the tongue (on the same side) and leads to atrophy and paralysis of the side affected. When the patient is asked to "stick out your tongue," the tongue deviates to the side that has the lesion.

A cortical lesion or a white matter lesion affecting the cortico-bulbar system is an upper motor lesion. In this case, there is weakness on the opposite side and perhaps mild atrophy, and the tongue protrusion deviates to the side opposite the lesion.

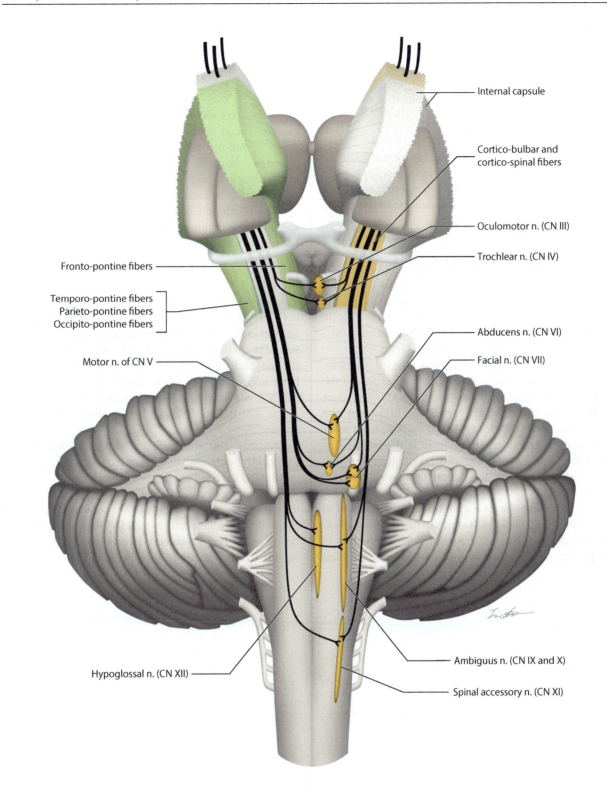

Internal capsule

Cortico-bulbar and cortico-spinal fibers

Oculomotor n. (CN III)

Trochlear n. (CN IV)

Fronto-pontine fibers

Temporo-pontine fibers
Parieto-pontine fibers
Occipito-pontine fibers

Abducens n. (CN VI)

Facial n. (CN VII)

Motor n. of CN V

Ambiguus n. (CN IX and X)

Hypoglossal n. (CN XII)

Spinal accessory n. (CN XI)

FIGURE 5.10: Cortico-Bulbar Fibers—Nuclei of the Brainstem

FIGURE 5.11—RUBRO-SPINAL TRACT

VOLUNTARY DIRECT/INDIRECT MOTOR CONTROL

The **red nucleus** is a prominent nucleus of the midbrain (see Figure 3.3). It takes its name from a reddish color seen in fresh dissections of the brain, presumably the result of its high vascularity. The nucleus (see Figure 4.2C and Appendix Figure A.3) has two portions, a small-celled upper division and a portion with large neurons more ventrally located. The rubro-spinal pathway originates, at least in humans, from the larger cells.

The red nucleus receives its input from the motor areas of the cerebral cortex (cortico-bulbar) and from the cerebellum (see Figure 5.17). The cortical input is directly onto the projecting cells, thus forming a potential two-step pathway from motor cortex to spinal cord.

The rubro-spinal tract is also a crossed pathway, with the decussation occurring in the ventral part of the midbrain (see Figure 6.9 and Figure 6.12). The tract descends within the central part of the brainstem (the tegmentum) and is not clearly distinguishable from other fiber systems. The fibers then course in the lateral portion of the white matter of the spinal cord, just anterior to and intermingled with the lateral cortico-spinal tract (see Figure 6.10).

The rubro-spinal tract is a well-developed pathway in some animals. In monkeys, it seems to be involved in flexion movements of the limbs. Stimulation of this tract in cats produces an increase in tone of the flexor muscles.

The extent of this tract and its function in humans are discussed in the following paragraphs.

NEUROLOGICAL NEUROANATOMY

The location of this tract within the brainstem is shown at cross-sectional levels of the upper midbrain, the mid-pons, the mid-medulla, and spinal cord level C8. The tract is said to continue throughout the length of the spinal cord in primates but probably extends only into the cervical spinal cord in humans (as shown).

The fibers of CN III (oculomotor) exit through the medial aspect of this nucleus at the level of the upper midbrain (see Appendix Figure A.3).

CLINICAL ASPECT

The functional significance of this pathway in humans is not well known. The number of large cells in the red nucleus in the human is significantly less than in the monkey. Motor deficits associated with a lesion involving only the red nucleus or only the rubro-spinal tract have not been adequately described. Although the rubro-spinal pathway may play a role in some flexion movements, it seems that the cortico-spinal tract predominates in the human.

Animal (cat) studies have been done with lesions either above or below the level of the red nucleus to try to understand the resultant motor deficits.

In humans, massive lesions above or below the level of the red nucleus region usually (if the person survives) lead to a comatose state accompanied by abnormal reflex posturing of the limbs—named decorticate and decerebrate rigidity (further discussed with Figure 5.13).

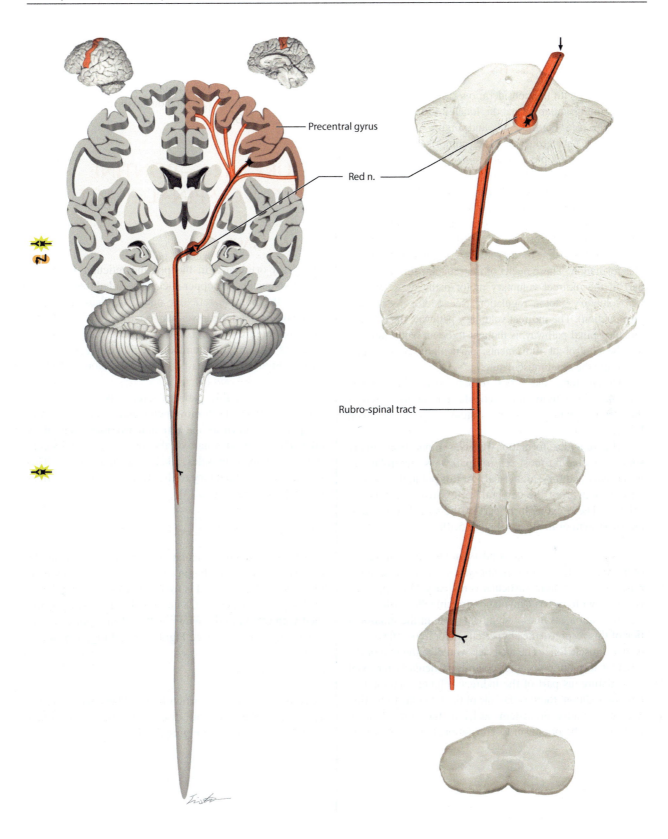

Precentral gyrus

Red n.

Rubro-spinal tract

FIGURE 5.11: Rubro-Spinal Tract—Voluntary Direct/Indirect Motor Control

FIGURE 5.12—RETICULAR FORMATION: INTRODUCTION

Interspersed with the consideration of the functional systems is the reticular formation, located in the core of the brainstem (see Figure 3.6A and Figure 3.6B and the Appendix). This group of nuclei comprises a rather old system with multiple functions—some generalized and some involving the sensory or the motor systems. Some sensory pathways have collaterals to the reticular formation, and some do not. Its nuclei participate in a number of functions, some quite general (e.g., "arousal") and others more specific (e.g., respiratory control).

In addition, some motor pathways originate in the reticular formation. These nuclei of the reticular formation are part of the indirect voluntary motor pathway, as well as non-voluntary motor regulation (see Motor Systems—Introduction). The indirect voluntary pathway, the cortico-reticulo-spinal pathway, is thought to be an older pathway for the control of movements, particularly of proximal joints and the axial musculature.

Muscle tone and reflex responsiveness are greatly influenced by activity in the reticular formation as part of the non-voluntary motor system (discussed with Figure 5.12B).

The reticular formation receives input from many sources, including most sensory pathways (anterolateral, trigeminal, auditory, and visual). At this point, the focus is on the input from the cerebral cortex, from both hemispheres. These axons form part of the so-called cortico-bulbar system of fibers (see Figure 5.10).

Note to the Learner: Understanding the complexity of the various parts of the motor system and the role of the reticular formation in particular is not easy. One approach is to start with the basic myotatic (stretch) reflex—the reticular formation assumes a significant role in the modification of this response (i.e., hyperreflexia or hyporeflexia), as well as muscle tone. The next step would be the role of the reticular formation in motor control, particularly for axial musculature, as part of the indirect voluntary motor system. In addition, there is the role of the reticular formation and other motor brainstem nuclei in the non-voluntary response of the organism to gravitational changes. It now becomes important to understand that the cortex has an important role in controlling this system.

There are two pathways from the reticular formation to the spinal cord—one originates in the pontine region (see below—Figure 5.12A) and one in the medullary region (see Figure 5.12B).

FIGURE 5.12A—PONTINE RETICULO-SPINAL TRACT

MEDIAL TRACT

This tract originates in the pontine reticular formation from two nuclei. The upper nucleus is called the **oral portion** of the pontine reticular nuclei (nucleus reticularis pontis oralis), and the lower part is called the **caudal portion** (see Figure 3.6B). The tract descends to the spinal cord and is located in the medial region of the white matter (see Figure 6.10); this pathway is therefore called the **medial reticulo-spinal tract**.

Functionally, this pathway exerts its action on the extensor muscles, both movements and tone. The area in the pons is known as the reticular extensor facilitatory area. The fibers terminate on the anterior horn cells controlling the axial muscles, likely via interneurons. This system is complementary to that from the lateral vestibular nucleus (see Figure 5.13).

NEUROLOGICAL NEUROANATOMY

The location of the reticular formation is shown in the brainstem. The tract begins in the pons and descends through the medulla and is found at cross-sectional levels of the spinal cord levels C8 and L3. The tract is intermingled with others in the white matter of the spinal cord in the anterior funiculus (see Figure 3.9 and Figure 6.10).

CLINICAL ASPECT

Lesions involving the cortico-bulbar fibers including the cortico-reticular fibers are discussed with the medullary reticular formation (see Figure 5.12B).

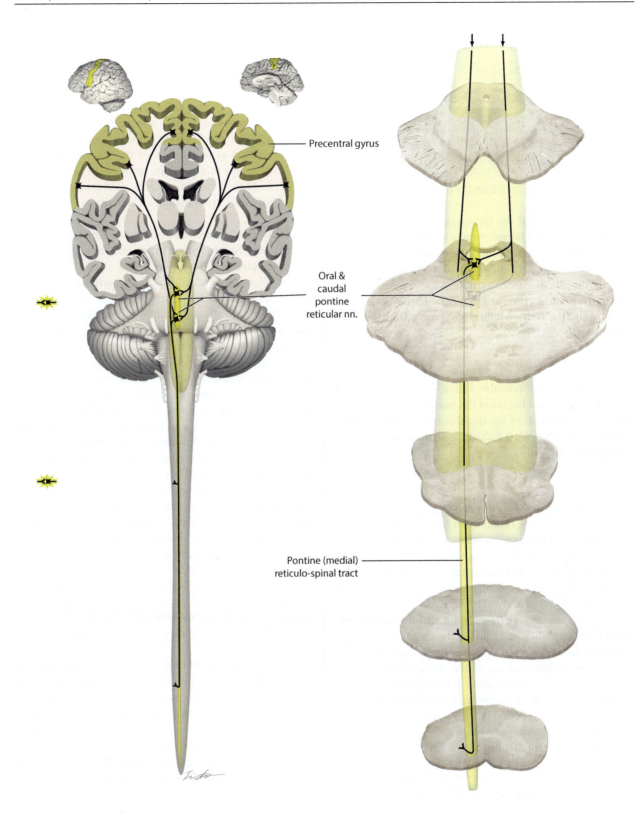

Precentral gyrus

Oral & caudal pontine reticular nn.

Pontine (medial) reticulo-spinal tract

FIGURE 5.12A: Pontine Reticulo-Spinal Tract—Medial Tract

FIGURE 5.12B—MEDULLARY RETICULO-SPINAL TRACT

LATERAL TRACT

This tract originates in the medullary reticular formation, mainly from the **nucleus gigantocellularis** (meaning very large cells; see Figure 3.6B). The tract descends more laterally in the spinal cord than the pontine pathway and is thus named the **lateral reticulo-spinal tract** (see Figure 6.10); some of the fibers cross and descend. The tract lies beside the (lateral) vestibulo-spinal pathway.

The pathway also has its greatest influence on axial musculature. This part of the reticular formation is functionally the reticular extensor inhibitory area, opposite to that of the pontine reticular formation. This area depends for its normal activity on influences coming from the cerebral cortex.

NEUROLOGICAL NEUROANATOMY

The location of the reticular formation is shown in the brainstem. The tract begins in the medulla and is found at cross-sectional levels of the spinal cord levels C8 and L3. The tract is intermingled with others in the white matter of the spinal cord.

CLINICAL ASPECT—SPASTICITY

A lesion destroying the cortico-bulbar fibers, an **upper motor neuron lesion**, results in an increase in the tone of the extensor/anti-gravity muscles that develops over a period of days. This increase in tone, called **spasticity**, is tested by passive flexion and extension of a limb; this test is velocity dependent, meaning that the joint of the limb has to be moved quickly. The anti-gravity muscles are affected in spasticity; in humans, for reasons that are difficult to explain, these muscles are the flexors of the upper limb and the extensors of the lower limb. There is also an increase in responsiveness of the stretch reflex, called **hyperreflexia**, as tested using the deep tendon reflex (DTR) discussed in Motor Systems—Introduction, which also develops over a period of a several days.

There are two hypotheses for the increase in the stretch (monosynaptic) reflex responsiveness:

- **Denervation supersensitivity**: One possibility is a change of the level of responsivity of the neurotransmitter receptors of the motor neurons themselves caused by the loss of the descending input that leads to an increase in excitability.

- **Collateral sprouting**: Another possibility is that axons adjacent to an area that has lost synaptic input will sprout branches and occupy the vacated synaptic sites of the lost descending fibers. In this case, the sprouting is thought to be of the incoming muscle afferents (called 1A afferents, from the muscle spindles).

There is experimental evidence (in animals) for both mechanisms. Spasticity and hyperreflexia usually occur in the same patient.

Another feature accompanying hyperreflexia is **clonus**. This can be elicited by grasping the foot and jerking the ankle upward; in a person with hyperreflexia, the response is a short burst of flexion-extension responses of the ankle, which the tester can feel and that can also be seen.

Lesions involving parts of the motor areas of the cerebral cortex, large lesions of the white matter of the hemispheres or of the posterior limb of the internal capsule, and certain lesions of the upper brainstem all may lead to a similar clinical state in which a patient is paralyzed or has marked weakness, with spasticity and hyperreflexia (with or without clonus) on the contralateral side some days after the time of the damage. The cortico-spinal tract would also be involved in most of these lesions, with loss of voluntary motor control and with the appearance of the extensor plantar response in most cases immediately after the lesion (see the Clinical Aspect discussion with Figure 5.9).

A similar situation can and does occur following large lesions of the spinal cord in which all the descending motor pathways are disrupted, both voluntary and non-voluntary. Destruction of the whole cord would lead to paralysis below the level of the lesion (paraplegia), bilateral spasticity, and hyperreflexia (with or without clonus), a severely debilitating state.

It is most important to distinguish this state from that seen in a Parkinsonian patient who has a change of muscle tone called **rigidity** (discussed with Figure 5.14), with no change in reflex responsiveness and a normal plantar response.

This state should be contrasted with a **lower motor neuron lesion** of the anterior horn cell, with *hypotonia* and *hyporeflexia,* as well as *weakness* (e.g., polio; discussed with Figure 5.7).

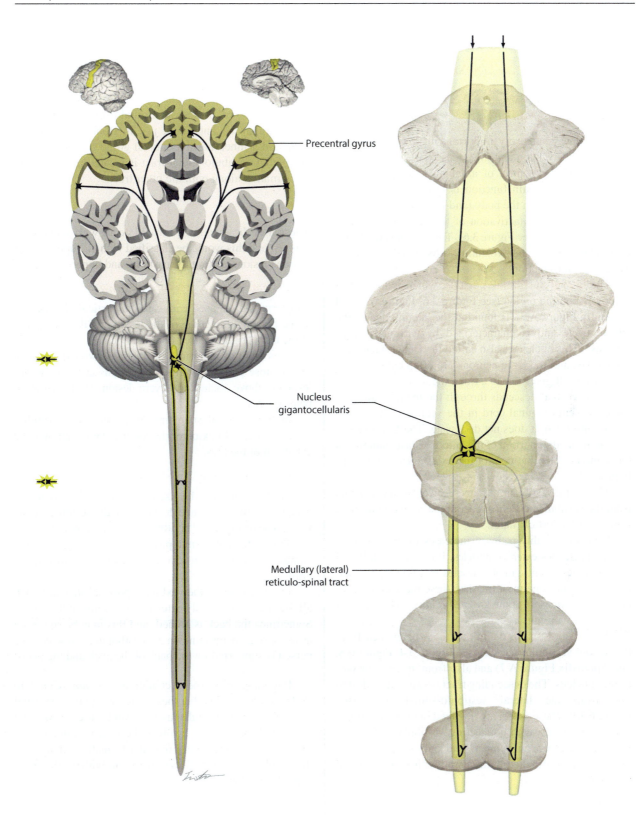

Precentral gyrus

Nucleus
gigantocellularis

Medullary (lateral)
reticulo-spinal tract

FIGURE 5.12B: Medullary Reticulo-Spinal Tract—Lateral Tract

FIGURE 5.13—DESCENDING VESTIBULO-SPINAL TRACTS

LATERAL AND MEDIAL

These pathways are very important in providing a link between the vestibular influences (i.e., gravity and balance) and the control of axial musculature, via the spinal cord. The main function is to provide corrective muscle activity when the body (and head) tilts or changes orientation in space (activation of the vestibular system, cranial nerve VIII; see Figure 3.6). This motor activity has been classified as non-voluntary (discussed in Motor Systems—Introduction).

The **lateral vestibular nucleus**, which is located in the lower pontine region (see Figure 3.6, Figure 6.8, and Appendix Figure A.7), is found at the lateral edge of the 4th ventricle and is characterized by extremely large neurons. (This nucleus is also called the Deiter nucleus in some texts, and the large neurons are often called by the same name.) It gives rise to a tract, the **lateral vestibulo-spinal tract**, that descends through the medulla and traverses the entire spinal cord in the ventral white matter (see Figure 6.10). It does not decussate. The fibers terminate in the medial portion of the anterior horn, namely, on those motor cells that control the axial musculature (see Figure 5.7).

The lateral vestibular nucleus receives its major inputs from the vestibular system and from the cerebellum; there is no cerebral cortical input.

Functionally, this pathway increases extensor muscle tone and activates extensor muscles. It is easier to think of these muscles as anti-gravity muscles in a four-legged animal; in humans, one must translate these muscles as functionally the extensors of the lower extremity and the flexors of the upper extremity.

The **medial vestibulo-spinal tract** originates from the medial vestibular nucleus (see Figure 3.4, Figure 6.8, and Appendix Figure A.7) and also from the inferior vestibular nucleus. The descending tract is situated medial to that from the lateral vestibulo-spinal tract (see Figure 6.10). These nuclei are involved in other functions, particularly to do with connecting vestibular influences with head and eye movements and only projects to the upper cervical spinal cord (discussed with Figure 6.8 and Figure 6.9).

The vestibular nuclei are found at the lower pontine level, and are seen through to the mid-medulla; the tracts descend throughout the spinal cord, as seen at C8 and L3. In the spinal cord, the lateral vestibulo-spinal tract is positioned somewhat anteriorly, just in front of the ventral horn, and it innervates the medial group of motor nuclei; the medial vestibulo-spinal tract is located in the ventral white matter, along with the other tracts.

CLINICAL ASPECT

A lesion of these pathways would occur with spinal cord injuries, and this would be classified with an "upper motor neuron" lesion, leading to spasticity and hyperreflexia.

Decorticate Rigidity (Flexor Posturing)

This term is to be used with caution, particularly because of the presumed location of the lesion and the wrong use of the term "rigidity" (see comment below).

This state of abnormal posturing is sometimes seen in comatose patients with severe lesions at the midbrain level or above including severe lesion of the cerebral hemispheres.

In this postural state known as decorticate rigidity there is a state of flexion of the forearm (at the elbow) and extension of the legs.

Decerebrate Rigidity (Extensor Posturing)

Again this term is to be used with caution, particularly because of the presumed location of the lesion and the wrong use of the term "rigidity" (see comment below).

This state of abnormal posturing is sometimes seen in comatose patients with severe lesions lower down in the brainstem.

In this condition they exhibit a postural state in which all four limbs are rigidly extended (including at the wrist). Sometimes the back is arched, and this may be so severe as to cause a posture known as opisthotonus, in which the person is supported by the back of the neck and the heels.

Physiologically, these conditions are *not* related to Parkinsonian rigidity, but they are related to the abnormal state of spasticity (see discussion with Figure 5.12B). The postulated mechanism involves the relative influence of the pontine and medullary reticular formation, along with the vestibulo-spinal pathway, with and without the input from the cerebral cortex.

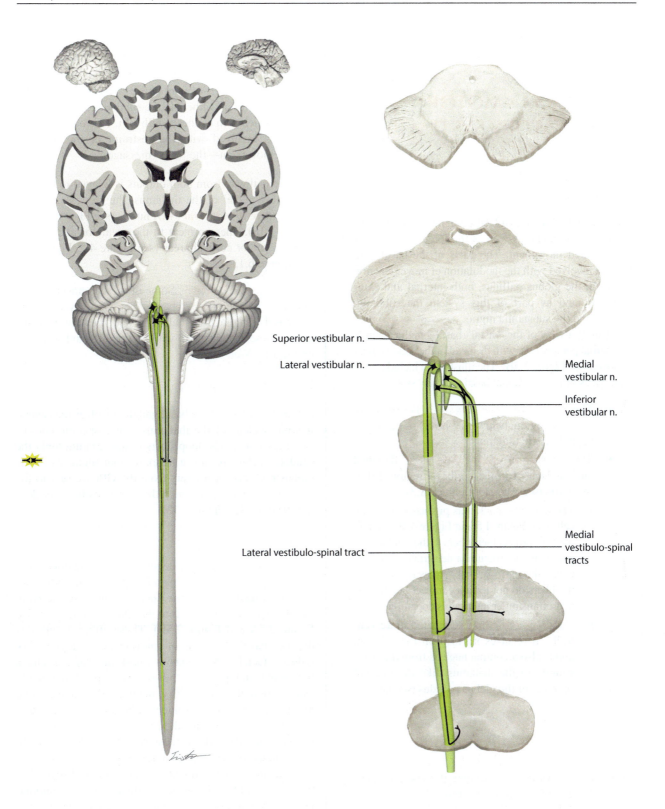

Superior vestibular n.

Lateral vestibular n.

Medial vestibular n.

Inferior vestibular n.

Lateral vestibulo-spinal tract

Medial vestibulo-spinal tracts

FIGURE 5.13: Descending Vestibulo-Spinal Tracts—Lateral and Medial

MOTOR MODULATION

FIGURE 5.14—MOTOR MODULATORY SYSTEM (1) BASAL GANGLIA

UPPER ILLUSTRATION

The basal ganglia are involved with the initiation and cessation of movement, as well as the force or amplitude of the intended movement. The parts of the basal ganglia that are involved with the modulation of movement are the putamen, the globus pallidus, both internal and external divisions, and as part of this system the subthalamic nucleus and the substantia nigra.

This is a view of the "distal" basal ganglia, seen from the medial perspective, with the head of the caudate nucleus removed (see Figure 2.5B). The illustration includes the two other parts of the functional basal ganglia "system":

- The **subthalamic nucleus (STh)** is situated in a small region below the level of the diencephalon, the thalamus.
- The **substantia nigra (SN)** is a flattened nucleus located in the midbrain region. It has two parts (see Appendix Figure A.3):
  - The **pars compacta** has pigment-containing cells (see Figure 1.5 and Figure 4.2C). These neurons project their fibers to the caudate and putamen (the striatum or neostriatum). This is called the nigro-striatal "pathway," although the fibers do not form a compact bundle; the neurotransmitter involved is dopamine.
  - The **pars reticulata** is situated more ventrally. It receives fibers from the striatum and is also an output nucleus from the basal ganglia to the thalamus, like the internal segment of the globus pallidus (see later).

LOWER ILLUSTRATION

BASAL GANGLIA CIRCUITRY

Information flows into the caudate (C) and putamen (P) from all areas of the cerebral cortex (in a topographic manner; see Figure 5.18), from the SN (dopaminergic from the pars compacta), and from the centromedian nucleus of the thalamus (see later).

There are two circuits from the putamen for motor modulation:

1. The information is processed and passed through to the globus pallidus, *internal* segment (GPi) (and the pars reticulata of the SN); these are the output

nuclei of the basal ganglia for the facilitation of the intended movement—the "go" message.

2. In the other circuit, the information goes first to the *external* segment of the globus pallidus (GPe), which sends fibers to the subthalamic nucleus (S), which then connects with GPi. This output system is involved with the restraint of the intended movement—the "stop" message.

The pathway from both outputs is relayed to the specific relay nuclei of the thalamus, the **ventral anterior (VA)** and **ventral lateral (VL)** nuclei (see Figure 4.3 and Figure 5.18). These project to the premotor and supplementary motor cortical areas (see Figure 4.5, Figure 5.8, and Figure 6.13).

Note to the Learner: Some of these connections are animated on the atlas Web site. The circuitry involving the basal ganglia, the thalamus, and the motor cortex areas is described in detail with Figure 5.18 (and animated on the atlas Web site).

ADDITIONAL DETAIL

Another sub-loop of the basal ganglia involves the centromedian nucleus of the thalamus, a non-specific nucleus (see Figure 4.3). The loop starts in the striatum (only the caudate nucleus is shown here), to both segments of the globus pallidus; then fibers from the GPi are sent to the centromedian nucleus, which then sends its fibers back to the striatum (see Figure 6.13).

CLINICAL ASPECT

Parkinson's disease: The degeneration of the dopamine-containing neurons of the pars compacta of the SN, with the consequent loss of their dopamine input to the basal ganglia (the striatum), leads to this clinical entity (see also Figure 2.5A and Figure 2.5B). Those afflicted with this disease have slowness of movement (bradykinesia), reduced facial expressiveness (mask-like facies), and a tremor at rest, typically a "pill-rolling" type of tremor. On examination, there is **rigidity**, manifested as an increased resistance to passive movement of both flexors and extensors that is not velocity dependent. (This is to be contrasted with spasticity; discussed with Figure 5.12B.) In addition, there is *no* change in reflexes.

The medical treatment of Parkinson's disease has limitations, although various medications and combinations (as well as newer drugs) can be used for many years. The medications are used to replenish dopamine for the receptors involved in "go" aspect of movements. Too little or too much medication has clear effects on the modulation of motor control. After many years, the medical management becomes more difficult or ceases to work.

For some of these and other select patients, a surgical approach for the alleviation of the symptoms of Parkinson's disease has been recommended.

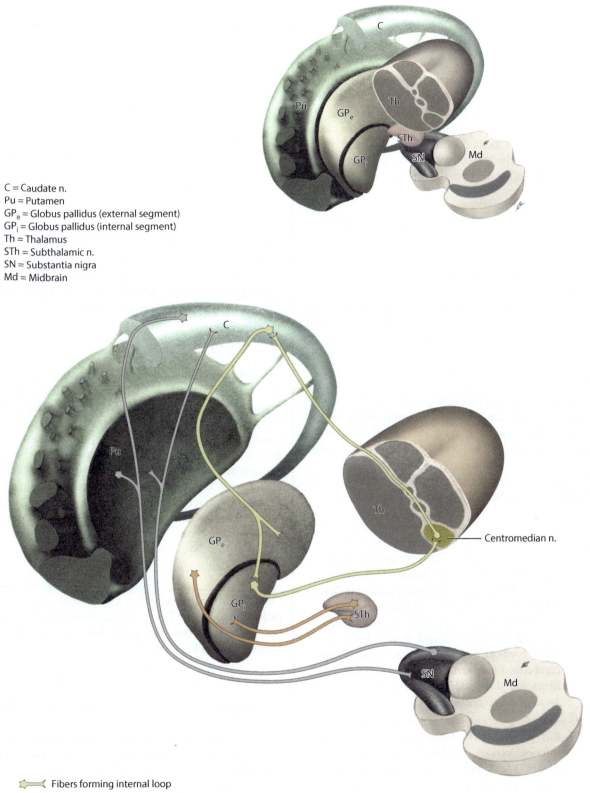

C = Caudate n.
Pu = Putamen
GP$_e$ = Globus pallidus (external segment)
GP$_i$ = Globus pallidus (internal segment)
Th = Thalamus
STh = Subthalamic n.
SN = Substantia nigra
Md = Midbrain

Centromedian n.

Fibers forming internal loop
Pallido-subthalamic and subthalamo-pallidal fibers
Striato-nigral and nigro-striatal fibers

FIGURE 5.14: Motor Modulatory System (i)—Basal Ganglia

MOTOR MODULATORY SYSTEM (ii)

The cerebellum regulates the ongoing movement and also compares the actual movement with the intended (planned) motor act. The part of the cerebellum primarily involved is the neocerebellum (see Figure 3.7 and Figure 3.8).

To understand the role of the cerebellum in the modulation of voluntary motor control, it is necessary to review the afferents to the cerebellum, the intra-cerebellar circuitry, and the efferents from the cerebellum.

FIGURE 5.15 — CEREBELLUM A: AFFERENTS

Information relevant to the role of the cerebellum in motor regulation comes from the cerebral cortex, the brainstem, and the muscle receptors in the periphery. The information is conveyed to the cerebellum mainly via the middle and inferior cerebellar peduncles.

INFERIOR CEREBELLAR PEDUNCLE

The inferior cerebellar peduncle goes from the medulla to the cerebellum. It lies behind the inferior olivary nucleus and can sometimes be seen on the ventral view of the brainstem (as in Figure 1.8). This peduncle conveys a number of fiber systems to the cerebellum. These are shown schematically in this diagram of the ventral view of the brainstem and cerebellum. They include the following:

- The **posterior (dorsal) spino-cerebellar** pathway is conveying proprioceptive information from most of the body. This is one of the major tracts of the inferior peduncle. These fibers, carrying information from the muscle spindles, relay in the dorsal nucleus of Clarke in the spinal cord (see Figure 5.1). They ascend ipsilaterally in a tract that is found at the edge of the spinal cord (see Figure 6.10). The dorsal spinocerebellar fibers terminate ipsilaterally; these fibers are distributed to the spino-cerebellar areas of the cerebellum.

- The **olivo-cerebellar** tract is also carried in this peduncle. The fibers originate from the inferior olivary nucleus (see Figure 1.8, Figure 3.1, Appendix Figure A.8, Appendix Figure A.9, and Appendix Figure A.10), cross

in the medulla, and are distributed to all parts of the cerebellum. These axons have been shown to be the climbing fibers to the main dendritic branches of the Purkinje neuron.

- Other cerebellar afferents from other nuclei of the brainstem, including the reticular formation, are conveyed to the cerebellum via this peduncle. Most important are those from the medial (and inferior) vestibular nuclei to the vestibulocerebellum. Afferents from the visual and auditory system are also known to be conveyed to the cerebellum.

ADDITIONAL DETAIL

The homologous spino-cerebellar tract for the upper limb is the **cuneo-cerebellar** tract. These fibers relay in the accessory (external) cuneate nucleus of the lower medulla (see Appendix Figure A.9 and Appendix Figure A.10). This pathway is not shown in the diagram.

MIDDLE CEREBELLAR PEDUNCLE

All parts of the cerebral cortex contribute to the massive **cortico-pontine** system of fibers (also described with Figure 3.8, Figure 5.10, and Figure 6.12). These fibers descend via the anterior and posterior limbs of the internal capsule, then the inner and outer parts of the cerebral peduncle, and terminate in the pontine nuclei. The fibers synapse and cross and then project mainly to the neocerebellum via the middle cerebellar peduncle. This input provides the cerebellum with the cortical information relevant to motor commands and the planned (intended) motor activities.

SUPERIOR CEREBELLAR PEDUNCLE

Only one afferent tract enters via the superior cerebellar peduncle (see later). This peduncle carries the major efferent pathway from the cerebellum (discussed with Figure 5.17 and Figure 5.18).

ADDITIONAL DETAIL

One group of cerebellar afferents, those carried in the **ventral (anterior) spino-cerebellar tract**, enters the cerebellum via the superior cerebellar peduncle. These fibers cross in the spinal cord, ascend (see Figure 6.10), enter the cerebellum, and cross again, thus terminating on the same side from which they originated.

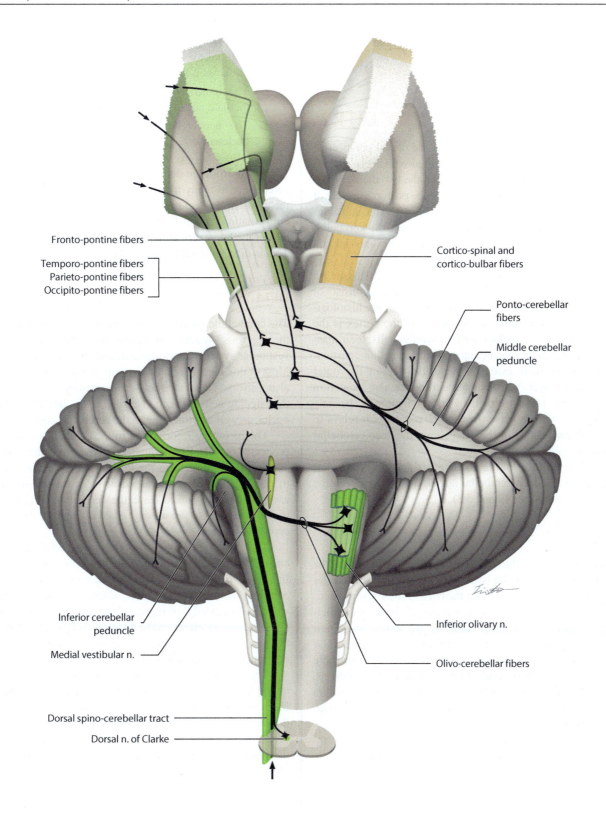

Fronto-pontine fibers

Temporo-pontine fibers
Parieto-pontine fibers
Occipito-pontine fibers

Cortico-spinal and
cortico-bulbar fibers

Ponto-cerebellar
fibers

Middle cerebellar
peduncle

Inferior cerebellar
peduncle

Medial vestibular n.

Inferior olivary n.

Olivo-cerebellar fibers

Dorsal spino-cerebellar tract

Dorsal n. of Clarke

FIGURE 5.15: Motor Modulatory System (ii)—Cerebellum A: Afferents

FIGURE 5.16— MOTOR MODULATORY SYSTEM (ii) CEREBELLUM B: INTRACEREBELLAR CIRCUITRY

The cerebellum is presented from the dorsal perspective (as in Figure 1.9, Figure 6.11, and Figure 6.12). To review, the 3rd ventricle is situated between the two diencephala; the pineal gland is seen attached to the posterior aspect of the thalamus. Below are the colliculi, superior and inferior. On the right side of the illustration, the cerebellar hemisphere has been cut away, revealing the "interior" on this side.

The cerebellum is organized with cortical tissue on the outside, the cerebellar cortex. The cortex consists of three layers, and all areas of the cerebellum are histologically alike. The most important cell of the cortex is the Purkinje neuron, which forms a layer of cells; their massive dendrites receive the input to the cerebellum, particularly along their extensive dendritic tree in the outer (granular) layer. Various interneurons are also located in the cortex. The axon of the Purkinje neuron is the only axonal system to leave the cerebellar cortex. It relays to the efferent nuclei of the cerebellum, the deep cerebellar nuclei.

Deep within the cerebellum are the intracerebellar nuclei or the deep cerebellar nuclei, now shown from the posterior view (see Figure 3.8).

CIRCUITRY OVERVIEW

All (excitatory) afferents to the cerebellum go to both the deep cerebellar nuclei (via collaterals) and the cerebellar cortex. After processing in the cortex, the Purkinje neuron sends its axon on to the neurons of the deep cerebellar nuclei—all Purkinje neurons are inhibitory. Their influence modulates the activity of the deep cerebellar neurons, which are tonically active (described in more detail later). The output of the deep cerebellar neurons, which is excitatory, influences neurons in the brainstem and cerebral cortex via the thalamus (discussed with Figure 5.17).

The connections of the cortical areas with the intracerebellar nuclei follow the functional divisions of the cerebellum:

- The vestibulocerebellum is connected to the fastigial nucleus, as well as to the lateral vestibular nucleus.
- The spinocerebellum connects with the intermediate nuclei (the globose and emboliform).
- The neocerebellum connects to the dentate nucleus.

Axons from the deep nuclei neurons project from the cerebellum to many areas of the central nervous system, including brainstem motor nuclei (e.g., vestibular, reticular formation) and thalamus (to motor cortex). In this way, the cerebellum exerts its influence on motor performance. This is discussed with Figure 5.17.

DETAILS OF CEREBELLAR CIRCUITRY

The cerebellum receives information from many parts of the nervous system, including the spinal cord, the vestibular system, the brainstem, and the cerebral cortex. Most of this input is related to motor function, but some is also sensory. These afferents are excitatory and influence the ongoing activity of the neurons in the intracerebellar nuclei, as well as projecting to the cerebellar cortex.

The incoming information to the cerebellar cortex is processed by various interneurons of the cerebellar cortex and eventually influences the Purkinje neuron. This leads to either increased or decreased firing of this neuron. Its axon is the only one to leave the cerebellar cortex, and these axons project, in an organized manner, to the deep cerebellar nuclei.

The Purkinje neurons are inhibitory, and their influence modulates the activity of the deep cerebellar nuclei. Increased firing of the Purkinje neuron increases the ongoing inhibition onto these deep cerebellar nuclei, whereas decreased Purkinje cell firing results in a decrease in the inhibitory effect on the deep cerebellar cells, that is, this results in the increased firing of the deep cerebellar neurons (called disinhibition).

The cerebellar cortex projects fibers directly to the lateral vestibular nucleus. As would be anticipated, these fibers are inhibitory. The lateral vestibular nucleus could therefore, in some sense, be considered one of the intracerebellar nuclei. This nucleus also receives input from the vestibular system and then projects to the spinal cord (discussed with Figure 5.13).

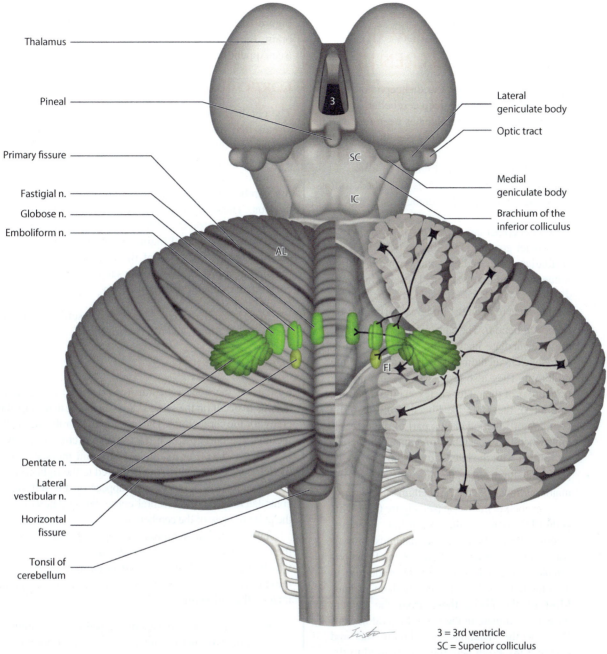

Thalamus

Pineal

Primary fissure

Fastigial n.

Globose n.

Emboliform n.

3

SC

IC

AL

FI

Lateral
geniculate body

Optic tract

Medial
geniculate body

Brachium of the
inferior colliculus

Dentate n.

Lateral
vestibular n.

Horizontal
fissure

Tonsil of
cerebellum

3 = 3rd ventricle
SC = Superior colliculus
IC = Inferior colliculus

AL = Anterior lobe
FI = Flocculonodular lobe

FIGURE 5.16: Motor Modulatory System (ii)—Cerebellum B: Intracerebellar Circuitry

FIGURE 5.17—MOTOR MODULATORY SYSTEM (ii) CEREBELLUM C: EFFERENTS

This is again a dorsal view of the diencephalon, brainstem, and cerebellum, with the deep cerebellar (intracerebellar) nuclei. The cerebellar tissue has been removed in the midline to reveal the 4th ventricle (as in Figure 1.9); the three cerebellar peduncles are also visualized from this posterior perspective (also in Figure 1.9).

The output from the cerebellum is described, following the functional division of the cerebellum:

- Vestibulocerebellum: Efferents from the fastigial nuclei go to brainstem motor nuclei (e.g., vestibular nuclei and reticular formation), thus influencing balance and gait. They exit in a bundle that is found adjacent to the inferior cerebellar peduncle (named the juxtarestiform body).

- Spinocerebellum: The emboliform and globose, the intermediate nuclei, also project to brainstem nuclei, including the red nucleus of the midbrain. They also project to the appropriate limb areas of the motor cortex via the thalamus (see later); these are the fibers involved in the comparator function of this part of the cerebellum, whereby the cerebellum adjusts the actual movement with the intended movement.

- Neocerebellum: The dentate nucleus is the major outflow from the cerebellum via the **superior cerebellar peduncle** (see Figure 6.11). This peduncle connects the cerebellar efferents, through the midbrain, to the thalamus on their way to the motor cortex. Some of the fibers terminate in the red nucleus of the midbrain, in addition to those from the intermediate nucleus. Most of the fibers, those from the dentate nucleus, terminate in the ventral lateral nucleus (VL) of the thalamus (see Figure 4.3 and Figure 5.18). From here they are relayed to the motor cortex, predominantly area 4, and also the premotor cortex, area 6. The neocerebellum is involved in motor coordination and planning. (This is to be compared with the influence of the basal ganglia on motor activity; see Figure 5.18.)

DETAILED PATHWAY

The outflow fibers of the superior cerebellar peduncles originate mainly from the dentate nucleus, with some from the intermediate nucleus (as shown). The axons start laterally and converge toward the midline (see Figure 5.3), passing in the roof of the upper half of the 4th ventricle (see Figure 1.9 and Figure 3.3). The fibers continue to "ascend" through the upper part of the pons (see Appendix Figure A.5 and Appendix Figure A.6). In the lower midbrain, there is a complete decussation of the peduncles (see Appendix Figure A.4).

CORTICAL LOOP

The cerebral cortex is linked to the neocerebellum by a circuit that forms a loop. Fibers are relayed from the cerebral cortex via the pontine nuclei to the cerebellum. The ponto-cerebellar fibers cross and go to the neocerebellum of the opposite side. After cortical processing in the cerebellar cortex, the fibers project to the dentate nucleus. These efferents cross (decussate) in the lower midbrain and project to the thalamus. From the thalamus, fibers are relayed mainly to the motor areas of the cerebral cortex. Because of the two (double) crossings, the messages are returned to the same side of the cerebral cortex from which the circuit began.

Note to the Learner: This is also described and illustrated in *The Integrated Nervous System*.

CLINICAL ASPECT

Lesions of the neocerebellum (of one side) cause motor deficits to occur on the same side of the body, that is, ipsilaterally for the cerebellum. The explanation for this lies in the fact that the cortico-spinal tract is also a crossed pathway (see Figure 5.9). For example, the errant messages from the left cerebellum which are delivered to the right cerebral cortex cause the symptoms to appear on the left side—contralaterally for the cerebral cortex but ipsilaterally from the point of view of the cerebellum.

The cerebellar symptoms associated with lesions of the neocerebellum are collectively called **dyssynergia**, in which the range, direction, and amplitude of voluntary muscle activity are disturbed. The specific symptoms include the following:

- Distances are improperly gauged when pointing, called dysmetria, and this includes past-pointing.

- Rapid alternating movements are poorly performed, called dysdiadochokinesis.

- Complex movements are performed as a series of successive movements, which is called a decomposition of movement.

- There is a tremor seen during voluntary movement, an **intention tremor** (this is in contrast to the Parkinsonian tremor, which is present at rest).

- Disturbances also occur in the normally smooth production of words, resulting in slurred and explosive speech.

In addition, cerebellar lesions in humans are often associated with hypotonia and sluggish deep tendon reflexes.

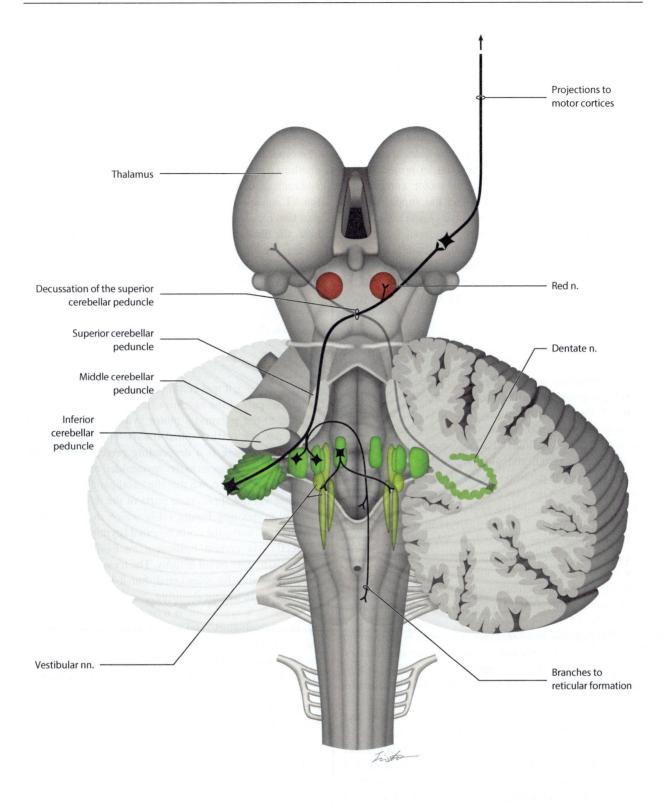

Thalamus

Decussation of the superior
cerebellar peduncle

Superior cerebellar
peduncle

Middle cerebellar
peduncle

Inferior
cerebellar
peduncle

Vestibular nn.

Projections to
motor cortices

Red n.

Dentate n.

Branches to
reticular formation

FIGURE 5.17: Motor Modulatory System (ii)—Cerebellum C: Efferents

FIGURE 5.18—MOTOR MODULATORY SYSTEMS (III) THALAMUS: MOTOR CIRCUITS

The specific relay nuclei of the thalamus that are linked with the motor systems, the basal ganglia, and the cerebellum are the **ventral lateral (VL)** and the **ventral anterior (VA)** nuclei (see Figure 4.3). These nuclei project to the different cortical areas involved in motor control, the motor strip, the premotor area, and the supplementary motor area (as shown in the upper insets—see Figure 5.8).

These thalamic nuclei also receive input from these cortical areas, in line with the reciprocal connections of the thalamus and cortex (discussed with Figure 6.13). One of the intralaminar nuclei, the centromedian nucleus, is also linked with the circuitry of the basal ganglia (discussed with Figure 5.14).

BASAL GANGLIA

The neostriatum receives input from wide areas of the cerebral cortex, as well as from the dopaminergic neurons of the substantia nigra. The putamen, which is mainly involved with motor regulation, then connects with the globus pallidus. The major outflow from the basal ganglia, from the globus pallidus, follows two slightly different pathways to the thalamus, as pallido-thalamic fibers. One group of fibers passes around and the other passes through the fibers of the internal capsule which is (represented on the diagram by large stippled arrows). These merge and end in the VA and VL nuclei of the thalamus. (The VA nucleus is not seen on this section through the thalamus.) The other outflow from the basal ganglia via the pars reticulata of the substantia nigra generally follows the same projection to these thalami nuclei (not shown).

The pathway from thalamus to cortex is excitatory. The basal ganglia influence is to modulate the level of excitation of the thalamic nuclei. Too much inhibition leads to a situation in which the motor cortex has insufficient activation, and the prototypical syndrome for this is Parkinson's disease (discussed with Figure 2.5A and Figure 5.14). Too little inhibition leads to a situation in which the motor cortex receives too much stimulation and the prototypical syndrome for this is Huntington's chorea (discussed with Figure 2.5A). The analogy that has been used to understand these diseases is to a motor vehicle, in which a balance is needed between the brake and the gas pedal for controlled forward motion in traffic.

The **motor** areas of the cerebral cortex that receive input from these two subsystems of the motor system are shown in the small insets at the top, both on the dorsolateral surface and on the medial surface of the hemispheres (see Figure 1.3, Figure 1.7, and Figure 4.5). The cortical projection from these thalamic nuclei is to the premotor and supplementary motor areas, as shown, cortical areas concerned with motor regulation and planning (see Figure 5.8 and Figure 6.13).

CEREBELLUM

The other part of the motor regulatory systems, the cerebellum, also projects (via the superior cerebellar peduncles) to the thalamus. The major projection is to the VL nucleus, but to a different portion of it than the part that receives the input from the basal ganglia. From here, the fibers project to the motor areas of the cerebral cortex, predominantly the precentral gyrus and the premotor area, areas 4 and 6, respectively (see Figure 4.3, Figure 5.8, and Figure 6.13).

Note to the Learner: These connections are animated on the atlas Web site (www.atlasbrain.com).

CLINICAL ASPECT

Many years ago, it was commonplace to refer to the basal ganglia as part of the **extrapyramidal motor system** (in contrast to the pyramidal motor system [discussed with Figure 5.9, the cortico-spinal tract]). This could lead to the idea that there is a descending projection from the basal ganglia, analogous to the "pyramidal" (cortico-spinal) pathway. It is now known that the basal ganglia exert their influence via the appropriate parts of the thalamus to the cerebral cortex (see Figure 6.13), which then acts either directly (i.e., using the cortico-spinal [pyramidal] tract) or indirectly via certain brainstem nuclei (cortico-bulbar pathways) to alter motor activity. The term extrapyramidal should probably be abandoned, but it is still frequently encountered in a clinical setting.

Tourette's syndrome is a motor disorder manifested by tics, involuntary sudden movements; occasionally, these individuals have bursts of involuntary language that rarely contains vulgar expletives. This disorder starts in childhood and usually has other associated behavioral problems, including difficulty with attention. There is growing evidence that this disorder is centered in the basal ganglia. The condition may persist into adulthood.

Note to the Learner: A hypothetical clinical case with this syndrome is discussed in the *Integrated Nervous System* text.

ADDITIONAL CLINICAL ASPECT

The motor abnormality associated with an isolated lesion of the subthalamic nucleus is called **hemiballismus**. The person is seen to have sudden flinging movements of a limb, on the side of the body opposite to the lesion. The usual cause of hemiballismus is a vascular lesion.

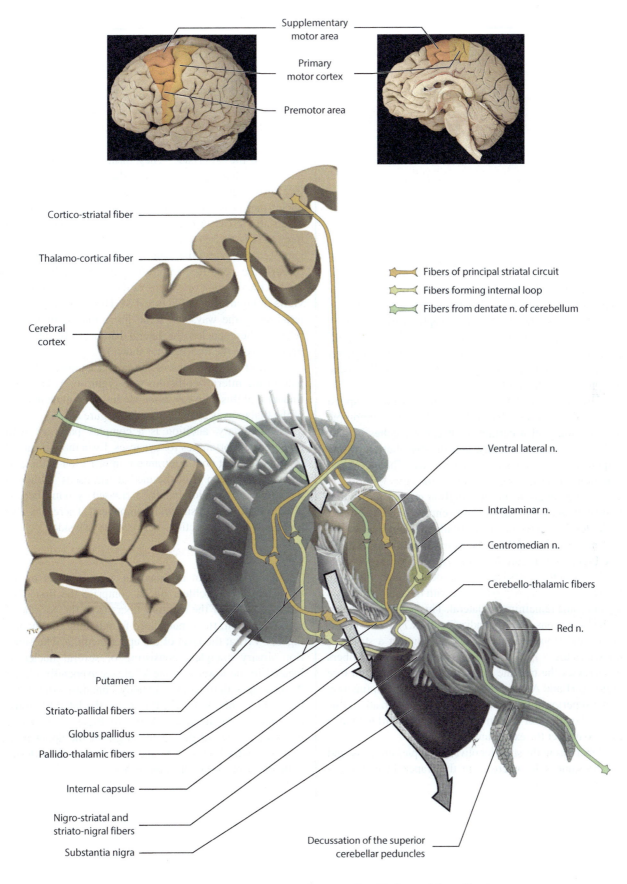

FIGURE 5.18: Motor Modulatory Systems (iii)—Thalamus: Motor Circuits

Chapter 6

Special Senses and Pathways Summary

FIGURE 6.1—AUDITION 1

AUDITORY PATHWAY 1: BRAINSTEM

The auditory pathway is somewhat more complex than the somatosensory pathways, first because it is bilateral and second because there are more synaptic stations (nuclei) along the way, with numerous connections across the midline. It also has a unique feature—a feedback pathway from the central nervous system (CNS) to cells in the receptor organ, the cochlea.

The specialized hair cells in the cochlea respond maximally to certain frequencies (pitch) in a tonotopic manner; tones of a certain pitch cause patches of hair cells to respond maximally, and the distribution of this response is continuous along the cochlea. The peripheral ganglion for these sensory fibers is the **spiral ganglion**. The central fibers from the ganglion project to the first brainstem nuclei, the dorsal and ventral **cochlear nuclei**, at the level of entry of the vestibulocochlear nerve (cranial nerve [CN] VIII) at the ponto-medullary junction (see Figure 1.8, Figure 6.11, and Appendix Figure A.8).

Many of the fibers leaving the cochlear nuclei synapse in the **superior olivary complex**, with the majority crossing but some remaining ipsilateral. This nucleus is located in the lower pons (see Appendix Figure A.7). Fibers crossing to and from the other side are found in a structure known as the **trapezoid body**, a compact bundle of fibers that crosses the midline in the lower pontine region (see Figure 6.11 and Appendix Figure A.7). The main function of the superior olivary complex is sound localization; this is based on the fact that an incoming sound will not reach the two ears at the exact same moment.

Fibers from the superior olivary complex either ascend on the same side or cross (in the trapezoid body) and ascend on the other side. They form a tract, the **lateral lemniscus**, that begins just above the level of these nuclei (see also Figure 6.11). Some fibers from the cochlear nuclei that cross in the intermediate acoustic stria and others that cross in the dorsal acoustic stria (not shown here, see Figure 3.3) join the lateral lemniscus. The lateral lemniscus carries the auditory information upward through the pons (see Figure 6.2 and Appendix Figure A.6) to the **inferior colliculus** of the midbrain (see dorsal views in Figure 1.9 and Figure 3.3). There are nuclei scattered along the way, and some fibers may terminate or relay in these nuclei; the lateral lemnisci are interconnected across the midline (not shown).

Almost all the axons of the lateral lemniscus terminate in the inferior colliculus (see Figure 6.2). The continuation of this pathway to the medial geniculate nucleus of the thalamus is also discussed in Figure 6.2.

In summary, audition is a complex pathway, with numerous opportunities for synapses. Even though named a "lemniscus," it does not transmit information in the efficient manner seen with the medial lemniscus. Although the pathway is predominantly a crossed system, there is also a significant ipsilateral component. There are also numerous interconnections between the two sides.

NEUROLOGICAL NEUROANATOMY

The auditory system is shown at various levels of the brainstem. The cochlear nuclei comprise the first CNS synaptic relay for the auditory fibers from the peripheral spiral ganglion; these nuclei are found along the incoming CN VIII at the level of the upper medulla. The superior olivary complex, consisting of several nuclei, is located in the lower pontine level (see Appendix Figure A.7), along with the trapezoid body, containing the crossing auditory fibers. By the mid-pons, the lateral lemniscus can be recognized (see Appendix Figure A.6). These fibers move toward the outer margin of the upper pons (see Appendix Figure A.5) and terminate in the inferior colliculus (see Appendix Figure A.4).

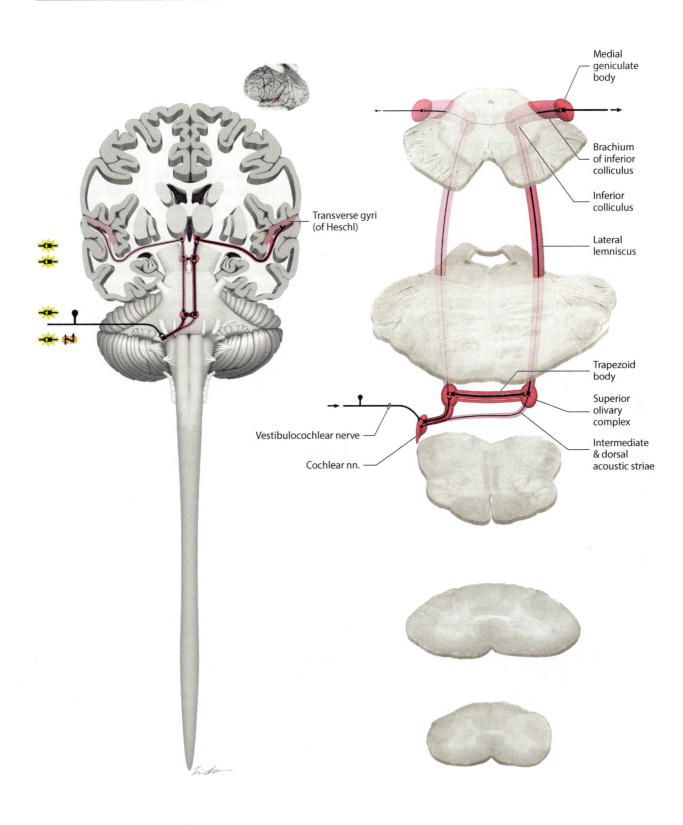

FIGURE 6.1: Audition 1—Auditory Pathway 1: Brainstem

FIGURE 6.2—AUDITION 2

AUDITORY PATHWAY 2: THALAMUS

This illustration shows the projection of the auditory system fibers from the level of the inferior colliculus, the lower midbrain, to the thalamus, and then to the cortex.

Auditory information is carried via the lateral lemniscus to the inferior colliculus (see Figure 6.1), after several synaptic relays. There is another synapse in this nucleus, making the auditory pathway overall somewhat different from and more complex than the medial lemniscal and different from the visual pathways (discussed in the next part of the chapter). The inferior colliculi are connected to each other by a small commissure (not labeled).

The auditory information is next projected to a specific relay nucleus of the thalamus, the **medial geniculate** (nucleus) body (**MGB**; see Figure 4.3). The tract that connects the two, the **brachium** of the inferior colliculus, can be seen on the dorsal aspect of the midbrain (see Figure 1.9 [not labeled] and Figure 3.3); this is shown diagrammatically in the present figure. The medial geniculate nucleus is likely involved with some analysis and integration of the auditory information.

From the medial geniculate nucleus the auditory pathway continues to the cortex. This projection, which courses beneath the lenticular (lentiform) nucleus of the basal ganglia, is called the sublenticular pathway, the inferior limb of the internal capsule, or simply the **auditory radiation**. The cortical areas involved with receiving this information are the **transverse gyri of Heschl**, situated on the superior temporal gyrus, within the lateral fissure. The location of these gyri is shown in the inset as the primary auditory areas (also seen in a photographic view in Figure 6.3).

More exact auditory analysis occurs in the cortex. Further elaboration of auditory information is carried out in the adjacent cortical areas. On the dominant side for language, these cortical areas are adjacent to Wernicke's language area (see Figure 4.5).

Sound frequency, known as **tonotopic** organization, is maintained all along the auditory pathway, starting in the cochlea. This can be depicted as a musical scale with high and low notes. The auditory system localizes the direction of a sound in the superior olivary complex (discussed with Figure 6.1); this is done by analyzing the difference in the timing that sounds reach each ear and by the difference in sound intensity reaching each ear. The loudness of a sound would be represented physiologically by the number of receptors stimulated and by the frequency of impulses, as in other sensory modalities.

NEUROLOGICAL NEUROANATOMY

This view of the brain includes the midbrain level and the thalamus, with the lentiform nucleus lateral to it. The lateral ventricle is open (cut through its body) and the thalamus is seen to form the floor of the ventricle; the body of the caudate nucleus lies above the thalamus and on the lateral aspect of the ventricle.

The auditory fibers leave the inferior colliculus and course via the brachium of the inferior colliculus to the medial geniculate nucleus of the thalamus. From here the auditory radiation courses below the lentiform nucleus to the auditory gyri on the superior surface of the temporal lobe within the lateral fissure. The gyri are shown in small "figurine" (above the main part of the illustration) and in Figure 6.3.

This diagram also includes the lateral geniculate body (nucleus), which subserves the visual system and its projection, the optic radiation (to be discussed with Figure 6.4 and Figure 6.6).

The temporal lobe structures are also shown, including the inferior horn of the lateral ventricle, the tail of the caudate nucleus, the hippocampal formation, and adjoining structures relevant to the limbic system (see Figure 9.5A in Section 4).

ADDITIONAL DETAIL

The auditory pathway has a feedback system, from the higher levels to lower levels (e.g., from the inferior colliculus to the superior olivary complex). The final link in this feedback is somewhat unique in the mammalian CNS because it influences the cells in the receptor organ itself. This pathway, known as the **olivo-cochlear bundle**, has its cells of origin in the vicinity of the superior olivary complex. It has both a crossed and an uncrossed component. Its axons reach the hair cells of the cochlea by traveling in CN VIII. This system changes the responsiveness of the peripheral hair cells.

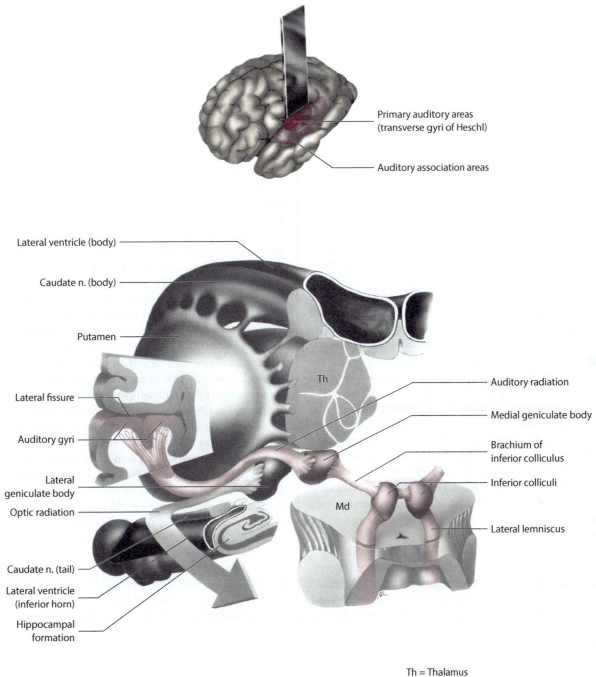

Primary auditory areas
(transverse gyri of Heschl)

Auditory association areas

Lateral ventricle (body)

Caudate n. (body)

Putamen

Th

Auditory radiation

Lateral fissure

Medial geniculate body

Auditory gyri

Brachium of
inferior colliculus

Lateral
geniculate body

Inferior colliculi

Optic radiation

Md

Lateral lemniscus

Caudate n. (tail)

Lateral ventricle
(inferior horn)

Hippocampal
formation

Th = Thalamus
Md = Midbrain

FIGURE 6.2: Audition 2—Auditory Pathway 2: Thalamus

FIGURE 6.3—AUDITION 3

AUDITORY PATHWAY 3: AUDITORY GYRI (PHOTOGRAPH)

This photographic view of the left cerebral hemisphere is shown from the lateral perspective (see Figure 1.3 and Figure 4.5). The lateral fissure has been opened, and this exposes two gyri that are oriented transversely, the **auditory gyri**. These gyri are the areas of the cortex that receive the incoming auditory sensory information first. They are also known as the **transverse gyri of Heschl** (also shown in Figure 6.2).

The lateral fissure forms a complete separation between this part of the temporal lobe and the frontal and parietal lobes above. Looked at descriptively, the auditory gyri occupy the superior aspect of the temporal lobe, within the lateral fissure.

Cortical representation of sensory systems reflects the particular sensation (modality). The auditory gyri are organized according to pitch, thus giving rise to the term **tonotopic** localization. This is similar to the representation of the somatosensory system on the postcentral gyrus (somatotopic localization; the sensory "homunculus").

Further opening of the lateral fissure reveals some cortical tissue that is normally completely hidden from view. This area is the **insula** or insular cortex (see Figure 1.4). The insula typically has five short gyri, and these are seen in the depth of the lateral fissure. It is important not to confuse the auditory gyri and insula. The position of the insula in the depth of the lateral fissure is also shown in a dissection of white matter bundles (see Figure 2.3) and in the coronal slice of the brain (see Figure 2.9A).

The lateral fissure has within it a large number of blood vessels, branches of the middle cerebral artery, which have been removed (see Figure 8.4). These branches emerge and then become distributed to the cortical tissue of the dorsolateral surface, including the frontal, temporal, parietal, and occipital cortex. Other small branches to the internal capsule and basal ganglia are given off within the lateral fissure (discussed with Figure 8.6).

CLINICAL ASPECT

The loss or decrease of hearing on one side may result from problems in the external ear (due to infection or excess ear wax) or middle ear (due to fluid, or infection, or pathology of the ossicles) that interfere with the transmission of the sound waves.

Diminished hearing, particularly for the higher frequencies, is common with advancing age and is often accompanied by a "ringing" noise, called tinnitus.

A tumor of cranial nerve (CN) VIII (within the internal auditory canal), called a schwannoma, also called an acoustic neuroma, is not rare and causes hearing loss on the side of the lesion. Because of its location, as the tumor grows it begins to compress the adjacent nerves (including CN VII). Eventually, if left unattended, additional symptoms result from further compression of the brainstem and an increase in intracranial pressure. Modern imaging techniques allow early detection of this tumor. Surgical removal, however, still requires considerable skill so as not to damage CN VIII itself (which would produce a loss of hearing) or CN VII (which would produce a paralysis of facial muscles) and adjacent neural structures; focused radiotherapy has also been used to destroy the tumour.

Because the auditory system has a bilateral pathway to the cortex, a lesion of the pathway or cortex on one side does not lead to a total loss of hearing (deafness) on the same side or the opposite ear. Nonetheless, the pathway still has a strong crossed aspect; the auditory spectrum for speech is directed to the dominant hemisphere.

Note to the Learner: The auditory system is further described in the *Integrated Nervous System* with a clinical case involving a tumor of CN VIII; animation of the pathway is found on the Web site for the *Integrated* text.

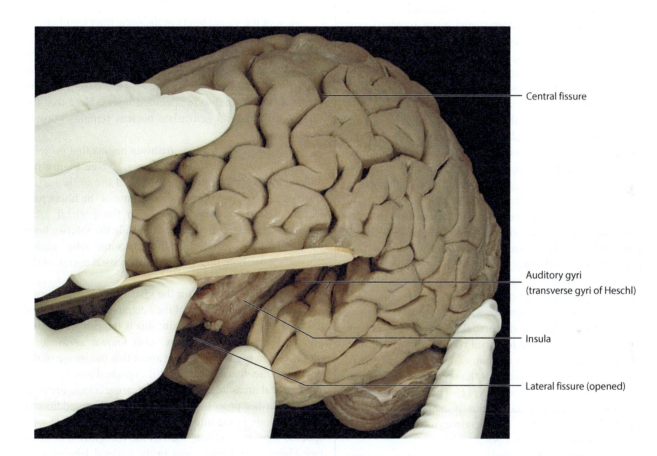

Central fissure

Auditory gyri
(transverse gyri of Heschl)

Insula

Lateral fissure (opened)

FIGURE 6.3: Audition 3—Auditory Pathway 3: Auditory Gyri (photograph)

VISUAL SYSTEM

FIGURE 6.4—VISION 1

VISUAL PATHWAY 1: VISUAL FIELD TO CORTEX

The visual image exists in the **visual field**, the outside world. This image is projected onto the retina and is known as the **retinal field**. The visual fields are also divided into temporal (lateral) and nasal (medial) portions. The temporal visual field of one eye is projected onto the nasal part of the retina of the ipsilateral eye and onto the temporal part of the retina of the contralateral eye. The primary purpose of the visual apparatus (e.g., muscles) is to align the visual image on corresponding points of the retina of both eyes. Because of the lens of the eye, the upper visual field projects to the lower retina (and the converse for the lower visual field); therefore, the image on the retina is "upside down."

Visual processing begins in the retina with the photoreceptors, the highly specialized receptor cells, the **rods and cones**. The central portion of the visual field projects onto the **macular** area of the retina, composed of only cones, which is the area required for discriminative vision (e.g., reading) and color vision. Rods are found in the peripheral areas of the retina and are used for peripheral vision and seeing under conditions of low level illumination. These receptors synapse with the **bipolar neurons** located in the retina, the first actual neurons in this system (functionally equivalent to dorsal root ganglion neurons). These connect with the **ganglion cells** (still in the retina), whose axons leave the retina at the optic disc to form the **optic nerve (cranial nerve [CN] II)**. The optic nerve is in fact a tract of the central nervous system (CNS) because its myelin is formed by oligodendrocytes (the glial cell that forms and maintains CNS myelin).

After exiting the eyeball, the **optic nerves** course through the orbit and exit through the optic foramen to enter the interior of the skull. In the area above the pituitary gland, the nerves undergo a partial crossing (decussation) of fibers in a structure called the **optic chiasm**.

Note to the Learner: There is no synapse in the optic chiasm.

The fibers from the nasal retina on one side cross the midline and join with those from the temporal retina from the other eye (which do not cross) to form the **optic tract**. Thus, the image of the visual world that started in different parts of the retina of the two eyes is now brought back together in the optic tract (described in detail later). The result of this rearrangement is to bring together the visual information from the visual field of one side from both eyes to the opposite side of the brain.

Using a specific example, the visual object on the left side, the visual field, projects to the nasal retina for the left eye and the temporal retina for the right eye.

Note to the Learner: Making a sketch diagram of the visual system using colored pens or pencils is a simple and effective way of understanding the visual pathway.

Most of the visual fibers in the optic tract terminate in the **lateral geniculate nucleus (LGN or LGB)**, a specific relay nucleus of the thalamus (see Figure 4.3). The lateral geniculate is a layered nucleus (this is a unique arrangement for a thalamic nucleus—see Figure 6.6); the fibers from each eye synapse in specified layers. Note that the image in the lateral geniculate nucleus remains "upside down."

After processing, a new pathway begins that projects to the **primary visual cortex, area 17** (see Figure 6.5 and Figure 6.6). The projection, called the **optic radiation**, consists of two portions with some of the fibers projecting directly posteriorly deep in the parietal lobe, whereas others sweep forward alongside the inferior horn of the lateral ventricle in the temporal lobe, called **Meyer's loop**; both then project to the visual cortex of the occipital lobe.

The final destination for the visual fibers is the cortex along the calcarine fissure of the occipital lobe, located on the medial surface of the brain; this is the primary visual area, V1 or calcarine cortex, also known as **area 17** (described in Figure 6.6). Note again that the image of the objects in the visual cortex is still "upside down."

Cortical areas adjacent to the calcarine cortex, areas 18 and 19, further process the visual information; additional visual regions in the inferior aspect of the brain deal with specific aspects of vision, such as the recognition of faces (see Figure 1.5). Other areas in the parietal lobe process visuo-spatial information (discussed with Figure 6.13).

CLINICAL ASPECT

Note that the visual pathway—from cornea to the calcarine cortex—extends through the whole brain (excluding the frontal lobe), hence its importance in the assessment of nervous system integrity.

The visual pathway is easily testable, even at the bedside. Students should be able to draw the visual field defect in both eyes that would follow a lesion of the optic nerve, at the optic chiasm, and in the optic tract.

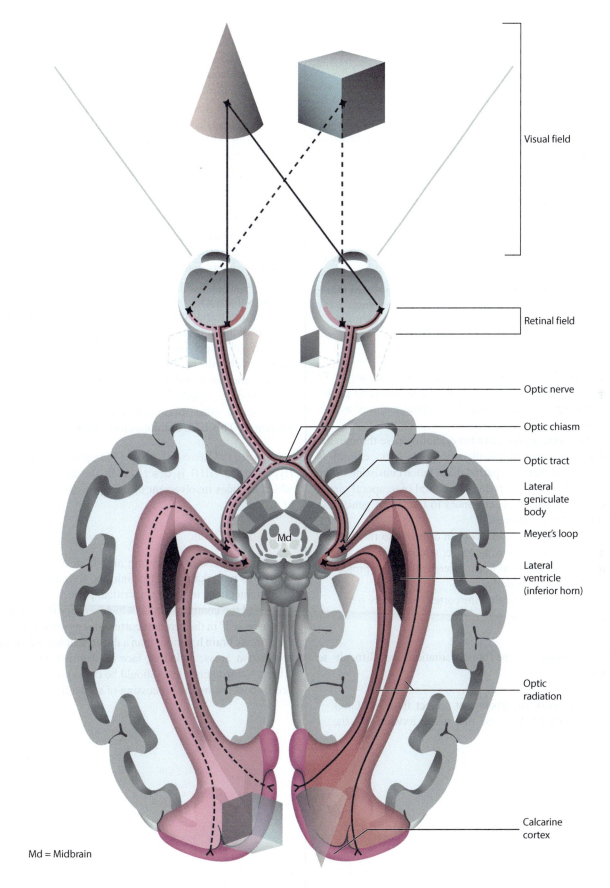

Visual field

Retinal field

Optic nerve

Optic chiasm

Optic tract

Lateral
geniculate
body

Meyer's loop

Lateral
ventricle
(inferior horn)

Optic
radiation

Calcarine
cortex

Md

Md = Midbrain

FIGURE 6.4: Vision 1—Visual Pathway 1: Visual Field to Cortex

FIGURE 6.5—VISION 2

VISUAL PATHWAY 2 (RADIOLOGICAL RECONSTRUCTION)

This is an unusual but instructive image—it is a reconstruction, from several T1 magnetic resonance images, of the visual pathway. Because this pathway, from orbit to cortex, does not exist on one plane, several "cuts" were used to piece together the whole pathway including the optic radiation from lateral geniculate to calcarine cortex.

UPPER ILLUSTRATION

This illustration, using a T1 magnetic resonance image of the medial aspect of the brain, shows the approximate path taken by the visual pathway, from the orbit to thalamus and the visual radiation to the occipital cortex.

LOWER ILLUSTRATION

This radiological image is a reconstruction, using T1 magnetic resonance images of the brain in several planes (see the upper illustration), of the visual pathway as if it exists in a single plane.

The optic nerves leave the eyeball, course through the orbit, and enter the skull, where the optic chiasm is located, and there is a partial decussation consisting of the fibers from the nasal retina (the lateral or temporal visual fields). The optic tract continues to the lateral geniculate nucleus of the thalamus.

The optic radiation has been drawn onto the illustration on one side, showing Meyer's loop swinging anteriorly into the temporal horn, whereas the other projection goes directly posteriorly.

Both end in the calcarine cortex (see Figure 6.6).

CLINICAL ASPECT

Some lesions of the optic radiation are difficult to understand:

- Loss of the fibers that project from the lower retinal field, those that sweep forward into the temporal lobe (Meyer's loop, see Figure 6.4 and also Figure 6.6), results in a loss of vision in the upper visual field of both eyes on the side opposite the lesion, specifically the upper quadrant of both eyes.

- Loss of those fibers coming from the upper retinal field that project directly posteriorly, passing deep within the parietal lobe, results in the loss of the lower visual field of both eyes on the side opposite the lesion, specifically the lower quadrant of both eyes.

From the developmental perspective, the retina is an "extension" of the brain, and the optic nerve is in fact a central nervous system tract. In its path through the orbit, the optic nerve is ensheathed by the meninges of the brain, dura, arachnoid, and pia, with a typical subarachnoid space containing cerebrospinal fluid. Any increase of intracranial pressure may be reflected via this space onto the optic nerve and cause its compression, as well as compromising the blood vessels supplying the retina (the arteries and veins) running within the nerve.

The end result of this process will cause "blurring" of the optic disc, called **papilledema**, as seen with an ophthalmoscope (also discussed with the introduction to Section 3). This is an ominous clinical sign!

Note to the Learner: A case of Idiopathic Intracranial Hypertension (IIH) is described in the *Integrated* text, which includes involvement of the optic nerve and a loss of vision.

ADDITIONAL DETAIL

The work on visual processing and its development has offered us remarkable insights into the formation of synaptic connections in the brain, critical periods in development, and the complex way in which sensory information is "processed" in the cerebral cortex. It is now thought that the primate brain has more than a dozen specialized visual association areas, including face recognition, color, and others. Neuroscience texts should be consulted for further details concerning the processing of visual information.

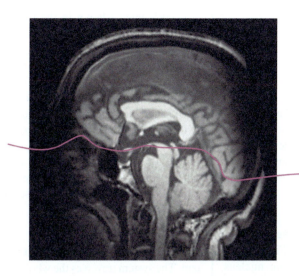

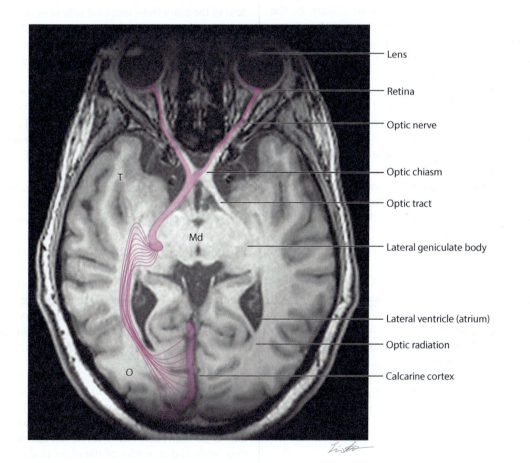

Lens

Retina

Optic nerve

Optic chiasm

Optic tract

Lateral geniculate body

Lateral ventricle (atrium)

Optic radiation

Calcarine cortex

T = Temporal lobe
O = Occipital lobe
Md = Midbrain

FIGURE 6.5: Vision 2—Visual Pathway 2 (radiological reconstruction)

FIGURE 6.6—VISION 3

VISUAL PATHWAY 3: VISUAL CORTEX (ILLUSTRATION AND PHOTOGRAPH)

We humans are visual creatures. We depend on vision for access to information (the written word), the world of images (e.g., photographs, television, online viewing), and the complex urban landscape. There are many cortical areas devoted to interpreting the visual world.

UPPER ILLUSTRATION

This axial image at the uppermost level of the midbrain includes the pulvinar of the thalamus (see Figure 4.3). The visual fibers leave the lateral geniculate nucleus (body) and form the optic radiation that terminates in **area 17**, the primary visual area, within and along the upper and lower gyri of the calcarine fissure of the occipital lobe (see Figure 1.7 and seen in the photographic image below). It is important to note the relationship of the optic radiation to the lateral ventricle (discussed with the Clinical Aspect to Figure 6.5). Note the layers of the lateral geniculate nucleus shown on the other side; this nucleus is laminated with different layers representing the visual fields of the ipsilateral and contralateral eyes.

The visual image remains "upside down" in the occipital cortex because the fibers from the lower retina—representing the upper visual field—are found in the lower portion of the calcarine cortex; the fibers in the upper portion of the calcarine cortex represent the lower visual field (see Figure 6.4).

Some fibers from the retina leave the optic tract and terminate in the **superior colliculi** (see Figure 1.9 and also Figure 1.5, Figure 3.2, and Figure 3.3), which are involved with coordinating eye movements (see Additional Detail below).

Visual fibers also end in the **pre-tectal "nucleus,"** an area in front of the superior colliculus, for the pupillary light reflex (reviewed with Figure 6.7).

Some other fibers terminate in the **suprachiasmatic** nucleus of the hypothalamus (located above the optic chiasm, not shown), which is involved in the regulation of the diurnal (circadian) rhythm.

LOWER ILLUSTRATION (PHOTOGRAPH)

The posterior portion of area 17, extending to the occipital pole, is where macular vision is represented (this is shown in the illustration with a darker coloration); the visual cortex in the more anterior portion of area 17 is the cortical region where the peripheral areas of the retina are represented (shown in the illustration with a lighter coloration).

The adjacent cortical areas, **areas 18 and 19** (on both the medial and lateral surfaces of the occipital lobe), are visual association areas; fibers are also relayed here via the pulvinar of the thalamus (see Figure 4.3 and Figure 6.13). There are many other cortical areas for elaboration of the visual information, including regions on the inferior aspect of the hemisphere for face recognition (see Figure 1.5) as well as areas of the temporal lobe (see Figure 6.13). Other areas in the parietal lobe process visuo-spatial information (discussed with Figure 6.13).

CLINICAL ASPECT

It is very important for the student to know the visual system. The system traverses the whole brain and cranial fossa, from front to back, and testing the complete visual pathway from retina to cortex is an opportunity to sample the intactness of the brain from temporal pole to occipital pole.

Diseases of CNS myelin, such as multiple sclerosis (MS), can and do affect the optic nerve and/or optic tract and cause visual loss. Sometimes this is the first manifestation of MS.

Visual loss involving the retina can occur, as in a stroke, since the retinal artery is a branch of the ophthalmic artery which is a branch of the internal carotid artery. Thrombi from the internal carotid bifurcation can embolize via this route, causing an infarction of some or all of the retina.

Visual loss can occur for other reasons, one of which is the loss of blood supply to the cortical areas. The visual cortex is supplied by the posterior cerebral artery (from the vertebro-basilar system; discussed with Figure 8.5). Part of the occipital pole, with the representation of the macular area of vision, may be supplied by the middle cerebral artery (from the internal carotid system; see Figure 8.4). In some cases, macular sparing is found after occlusion of the posterior cerebral artery, presumably because the blood supply to this area in this person was coming from the other vascular supply.

ADDITIONAL DETAIL

Fibers from the optic tract project to the **superior colliculus,** by-passing the lateral geniculate. The superior colliculus is an important center for visual reflexes. Fibers are then projected to nuclei of the extra-ocular muscles and neck muscles via a small pathway, the **tecto-spinal tract** which is found incorporated with the MLF, the medial longitudinal fasciculus (discussed with Figure 6.9). Additional areas involved with eye movements include the occipital cortex, the frontal eye field (see Figure 4.5) and the paramedian pontine reticular nucleus (see Figure 6.8).

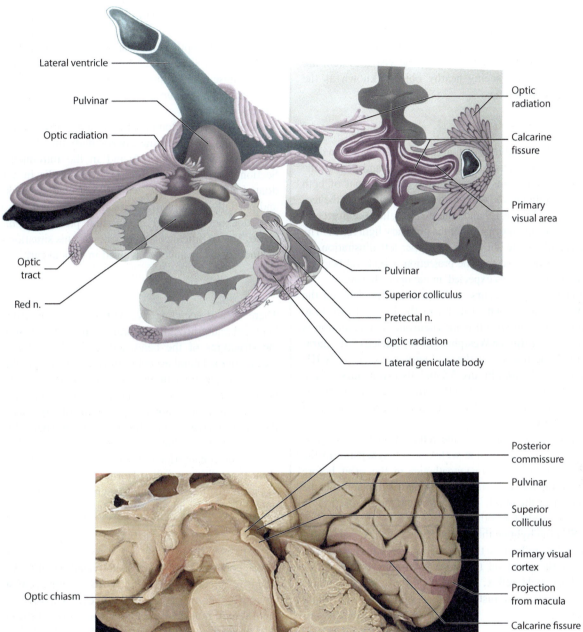

Lateral ventricle

Pulvinar

Optic radiation

Optic tract

Red n.

Optic radiation

Calcarine fissure

Primary visual area

Pulvinar

Superior colliculus

Pretectal n.

Optic radiation

Lateral geniculate body

Posterior commissure

Pulvinar

Superior colliculus

Primary visual cortex

Projection from macula

Calcarine fissure

Optic chiasm

FIGURE 6.6: Vision 3—Visual Pathway 3: Visual Cortex (illustration and photograph)

FIGURE 6.7—VISION 4

VISUAL REFLEXES: PUPILLARY AND ACCOMMODATION REFLEXES

A small but extremely important group of fibers from the optic tract projects to the pretectal area for the pupillary light reflex. Reflex adjustments of the visual system are also required for seeing nearby objects, known as the accommodation reflex.

PUPILLARY LIGHT REFLEX

Some of the visual information (from certain ganglion cells in the retina) is carried in the optic nerve and tract to the midbrain. A nucleus located in the area in front of both superior colliculi (the other name for the colliculi is the tectal area; see Figure 1.9 and Figure 3.3), called the **pretectal area**, is the site of synapse for the pupillary light reflex. The afferent fibers are shown in the upper left illustration for only one eye. Note that the projection is bilateral because some fibers cross, as expected, in the optic chiasm.

The synapse occurs in the pretectal nucleus on the same side, as shown in detail in the lower illustration. After synapsing with this interneuron, the connection is sent to the Edinger-Westphal (E-W) nucleus, the parasympathetic portion of the oculomotor nucleus (CN III) (see Figure 1.8 and Figure 3.5 and also Appendix Figure A.3). A collateral is sent to the other side, via the posterior commissure (labeled in Figure 6.6; also shown in Figure 9.5A).

The efferent fibers of the reflex from the **Edinger-Westphal** nucleus course with **CN III**, synapsing in the **ciliary ganglion** (parasympathetic) in the orbit before innervating the smooth muscle of the iris that controls the diameter of the pupil. This is shown in the illustration on the upper right for both sides.

Shining light on the retina of one eye, the usual way of testing this reflex (see the next paragraph) causes constriction of the pupil on the same side—this is the **direct** pupillary light reflex; the pupil of the other eye reacts almost at the same instance—this is the **consensual** light reflex.

Clinical Aspect

The pupillary light reflex is a critically important clinical sign, particularly in patients who are in a coma or following a head injury. It is essential to ascertain the status of the reaction of the pupil to light, ipsilaterally and on the opposite side. The student is encouraged to draw this pathway and to work out the clinical picture of a lesion involving the afferent visual fibers, the midbrain area, and a lesion affecting the efferent fibers (CN III).

In a disease such as multiple sclerosis, or with diseases of the retina, there can be a reduced sensory input via the optic nerve, and this can cause a condition called a *relative afferent pupillary defect*. A specific test for this is the **swinging light reflex**, which is performed in a dimly lit room. Both pupils will constrict when the light is shone on the normal side. As the light is shone in the affected eye, because of the diminished afferent input to the pretectal nucleus, the pupil of this eye will dilate in a paradoxical manner.

CN III, the oculomotor nerve, is usually involved in brain herniation syndromes, particularly uncal herniation (discussed with Figure 1.6 and in the introduction to Section 3; see also Figure 7.1). This results in a fixed dilated pupil on one side, with the affected eye abducted and partially depressed, a critical sign when one is concerned about increased intracranial pressure from any cause. The significance and urgency of this situation must be understood by anyone involved in critical care.

HORNER'S SYNDROME

There is a descending sympathetic pathway from the hypothalamus through the brainstem to supply some of the structures of the head and neck. These fibers are located in the lateral aspect of the medulla, exit to the cervical sympathetic ganglia and are distributed with the branches of the external carotid artery. The superior eyelid receives this innervation along with the dilator muscles of the pupil. Interruption of this pathway anywhere along its course results in a "drooping" of the upper eyelid (called **ptosis**) and constriction of the pupil (**meiosis/miosis**). The clinical condition is called Horner's syndrome (discussed as part of the lateral medullary syndrome, also known as the Wallenberg syndrome, with Figure 6.11).

ACCOMMODATION REFLEX

The accommodation reflex is activated when looking at a nearby object, as in reading. Three events occur simultaneously—convergence of both eyes (involving both medial recti muscles), a change (rounding) of the curvature of the lens, and pupillary constriction. This reflex requires the visual information to be processed at the cortical level. The descending cortico-bulbar fibers (see Figure 5.10) go to the oculomotor nucleus and influence both the motor portion (to the medial recti muscles), and also to the parasympathetic (Edinger-Westphal) portion (to the smooth muscle of the lens and the pupil, via the ciliary ganglion) to effect the reflex.

Afferent pathway

Efferent pathway

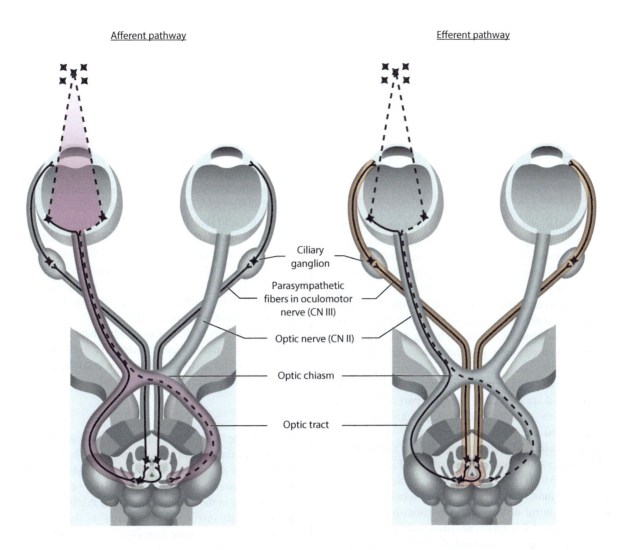

Ciliary
ganglion

Parasympathetic
fibers in oculomotor
nerve (CN III)

Optic nerve (CN II)

Optic chiasm

Optic tract

Midbrain detail

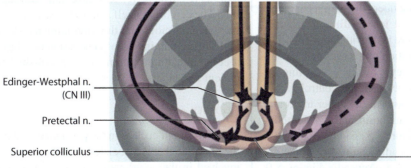

Edinger-Westphal n.
(CN III)

Pretectal n.

Superior colliculus

Posterior commissure

FIGURE 6.7: Vision 4—Visual Reflexes: Pupillary and Accommodation Reflexes

VESTIBULAR SYSTEM AND VISUOMOTOR

FIGURE 6.8—VESTIBULAR SYSTEM

VESTIBULAR NUCLEI AND VISUOMOTOR

The vestibular system carries information about our position in relation to gravity and changes in that position. The sensory system is located in the inner ear and consists of three **semicircular canals** (see Figure 3.8), as well as other sensory organs in a bony and membranous labyrinth. There is a peripheral ganglion (the spiral ganglion), and the central processes of these cells enter the brainstem as part of the vestibulocochlear nerve, CN VIII, at the cerebellar-pontine angle, just above the cerebellar flocculus (see Figure 1.8).

The vestibular information is carried to **four vestibular nuclei** that are located in the upper part of the medulla and lower pons: superior, lateral, medial, and inferior (see Figure 3.4 and also Appendix Figure A.7).

The **lateral vestibular nucleus** gives rise to the lateral vestibulo-spinal tract (as described in Figure 5.13; see also Figure 6.9). This is the pathway that serves to adjust the postural musculature to changes in relation to gravity; it is under control of the "older" portions of the cerebellum (see Figure 5.17). Fibers also descend from the **medial vestibular nucleus** as the medial vestibulo-spinal tracts (as described in Figure 5.13; see also Figure 6.9); these fibers are involved with postural adjustments to positional changes, using the axial musculature.

The **medial and inferior vestibular nuclei** give rise to both ascending and descending fibers that join a conglomerate bundle called the **medial longitudinal fasciculus (MLF)**, which is described more fully with Figure 6.9. The ascending fibers adjust the position of the eyes and coordinate eye movements of the two eyes by interconnecting the three cranial nerve nuclei involved in the control of eye movements—CN III (oculomotor) in the upper midbrain, CN IV (trochlear) in the lower midbrain, and CN VI (abducens) in the lower pons (see Figure 3.5 and Figure 6.13; see also Appendix Figure A.5, Appendix Figure A.6, Appendix Figure A.7, Appendix Figure A.8, Appendix Figure A.9, and Appendix Figure A.10). If one considers lateral gaze, a movement of the eyes to the side (in the horizontal plane), this requires the coordination of the lateral rectus muscle (abducens nucleus) of one side and the medial rectus (oculomotor nucleus) of the other side; this is called a **conjugate eye movement**. These fibers for coordinating the eye movements are carried in the MLF.

The descending fibers from the vestibular nuclei, also carried in the MLF (and further described in Figure 6.9), are likely concerned with the coordination of vision with the positioning of the neck.

PARAMEDIAN PONTINE RETICULAR FORMATION (PPRF)

There is a "gaze center" within the pontine reticular formation—the paramedian pontine reticular formation (PPRF)—for **saccadic** eye movements. These are extremely rapid (ballistic) movements of both eyes, yoked together, usually in the horizontal plane so that we can shift our focus extremely rapidly from one object to another. The fibers directing these movements originate from the different areas of the cortex, including from the frontal eye field (see Figure 4.5) and the occipital cortex, and also likely course in the MLF.

Several areas of the cortex and brainstem are involved in the coordination of eye movements, both pursuit and saccadic. Additional texts should be consulted to understand this subject fully.

CLINICAL ASPECT

Although the control of lateral gaze is relatively well understood, the control of vertical gaze—looking up and down—is not as definitively known. On the basis of lesions of the brainstem, the area involved in vertical gaze is thought to be in the region of the upper midbrain. Looking upward usually involves a "furrowing" of the forehead.

There is a somewhat rare condition, called a **locked-in syndrome**, caused by an occlusion of the vascular supply to the brainstem, usually the ventral pons. Most patients do not survive, but those who do suffer from complete paralysis of all musculature, including the cranial nerves, except vertical gaze and blinking of the eyelids. The cerebrum itself is not affected, which means that they are conscious and thinking humans who are "locked in"! This remaining bit of ocular function has allowed for a communication system to be established between the patient and others.

ADDITIONAL DETAIL

There is a small nucleus in the periaqueductal gray region of the midbrain that is associated with the visual system and is involved in the coordination of eye and neck movements. This nucleus is called the **interstitial nucleus (of Cajal)**. It is located near the oculomotor nucleus (see also Figure 6.9). This nucleus receives input from various sources and contributes fibers to the MLF (discussed with the Figure 6.9). Some have named this pathway the interstitio-spinal "tract."

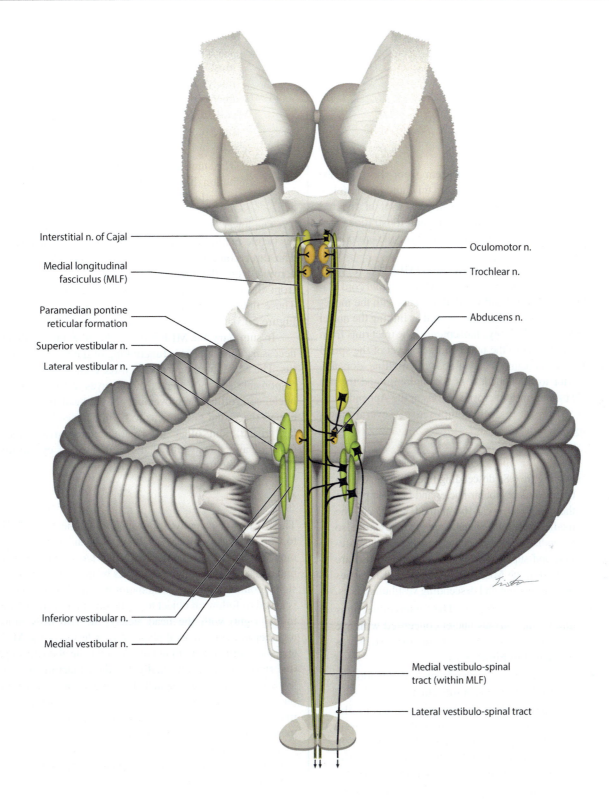

FIGURE 6.8: Vestibular System—Vestibular Nuclei and Visuomotor

FIGURE 6.9—MEDIAL LONGITUDINAL FASCICULUS

MEDIAL LONGITUDINAL FASCICULUS (MLF)—ASSOCIATED TRACTS AND EYE MOVEMENTS

This diagram shows the brainstem from the posterior perspective (as in Figure 1.9, Figure 6.11, and Figure 6.12). The upper small illustration is at the level of the red nucleus of the midbrain. The lower small illustrations is at the level of the cervical spinal cord. (Note the orientation of the spinal cord.)

The medial longitudinal fasciculus (MLF) is a composite pathway within the brainstem and upper spinal cord that links the visual world and vestibular events with the movements of the eyes and the neck, as well as linking the nuclei that are responsible for eye movements. The tract runs from the midbrain level to the upper thoracic level of the spinal cord. It has a constant location near the midline, dorsally, just anterior to the aqueduct of the midbrain and the 4th ventricle (see brainstem cross-sections in the Appendix).

The MLF is, in fact, composed of several tracts running together:

- **Vestibular fibers**: Of the four vestibular nuclei (see Figure 6.8), descending fibers originate from the medial vestibular nuclei and become part of the MLF; this can be named separately the medial vestibulo-spinal tract. There are also ascending fibers that come from the medial, inferior, and superior vestibular nuclei that also are carried in the MLF. Therefore, the MLF carries both ascending and descending vestibular fibers.

- **Visuomotor fibers**: The interconnections among the various nuclei concerned with eye movements are carried in the MLF (as described in Figure 6.8).

- **Vision-related fibers**: Visual information is received by various brainstem nuclei.
  - The superior colliculus is a nucleus for the coordination of visual-related reflexes, including eye movements (see Figure 1.9). The superior colliculus coordinates the movements of the eyes and the turning of the neck in response to visual information. It also receives input from the visual association cortical areas, areas 18 and 19 (see Figure 1.7, Figure 4.5, Figure 6.6, and Figure 6.13). The descending fibers from the superior colliculus, called the **tecto-spinal tract** (recall that the colliculi are part of the tectal plate), are very closely associated with the MLF and can be considered part of this system (although in most books this tract is discussed separately). As shown in the upper inset, these fibers cross in the midbrain. (The superior colliculus [SC] of only one side is shown in order not to obscure the crossing fiber systems at that level.)
  - The small interstitial nucleus and its contribution have already been noted and discussed with Figure 6.8.

The lower inset shows the MLF in the ventral funiculus (white matter) of the spinal cord, at the cervical level. The three components of the tract are identified—those coming from the medial vestibular nucleus, the fibers from the interstitial nucleus, and the tecto-spinal tract. These fibers are mingled together in the MLF (see also Figure 6.13).

In summary, the MLF is a complex fiber bundle that is necessary for the proper functioning of the visual apparatus. The MLF interconnects the three cranial nerve nuclei responsible for movements of the eyes with the motor nuclei controlling the movements of the head and neck. It allows the visual movements to be influenced by vestibular, visual, and other information and carries fibers (upward and downward) that coordinate the eye movements with the turning of the neck.

CLINICAL ASPECT

The MLF is principally responsible for connecting CN III to CN VI (review Figure 3.5) to coordinate conjugate eye movements. A lesion of the MLF interferes with this normal conjugate movement of the eyes by interfering with the signal connecting these oculomotor nuclei. When a person is asked to follow an object (e.g., the tip of a pencil moving to the right) with the head steady, the two eyes move together in the horizontal plane. With a lesion of the MLF (e.g., demyelination in multiple sclerosis), the abducting eye (the right eye) moves normally, but the adducting eye (the left eye) fails to follow; nonetheless, adduction is preserved on convergence. Clearly, the nuclei and the nerves are intact; the lesion then is in the fibers coordinating the movement. This condition is known as **internuclear ophthalmoplegia** (commonly called an **INO**). Sometimes there is also monocular horizontal nystagmus (rapid side-to-side movements) of the abducting eye.

ADDITIONAL DETAIL

The diagram also shows the posterior commissure (not labeled). This small commissure carries fibers connecting the superior colliculi. In addition, it carries the important fibers for the consensual pupillary light reflex coordinated in the pre-tectal "nucleus" (as discussed in Figure 6.7).

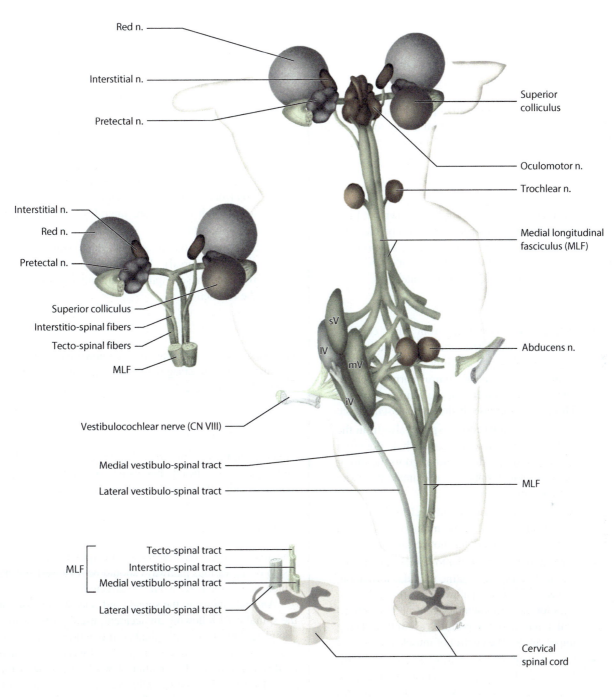

Red n.

Interstitial n.

Pretectal n.

Superior colliculus

Oculomotor n.

Trochlear n.

Medial longitudinal fasciculus (MLF)

Interstitial n.

Red n.

Pretectal n.

Superior colliculus

Interstitio-spinal fibers

Tecto-spinal fibers

MLF

sV

lV

mV

iV

Abducens n.

Vestibulocochlear nerve (CN VIII)

Medial vestibulo-spinal tract

Lateral vestibulo-spinal tract

MLF

Tecto-spinal tract

Interstitio-spinal tract

Medial vestibulo-spinal tract

Lateral vestibulo-spinal tract

MLF

Cervical spinal cord

sV = Superior vestibular n.
lV = Lateral vestibular n.
mV = Medial vestibular n.
iV = Interior vestibular n.

FIGURE 6.9: Medial Longitudinal Fasciculus (MLF)—Associated Tracts and Eye Movements

PATHWAYS SUMMARY

FIGURE 6.10—SPINAL CORD TRACTS: CROSS-SECTIONS

Note to the Learner: This is an opportune time to review all the illustrations of the spinal cord in Section 1 and in Section 2. The blood supply to the spinal cord is discussed with Figure 8.7 and Figure 8.8.

UPPER ILLUSTRATION: TRACTS—C8 LEVEL

The major tracts of the spinal cord are shown on this diagram, the descending (motor-related) tracts on the left side and the ascending (sensory-related) ones on the right side. In fact, both sets of pathways are present on both sides. Some salient features of each are presented here.

DESCENDING TRACTS

- Lateral cortico-spinal, from the cerebral (motor) cortex: These fibers for direct voluntary control supply mainly the lower motor neurons in the lateral ventral horn to control fine motor movements of the hand and fingers. This pathway crosses in the lowermost medulla.
- Anterior (ventral) cortico-spinal, also from the motor cortex: These fibers, which do not cross in the pyramidal decussation, go to the motor neurons that supply the proximal and axial musculature.
- Rubro-spinal, from the red nucleus: This tract crosses at the level of the midbrain. Its role in human motor function is not certain.
- Lateral and medial reticulo-spinal tracts, from the medullary and pontine reticular formation, respectively: These pathways are the additional ones for indirect voluntary control of the proximal joints and for posture, as well as being important for the control of muscle tone.
- Lateral vestibulo-spinal, from the lateral vestibular nucleus: Its important function is participating in the response of the axial muscles to changes in gravity. This pathway remains ipsilateral.
- Medial longitudinal fasciculus: This mixed pathway is involved in the response of the muscles of the eyes and of the neck to vestibular and visual input. It likely descends only to the cervical spinal cord level.

ASCENDING TRACTS

- Dorsal column tracts, consisting at this level of both the fasciculus cuneatus and fasciculus gracilis: These are the pathways for discriminative touch sensation, joint position, and "vibration" from the same side of the body, with the lower limb fibers medially (gracile) and the upper limb pathway laterally (cuneate).
- Anterolateral system, consisting of the anterior (ventral) spino-thalamic and lateral spino-thalamic tracts: These pathways carry pain and temperature, as well as crude touch information from the opposite side of the body, with the lower limb fibers more lateral and the upper limb fibers medially.
- Spino-cerebellar tracts, anterior (ventral) and posterior (dorsal): These convey information from the muscle spindles and other sources to the cerebellum.

SPECIAL TRACT

The dorsolateral fasciculus, better known as the tract of Lissauer, carries intersegmental information, particularly relating to pain afferents.

LOWER ILLUSTRATION

This diagram shows the tracts of the spinal cord—all on both sides (not labeled, but using the same functional colors)—with a three-dimensional perspective.

CLINICAL ASPECT

The loss of function that would be found following a lesion of the various pathways should be reviewed, noting which side of the body would be affected.

An acute injury to the spinal cord, such as severing of the spinal cord following an accident, usually results in a complete shutdown of all spinal cord functions, called **spinal shock**. After a period of about 3 to 4 weeks, the spinal cord reflexes return. In a matter of weeks, because of the loss of all the descending influences on the spinal cord, there is an increase in the reflex responsiveness (hyperreflexia) and a marked increase in tone (spasticity), along with the extensor plantar response.

A classic lesion of the spinal cord is the **Brown-Sequard syndrome**, which is a lesion of one half of the spinal cord on one side. Although rare, this is a useful lesion for a student to review the various deficits, sensory and motor, that would be found after such a lesion. In particular, it helps a student learn which side of the body would be affected because of the various crossing of the pathways (sensory and motor) at different levels (see Clinical Cases).

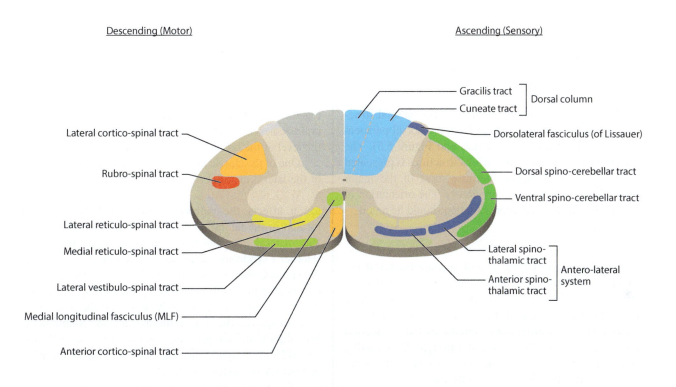

Descending (Motor)   Ascending (Sensory)

Gracilis tract ⎤
                ⎦ Dorsal column
Cuneate tract

Lateral cortico-spinal tract

Dorsolateral fasciculus (of Lissauer)

Rubro-spinal tract

Dorsal spino-cerebellar tract

Ventral spino-cerebellar tract

Lateral reticulo-spinal tract

Medial reticulo-spinal tract

Lateral spino-
thalamic tract ⎤
                ⎦ Antero-lateral
                   system
Anterior spino-
thalamic tract

Lateral vestibulo-spinal tract

Medial longitudinal fasciculus (MLF)

Anterior cortico-spinal tract

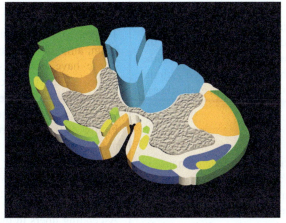

FIGURE 6.10: Spinal Cord Tracts: Cross-Sections

FIGURE 6.11—BRAINSTEM SENSORY SYSTEMS

ASCENDING TRACTS AND SENSORY NUCLEI

Note to the Learner: At this point, the sensory nuclei of the brainstem should be reviewed (see Figure 3.4 and Appendix Figure A.1) as well as the pathways (in Section 2). The blood supply of the brainstem is reviewed with Figure 8.1.

This diagrammatic presentation of the internal structures of the brainstem is shown from the dorsal perspective (as in Figure 1.9, Figure 3.2, and Figure 5.5). The information concerning the various structures is presented in an abbreviated manner because most of the major points have been reviewed previously. The orientation of the cervical spinal cord representation should be noted.

DORSAL COLUMN–MEDIAL LEMNISCUS

The dorsal columns (gracile and cuneate tracts) of the spinal cord terminate (synapse) in the nuclei gracilis and cuneatus in the lowermost medulla. Axons from these nuclei then cross the midline (decussate) as the internal arcuate fibers, forming a new bundle called the medial lemniscus. These fibers ascend through the medulla, change orientation in the pons, and move laterally, occupying a lateral position in the midbrain.

ANTEROLATERAL SYSTEM

This tract, having already crossed in the spinal cord, ascends and continues through the brainstem. In the medulla, it is situated posterior to the inferior olive. At the upper pontine level, this tract becomes associated with the medial lemniscus, and the two lie adjacent to each other in the midbrain region.

TRIGEMINAL PATHWAY

The sensory afferents for discriminative touch synapse in the principal nucleus of CN V; the fibers then cross at the level of the mid-pons and form a tract that joins the medial lemniscus. The pain and temperature fibers descend and form the descending trigeminal tract through the medulla with the nucleus adjacent to it. These fibers synapse and cross, over a wide area of the medulla and eventually join the other trigeminal tract. The two tracts form the trigeminal pathway, which joins with the medial lemniscus in the uppermost pons.

LATERAL LEMNISCUS

The auditory fibers (of CN VIII) enter the brainstem at the uppermost portion of the medulla. After the initial synapse in the cochlear nuclei, many of the fibers cross the midline, to form the trapezoid body. Some of the fibers synapse in the superior olivary complex. From this point, the tract known as the lateral lemniscus is formed. The fibers relay in the inferior colliculus.

CLINICAL ASPECT

This diagram allows the visualization of all the pathways together and thus assists in understanding lesions of the brainstem. The cranial nerve nuclei affected help locate the level of the lesion.

One of the classic lesions of the brainstem is an infarct of the lateral medulla (see Appendix Figure A.9), known as **Wallenberg (lateral medullary) syndrome**. This lesion affects the pathways and cranial nerve nuclei located in the lateral areas of the medulla, including the anterolateral tract and the descending sympathetic tract (Horner's syndrome, discussed with Figure 6.7), and also the spinocerebellar tract; the descending trigeminal system is also involved, and possibly the exiting fibers and the nuclei of CN IX and CN X. Based on the damage to these structures, the clinical findings include contralateral loss of pain and temperature, ipsilateral ataxia, and ipsilateral Horner's syndrome. The deficits of facial sensation depend on the extent of the damage to the descending (spinal) trigeminal and trigemino-thalamic tracts.

ADDITIONAL DETAIL

The superior cerebellar peduncles are shown in this diagram, although they are not part of the sensory systems. These have been described with the cerebellum (see Figure 5.17). This fiber pathway from the cerebellum to the thalamus decussates in the lower midbrain at the inferior collicular level (shown in cross-section; see Appendix Figure A.4); fibers are given off "en route" to the red nucleus.

MEDIAL LONGITUDINAL FASCICULUS (MLF) (NOT SHOWN)

Ascending fibers in the MLF include those from the vestibular nuclei and from CN VI to the visuomotor nuclei CN IV and III (see Figure 6.9).

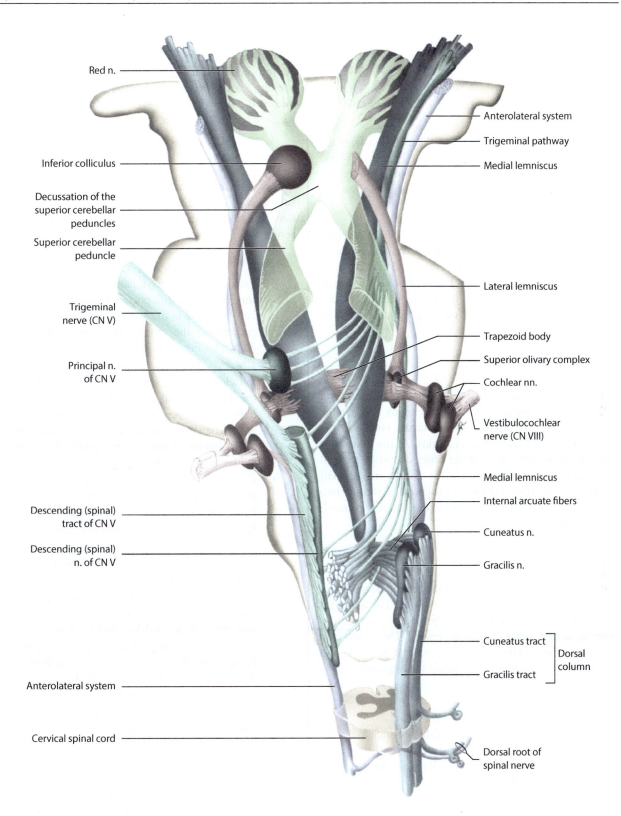

FIGURE 6.11: Brainstem Sensory Systems—Ascending Tracts and Sensory Nuclei

FIGURE 6.12—BRAINSTEM MOTOR SYSTEMS

DESCENDING TRACTS AND MOTOR NUCLEI

Note to the Learner: At this point, the motor nuclei of the brainstem should be reviewed (see Figure 3.5 and Appendix Figure A.1) as well as the motor pathways (in Section 2). The blood supply of the brainstem is reviewed with Figure 8.1.

The descending pathways that have been described are shown, using the posterior view of the brainstem (as in Figure 1.9, Figure 3.3, and Figure 5.5), along with those cranial nerve nuclei that have a motor component. These pathways are presented in summary form. The orientation of the cervical spinal cord representation should be noted.

CORTICO-SPINAL TRACT

These fibers course in the middle third of the cerebral peduncle, are dispersed in the pontine region between the pontine nuclei, and regroup as a compact bundle in the medulla, situated within the pyramids. At the lowermost part of the medulla, most of the fibers decussate to form the lateral cortico-spinal tract of the spinal cord. A small portion of the tract continues ipsilaterally, mostly into the cervical spinal cord region, as the anterior (ventral) cortico-spinal tract.

CORTICO-BULBAR FIBERS

The cortical fibers that project to the cranial nerve nuclei of the brainstem are shown in this diagram. The term also includes those cortical fibers that project to the reticular formation and other brainstem nuclei. These are also located in the middle third of the cerebral peduncle and are given off at various levels within the brainstem.

RUBRO-SPINAL TRACT

This tract from the lower portion of the red nucleus decussates in the midbrain region and descends through the brainstem. In the spinal cord, the fibers are located anterior to the lateral cortico-spinal tract.

CORTICO-PONTINE FIBERS

The cortico-pontine fibers are part of a circuit that involves the cerebellum. The cortical fibers arise from the motor areas as well as from widespread parts of the cerebral cortex. The fibers are located in the outer and inner thirds of the cerebral peduncle—the fronto-pontine fibers in the inner third and fibers from the other lobes in the outer third. They terminate in the nuclei of the pons proper, and the information is then relayed (after crossing) to the cerebellum via the massive middle cerebellar peduncle. The role of this circuit in motor control has been explained with the cerebellum.

MOTOR CRANIAL NERVES

The motor cranial nerve (CN) nuclei and their function have been discussed and their location within the brainstem described. Only topographical aspects are described here:

- **CN III—oculomotor** (to most extra-ocular muscles and parasympathetic): These fibers traverse the medial portion of the red nucleus before exiting in the fossa between the cerebral peduncles, the interpeduncular fossa (see Appendix Figure A.3).
- **CN IV—trochlear** (to the superior oblique muscle): The fibers from this nucleus cross in the posterior aspect of the lower midbrain before exiting posteriorly (see Appendix Figure A.5). The slender nerve then wraps around the lower border of the cerebral peduncles in its course anteriorly (see Figure 3.3).
- **CN V—trigeminal** (to muscles of mastication): The motor fibers pierce the middle cerebellar peduncle in the mid/upper pontine region, along with the sensory component.
- **CN VI—abducens** (to the lateral rectus muscle): The anterior course of the exiting fibers could not be depicted from this perspective.
- **CN VII—facial** (to muscles of facial expression): The fibers to the muscles of facial expression have an internal loop before exiting. The nerve loops over the abducens nucleus, to form a bump called the facial colliculus in the floor of the 4th ventricle (see Figure 3.3). The nerve of only one side is shown in this illustration.
- **CN IX—glossopharyngeal** and **CN X—vagus** (motor and parasympathetic): The fibers exit on the lateral aspect of the medulla, behind the inferior olive.
- **CN XI—spinal accessory** (to neck muscles): The fibers that supply the large muscles of the neck (sternomastoid and trapezius) originate in the upper spinal cord and ascend into the skull before exiting.
- **CN XII—hypoglossal** (to muscles of the tongue): These fibers actually course anteriorly, exiting from the medulla between the inferior olive and the cortico-spinal (pyramidal) tract.

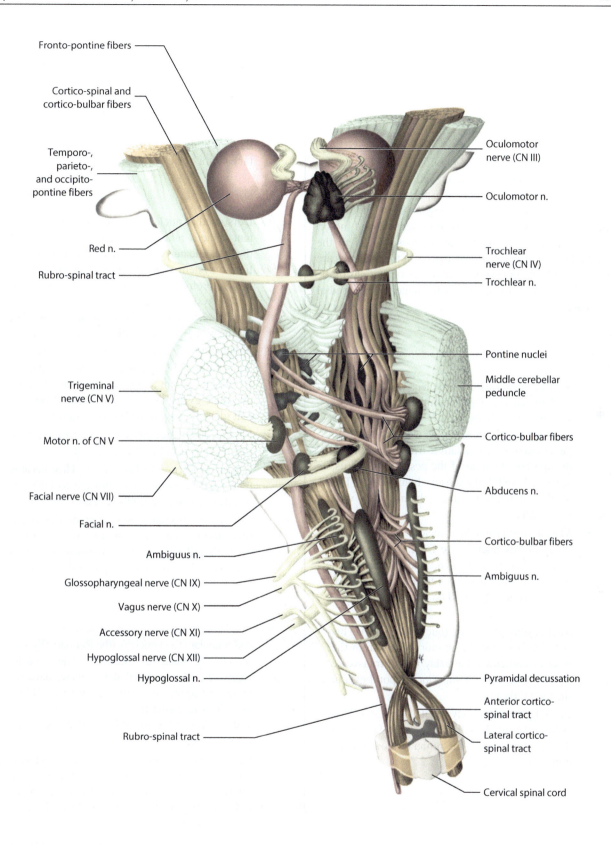

Fronto-pontine fibers

Cortico-spinal and cortico-bulbar fibers

Temporo-, parieto-, and occipito-pontine fibers

Red n.

Rubro-spinal tract

Trigeminal nerve (CN V)

Motor n. of CN V

Facial nerve (CN VII)

Facial n.

Ambiguus n.

Glossopharyngeal nerve (CN IX)

Vagus nerve (CN X)

Accessory nerve (CN XI)

Hypoglossal nerve (CN XII)

Hypoglossal n.

Rubro-spinal tract

Oculomotor nerve (CN III)

Oculomotor n.

Trochlear nerve (CN IV)

Trochlear n.

Pontine nuclei

Middle cerebellar peduncle

Cortico-bulbar fibers

Abducens n.

Cortico-bulbar fibers

Ambiguus n.

Pyramidal decussation

Anterior cortico-spinal tract

Lateral cortico-spinal tract

Cervical spinal cord

FIGURE 6.12: Brainstem Motor Systems—Descending Tracts and Motor Nuclei

FIGURE 6.13—THALAMUS AND CORTEX

THALAMIC NUCLEI: INPUTS AND CORTICAL CONNECTIONS AND FUNCTIONS

The thalamus was introduced previously (see Figure 2.6) with a schematic perspective, as well as an introduction to the nuclei and their functional aspects (see Figure 4.3). At this stage it is important to integrate knowledge of the thalamic nuclei with the inputs, both sensory and motor, and the connections (reciprocal) of these nuclei to the cerebral cortex. The functional aspects of the thalamus are reviewed, with color used to display the connections of the nuclei with the cortical areas (dorsolateral and medial aspects).

SPECIFIC RELAY NUCLEI

Sensory

- Ventral posterolateral nucleus (**VPL**): This nucleus receives input from the somatosensory systems of the body, mainly for discriminative touch and position sense, as well as the "fast" pain system for localization. The fibers relay to the appropriate areas of the post-central gyrus, areas 1, 2, and 3, the sensory homunculus. The hand, particularly the thumb, is well represented.
- Ventral posteromedial nucleus (**VPM**): The fibers to this nucleus are from the trigeminal system (TG) (i.e., the head and the face), and the information is relayed to the facial area of the post-central gyrus. The tongue and lips are well represented.
- Medial geniculate body (nucleus) (**MGB**): This is the nucleus for the auditory fibers from the inferior colliculus that relay to the transverse gyri of Heschl on the superior temporal gyri in the lateral fissure.
- Lateral geniculate body (nucleus) (**LGB**): This is the relay nucleus for the visual fibers from the ganglion cells of the retina to the calcarine cortex.

Motor

- Ventral anterior nucleus (**VA**) and ventral lateral nucleus (**VL**): Fibers to these nuclei originate in the globus pallidus and substantia nigra (pars reticulata), as well as the cerebellum, and are relayed to the motor and premotor areas of the cerebral cortex, as well as the supplementary motor cortex.

ASSOCIATION NUCLEI

- Dorsomedial nucleus (**DM**): This most important nucleus relays information from many of the thalamic nuclei, as well as from parts of the limbic system (hypothalamus and amygdala), to the prefrontal cortex (discussed with the Limbic system in Section 4—see Figure 10.1B). This connection may be extremely important for the processing of limbic (emotional) aspects of behavior.
- Anterior nuclei (**AN**): These nuclei are part of the limbic system and relay information to the cingulate gyrus; they are part of the Papez circuit (discussed with the limbic system).
- Lateral dorsal nucleus (**LD**): The function of this nucleus is not well established.
- Lateral posterior nucleus (**LP**): This nucleus relays to the parietal association areas of the cortex; again, it is not a well-understood nucleus.
- **Pulvinar**: This nucleus is part of the visual relay, to visual association areas of the cortex, areas 18 and 19 as well as to vision-associated regions of the temporal and parietal lobes lobe.

NON-SPECIFIC NUCLEI

- Intralaminar nucleus (**IL**), midline nucleus (**Mid**), and reticular (**Ret**) nucleus (the Ret partially surrounds the thalamus laterally): These nuclei receive input from other thalamic nuclei and from the ascending reticular activating system (ARAS), as well as receiving fibers from the "slow" pain system; they relay to widespread areas of the cerebral cortex.
- Centromedian nucleus (**CM**): This nucleus (one of the intralaminar nuclei) is part of an internal loop receiving from the globus pallidus and relaying to the neostriatum, the caudate and the putamen.

[TEXT CONTINUED]

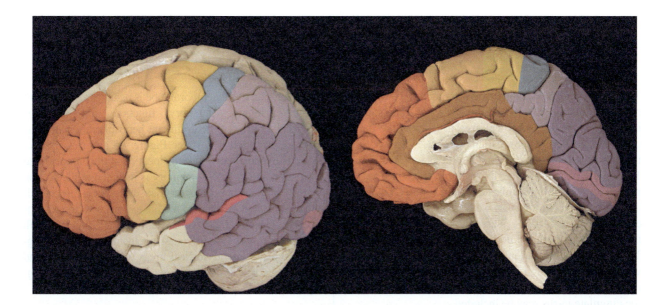

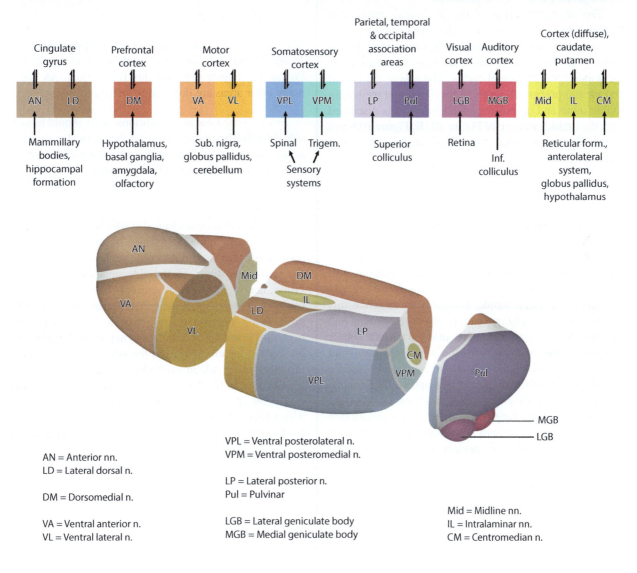

AN = Anterior nn.
LD = Lateral dorsal n.

DM = Dorsomedial n.

VA = Ventral anterior n.
VL = Ventral lateral n.

VPL = Ventral posterolateral n.
VPM = Ventral posteromedial n.

LP = Lateral posterior n.
Pul = Pulvinar

LGB = Lateral geniculate body
MGB = Medial geniculate body

Mid = Midline nn.
IL = Intralaminar nn.
CM = Centromedian n.

FIGURE 6.13: Thalamus and Cortex—Thalamic Nuclei: Inputs and Cortical Connections and Functions

ASSOCIATION CORTEX

PREFRONTAL CORTEX

The expansion of the frontal lobe is thought to be the most recent evolutionary change in higher apes and particularly in humans. The large cortical area in front of the motor cortices (see Figure 5.8), called the **prefrontal cortex**, includes that found on the dorsolateral aspect (see Figure 1.3, Figure 4.5, and this illustration), and also on the medial surface of the hemispheres (see Figure 1.7, Figure 4.5, and this illustration), as well as the inferior (orbital) cortex (see Figure 1.5)—all association cortex. It receives thalamic input from the dorsomedial nucleus (see Figure 10.1B and this illustration). In addition, the dopaminergic ventral tegmental area (VTA) projects to the medial and orbital cortical areas (see Figure 10.8). Note that parts of the cingulate gyrus may be included.

The characterization of the prefrontal cortex (the frontal lobe cortex anterior to the motor areas) as the executive director of activities of the brain, the CEO, is partially correct; that function seems to be localized to the dorsolateral prefrontal cortex. The role of the orbital and medial prefrontal cortex in receiving and mediating limbic input and overseeing social behavior is discussed with the limbic system (see Figure 10.1B, Figure 10.8 and in the discussion called "Synthesis" at the end of that section).

PARIETAL LOBE

The parietal lobe, aside from the postcentral (somatosensory) area (see Figure 4.5 and Figure 5.5), is functionally association cortex. It is situated between the various sensory areas— somatosensory and likely vestibular, as well as visual (and auditory). Its functional role, particularly the superior parietal lobule, is mainly to integrate this information especially in regard to object recognition and the guidance of movement. The inferior parietal lobule consists of two special gyri—the **supramarginal** and **angular** gyri (see Figure 1.3). There is clinical evidence that these gyri of the dominant hemisphere (in humans) are associated with mathematics (arithmetic), left-right orientation, and also language (writing). The constellation of deficits in these functions is known as the **Gerstmann syndrome**. Contralateral neglect is associated with parietal lesions in the non-dominant hemisphere (discussed below).

TEMPORAL LOBE

The evolution of the temporal lobe is again associated with higher primates. Auditory function and language (Wernicke's area of the dominant hemisphere) occupy the superior temporal gyrus (see Figure 4.5 and this illustration). The medial portion is the location of the hippocampal formation and its role in memory is included in the section on the limbic system. Olfactory sensory input is 'processed' in areas located also on the medial aspect, near the temporal pole. The remainder is association cortex, involved with the elaboration of visual processing, including object recognition (on the lateral cortical areas, this illustration), and facial and color recognition (on the inferior surface, merging with the occipital lobe, see Figure 1.5).

OCCIPITAL LOBE

The occipital association areas of the occipital lobe, areas 18 and 19, are adjacent to the primary visual area 17 (see Figure 1.5, Figure 6.6, and this illustration). As noted, many other areas are involved in visual processing.

Note to the Learner: Additional texts should be consulted for further perspectives on the cortical association areas and the clinical deficits connected with lesions in these areas (see Annotated Bibliography).

Clinical Aspect

One of the most unique clinical syndromes, **contralateral neglect**, occurs following a lesion of the inferior parietal area on the **non-dominant side**. Neglect may involve all the sensory systems on the side opposite the lesion—vision, hearing, and somatosensory. In addition, the person may deny having a deficit. A simple "office" test for this is to ask a person to copy (or draw) a clock with numbers; the person with contralateral neglect will not include one-half of the clock face, on the side opposite the lesion (usually the left side in a right-handed individual).

Other tests for the parietal lobe include stereognosis, two point discrimination and graphesthesia (discussed with Chapter 5 in the Introduction to Sensory Systems). For these tests to be valid, the primary modalities of sensation need to be normal.

MENINGES, CEREBROSPINAL FLUID, AND VASCULAR SYSTEMS

INTRODUCTION

The brain in adulthood is enclosed in a rigid container, the skull, and covered with connective tissue, the meninges, composed of three layers—dura, arachnoid and pia. The other contents within the skull include blood (arteries, veins and venous sinuses) and cerebrospinal fluid (CSF). The arterial blood supply (described in Chapter 8) should have sufficient pressure to maintain an adequate blood flow to the nervous tissue. The venous sinuses (to be described in this chapter), on the other hand, normally have a very low pressure. In addition, there is cerebrospinal fluid (CSF) both within the brain (within the ventricles) and surrounding the brain (within the meninges), and there is a continuous production, flow and reabsorption of CSF (described with Figure 7.8).

Note to the Learner: Two videos on the Web site of the atlas (www.atlasbrain.com) are relevant to this section— one video on the **Cerebrovascular System** and **Cerebrospinal Fluid** and the other on the **Interior of the Skull**. The suggestion is to view the videos, return to the text and illustrations, and then perhaps view the videos again.

INTRACRANIAL PRESSURE (ICP)

Many neurological disease processes exert their effect because of a rise in intracranial pressure (ICP).

Any increase in volume inside the skull—whether from brain swelling, a tumor, an abscess, hemorrhage (discussed with Figure 7.2), or an abnormal amount of CSF (discussed with Figure 7.8)—causes a rise in pressure inside the skull (i.e., ICP). This process may be acute, subacute, or chronic. Although brain tissue itself has no pain fibers, the blood vessels and meninges do; hence any pulling on the meninges may give rise to a severe headache.

Normal intracranial pressure in adults is less than 15 mm. Hg. (mercury), that is less than 20 cm. of water (H_2O). The pressure is normally measured with a manometer while doing a lumbar puncture, usually called an "LP" or sometimes a spinal tap (described with Figure 7.3).

CLINICAL ASPECT

A prolonged increase in ICP can be detected clinically by examining the optic disc with an ophthalmosocpe (also discussed with Figure 6.5). Increased ICP is reflected in pressure along the optic nerve with compression of the retinal veins just after they pass from the optic disc into the optic nerve. This compression results in swelling as the nerve leaves the eyeball at the optic disc with blurring of the margins of the disc, an abnormal finding called **papilledema**, and distension of the retinal veins. This examination is critical in the assessment of patients suspected of having increased ICP.

CAUTIONARY NOTE

A lumbar puncture (LP) should not be done if there is any indication that the ICP is elevated because of the danger of causing a tonsillar herniation syndrome (discussed with Figure 3.2; see also Figure 7.1). The assistance of neurosurgery and other devices would be needed to assess and monitor intracranial pressure.

PEDIATRIC PERSPECTIVE

The sutures holding the bones of the skull together are not yet fixed in newborns and infants; the anterior fontanelle (sometimes called the "soft spot") is quite noticeable during the first year of life. Therefore any increase in intracranial pressure will cause a bulging of the fontanelle and a separation of the sutures; part of a regular "well-baby" check up in fact involves measuring and charting the circumference of the infant's head. This issue is of importance in the context of hydrocephalus in infancy (discussed with Figure 7.8). The sutures of the skull normally close by 2 years of age.

Chapter 7

Meninges, Venous System, and Cerebrospinal Fluid Circulation

MENINGES

FIGURE 7.1—CRANIAL MENINGES 1

CEREBRAL HEMISPHERES: DORSAL VIEW (PHOTOGRAPH)

The cerebral hemispheres occupy the interior of the skull, the cranial cavity. The brain in this photograph is seen mainly from above and from the side—one hemisphere has the meninges removed and the other is still covered with meninges.

The meninges, the connective tissue coverings of the brain, consist of three layers—the dura, the arachnoid, and the pia—with spaces or potential spaces between the layers (see Figure 7.2).

The outermost layer of the meninges is the dura, a thick strong sheet of connective tissue. Within the cranial cavity, the dura and periosteum of the skull bones adhere to one another. The dural layer has additional folds within the skull that subdivide the cranial cavity and serve to keep the brain in place. The two major dural sheaths are:

- The falx cerebri in the sagittal plane, between the hemispheres (see Figure 7.4 and Figure 7.5; the falx has been removed from the interhemispheric fissure in Figure 2.2A).
- The tentorium cerebelli in the transverse plane, between the occipital lobe and the cerebellum (see Figure 7.5 and Figure 7.6; the cut edge of the tentorium can be seen in the hemisected brain as in Figure 1.7).

There is an opening in the tentorium for the brainstem, called the **tentorial notch** or incisura, located at the level of the upper midbrain (see Figure 7.5). The falx and tentorium, because of their attachment to the skull particularly at the sutures, do help to stabilize the brain hemispheres within the cranial cavity. These dural sheaths are further discussed with the venous sinuses (see Figure 7.4, Figure 7.5, and Figure 7.6).

In this illustration, the falx cerebri is still in place within the interhemispheric fissure (see also Figure 2.2A). At the attachment of the falx to the skull, there is a large venous sinus—the superior sagittal sinus (see Figure 7.2 and Figure 7.4).

The **subarachnoid space**, between the arachnoid and pia, is filled with cerebrospinal fluid (**CSF**) (see Figure 7.2). Therefore, the brain is actually "floating" inside the skull. Both the meninges and the CSF offer a certain measure of protection to the underlying brain tissue beyond that of the skull itself.

CLINICAL ASPECT

Any space-occupying lesion (e.g., sudden hemorrhage, slow-growing tumor) sooner-or-later causes a displacement of brain tissue from one compartment to another within the skull. This is called a **brain herniation** syndrome, and it typically occurs:

- Under the falx cerebri itself—**subfalcine herniation** (see Figure 7.4).
- Through the tentorial notch—**uncal herniation** (discussed with Figure 1.6).
- Through the foramen magnum—**tonsillar herniation** (discussed with Figure 3.2 and in the introduction to this section).

This pathological displacement itself causes damage to the brain.

These shifts are life-threatening and require emergency management.

Note to the Learner: This would be an opportune time to review the signs and symptoms associated with these clinical emergencies, such as testing of the pupillary light reflex and the pathways involved.

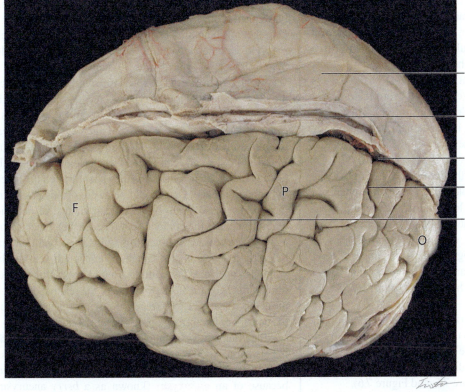

Dura

Superior sagittal
sinus (opened)

Interhemispheric fissure

Parieto-occipital fissure

Central fissure

F = Frontal lobe
P = Parietal lobe
O = Occipital lobe

FIGURE 7.1: Cranial Meninges 1—Cerebral Hemispheres: Dorsal View (photograph)

FIGURE 7.2—CRANIAL MENINGES 2

SCALP AND CRANIAL MENINGEAL LAYERS

The major layers of the scalp are shown in these illustrations (and labeled in the upper illustration), notably the skin (with hair), and the aponeurosis (the flattened tendon that connects the bellies of the frontalis and occipitalis muscles over the top of the cranium). (This illustration has been modified from that found in the *Integrated* text.)

The **skull** itself has outer and inner layers, called tables, with the middle layer where there is bone marrow for the formation of the blood cells (see Figure 2.9B and Figure 3.2).

The **dura** is composed of 2 layers—an outer periosteal layer which is firmly attached to the bony periosteum, and an inner meningeal layer. There is a potential space on the inner surface of the skull between the bone and the outer (periosteal) layer of the dura, the **epidural space**. This is where the artery that supplies most of the dura is located, the **middle meningeal artery** (from the external carotid artery, shown in the lower illustration). This artery typically produces grooves in the temporal bone on the interior of the skull (see the video in the atlas Web site [www.atlasbrain.com]).

The two dural sheaths that separate the parts of the brain from each other and hold the cerebral hemispheres in place, the falx (shown in the lower illustration) and the tentorium (not seen in these illustrations), attach to the bone (see Figure 7.1; described with the venous sinuses in Figure 7.4, Figure 7.5, and Figure 7.6). At the upper edge of the falx and at the lateral margins of the tentorium, where these dural sheaths are attached, the two dural layers split and form—within the dura—the cerebral **venous sinuses**. Therefore, we have the superior sagittal sinus at the upper edge of the falx (as shown in the lower illustration; also see Figure 7.1) and the transverse sinuses at the margins of the tentorium cerebelli (further discussed with the venous circulation; see Figure 7.4, Figure 7.5, and Figure 7.6).

The next layer is the **arachnoid**. There is a potential space between the dura and arachnoid, and the cerebral veins pass through this "space." These <u>bridging veins</u>, as they are called (shown in the lower illustration), course from the surface of the brain into the venous sinuses, particularly the superior sagittal sinus. There is a potential space between the dura and arachnoid, called the **subdural space**, where bleeding may occur, usually from venous sources (see later).

The innermost layer, the **pia**, lies on the surface of the brain and follows all its folds. The **subarachnoid space**, between the arachnoid and the pia, contains the cerebrospinal fluid (CSF). Large arteries and veins are also found in this space. The arachnoid granulations (shown in the lower illustration) convey the CSF from the subarachnoid space back to the venous circulation (discussed with Figure 7.8).

CLINICAL ASPECT

Bleeding inside the skull occurs in somewhat typical situations and in predictable locations.

A forcible traumatic injury to the side of the head may lead to a fracture of the temporal bone and disruption of the middle meningeal artery. This causes the typical syndrome associated with an **epidural hemorrhage**. This life-threatening arterial bleeding often follows a typical course: trauma to the side of the head with or without a loss of consciousness, then a lucid period, and then rapid neurological deterioration that, if not recognized soon enough and managed appropriately, usually leads to brain herniation (discussed in the introduction to this section and with Figure 7.1) and possibly death.

The bridging veins may be disrupted by trauma to the head, in which case blood leaks into the potential space between the dura and arachnoid, to produce a **subdural hemorrhage**. Subdural (venous) hemorrhages usually occur slowly over time (subacute or chronic), but they may also manifest acutely. Typically, they occur more commonly in very young and very old people. Apparently, elderly people are more vulnerable to this type of bleeding because of the extra space available for the brain to "move about" as a result of age-related cerebral shrinkage. Any type of head trauma, even the most mild (e.g., bumping your head getting into or out of a car), can cause these bridging veins to be disrupted ("sheared" or torn) as they pass through the potential subdural space. One should be aware of this possibility in an older person who has a sudden change of behavior (with or without the complaint of a headache). Once recognized and assessed with neuroimaging, this condition may be easily treatable and, if caught soon enough, can resolve without sequelae. CT scanning can be very useful in determining the age of a bleed by virtue of the relative density of the hemorrhage. The more recent the hemorrhage, the higher the density of hemorrhage on CT scan (Hounsfield Units).

Bleeding may occur into the CSF as a result of the bursting of a large blood vessel coursing within this space. This is called a **subarachnoid hemorrhage**. Typically, this occurs because of an **aneurysm** (known as a *berry* aneurysm), which is a weakening of the blood vessel wall, most often involving the arteries of the circle of Willis (discussed with Figure 8.2). Meningeal irritation causes pain, and the person complains of an ultra-severe headache; because this is arterial bleeding, loss of consciousness may occur very quickly.

Bleeding also occurs within brain itself, a **brain hemorrhage**, destroying brain tissue. The clinical picture depends on the size and location of the bleed. It is estimated that about 15% of cerebrovascular "accidents" (CVA) are due to hemorrhage. CT scanning is mandatory for determining whether a bleed has occurred.

As discussed in the introduction to this section, any increase of volume inside the skull leads to an increase in intracranial pressure (ICP) and the possibility of a brain herniation syndrome. Note again the necessity of examining the optic discs for papilledema.

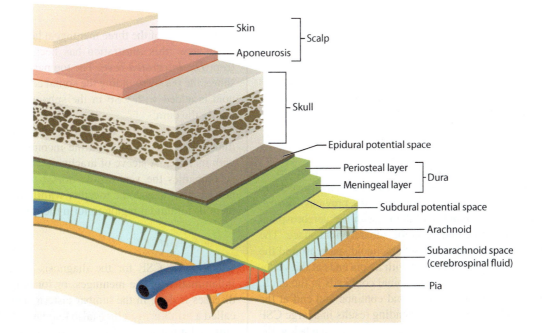

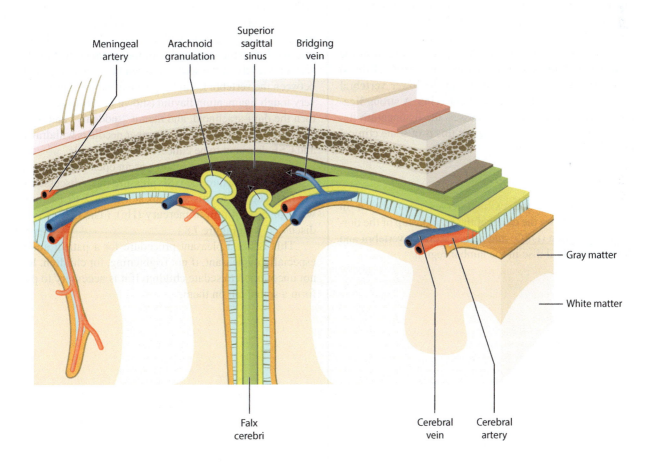

FIGURE 7.2: Cranial Meninges 2—Scalp and Cranial Meningeal Layers

FIGURE 7.3—SPINAL MENINGES

VERTEBRAL CANAL

The meninges continue around the spinal cord within the vertebral canal. Three views of the spinal meninges are shown:

LONGITUDINAL VIEW

The vertebral canal and spinal cord are shown as in Figure 1.10 (also Figure 1.11). The meningeal layers are illustrated with color coding (as in the previous illustration)—pia, arachnoid, and dura—with the cerebrospinal fluid (CSF colored blue), in the subarachnoid space. The CSF continues within the subarachnoid space around the spinal cord. The spinal cord with its pia ends at the vertebral level of L2—where the spinal cord ends in the adult, whereas the dura and arachnoid continue and end at the level of S2. This differential ending results in a large CSF space, known as a cistern, in the vertebral canal below the level of the spinal cord—called the **lumbar cistern** (see Figure 1.2, Figure 1.10, and Figure 1.11). This site is used for sampling of CSF, called a spinal tap or lumbar puncture (see Figure 7.8 and discussed later).

Note that the spinal cord dura is separated from the periosteum of the vertebra (and the intervertebral discs) by a space that is filled with fat in the lower vertebral region and that contains a plexus of veins. A strong ligament between the spinous processes, the ligamentum flavum, is also located in this region; this is of some clinical importance (see below).

SCHEMATIC

The schematic of the exiting nerves (also shown in Figure 1.12) shows the detail of the relationship of the dorsal root ganglion (DRG) and nerves with the vertebra and the intervertebral disc in the lumbar region.

AXIAL VIEW

This is a view of the three meningeal layers depicted in an axial view (this illustration has been slightly modified from *The Integrated Text*). The top part of the illustration shows the pia surrounding the spinal cord, then the arachnoid is added with CSF in the subarachnoid space, and finally the dura.

Note particularly the exiting nerve roots ventrally (motor, see Figure 8.7) and the incoming (sensory) root with its DRG. A sleeve of arachnoid and dura, with CSF, accompanies the ventral and dorsal roots of the spinal nerves, until they come together within the intervertebral (neural) foraminal region to form the mixed spinal nerve.

CLINICAL ASPECT

Sampling of CSF for the diagnosis of meningitis, for inflammation of the meninges, or for other neurological diseases is done in the lumbar cistern. This procedure is called a **lumbar puncture** (also known as an LP, or spinal tap), and it must be performed using sterile technique.

The patient is positioned on her or his side, in the so-called fetal position, and the area of the lower back is cleansed. After appropriate local anesthesia, a trochar (which is a large needle with a smaller needle inside) is inserted *below* the termination of the spinal cord at the L2 vertebral level), in the space between the vertebra, usually between the vertebra L4 to L5. The trochar must pierce the very tough ligamentum flavum (shown in this illustration), then the dura-arachnoid, and then it "suddenly" enters the lumbar cistern. At this point the inner needle is withdrawn, and CSF drips out to be collected in sterile vials.

A pressure apparatus known as an open manometer is also used to measure the CSF pressure (also discussed in the Introduction to this section), normally 7 to 19 cm of water *or* 5 to 14 mm of mercury (Hg). (The CSF is fully discussed with Figure 7.8.)

This is not a pleasant procedure for a patient and is especially unpleasant, if not frightening, for children. It is not uncommon to sedate children if it is necessary to perform a spinal tap on them.

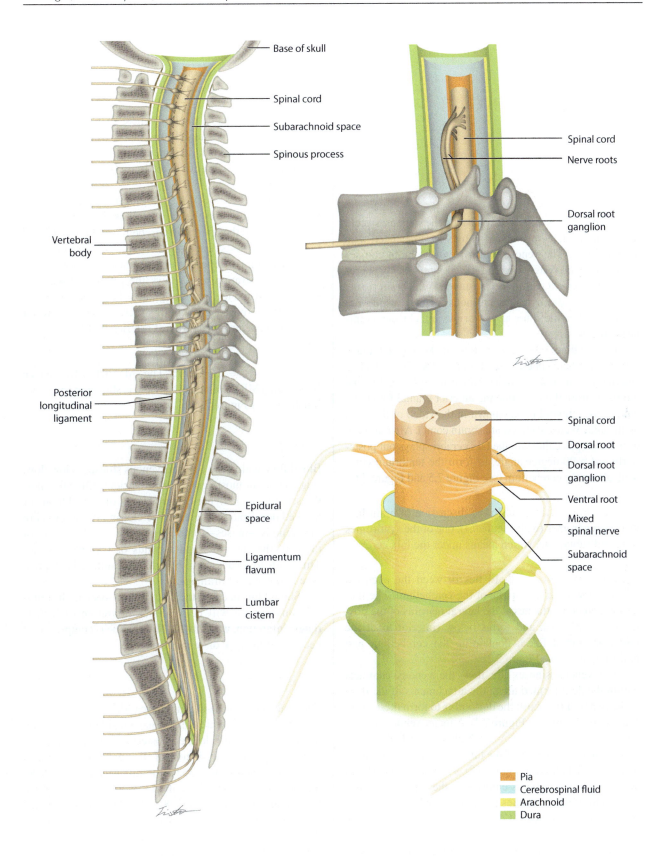

Base of skull

Spinal cord

Subarachnoid space

Spinous process

Vertebral body

Posterior longitudinal ligament

Epidural space

Ligamentum flavum

Lumbar cistern

Spinal cord

Nerve roots

Dorsal root ganglion

Spinal cord

Dorsal root

Dorsal root ganglion

Ventral root

Mixed spinal nerve

Subarachnoid space

Pia
Cerebrospinal fluid
Arachnoid
Dura

FIGURE 7.3: Spinal Meninges—Vertebral Canal

VENOUS SYSTEM

FIGURE 7.4—VENOUS SINUSES 1

VENOUS CIRCULATION MID-SAGITTAL (PHOTOGRAPHIC VIEW WITH OVERLAY)

At this point it is timely to consider the venous system inside the skull, as the venous sinuses are associated with the meninges. The venous return from the brain courses via a superficial system over the surface of the hemispheres and a deep system within the substance of the brain tissue. Both systems drain into the venous sinuses, which are formed within the dura at its various attachment points inside the skull.

The midline **falx cerebri** lies in the sagittal plane between the hemispheres (see Figure 2.2A). It is attached anteriorly to a spike of bone, called the crista galli, that protrudes from the nose into the anterior cranial fossa; the bone is the ethmoid bone (shown in the video of the skull on the atlas Web site [www.atlasbrain.com]). After arching over the corpus callosum, the falx cerebri splays out posteriorly—at a 90-degree angle, to form the tentorium cerebelli, above the cerebellum (see Figure 7.5 and Figure 7.6).

Note to the Learner: In the discussion of brain herniation syndromes (with Figure 7.1), one of the syndromes is named subfalcine, which means under the falx cerebri.

The **tentorium cerebelli** lies between the occipital lobe and the superior surface of the cerebellum. The tentorium cerebelli attaches at the lateral margins of the posterior cranial fossa and forms below (inferior to) it the posterior cranial fossa, containing the cerebellum and brainstem (see also Figure 3.2).

The venous sinuses are then the venous channels within the dura formed along the attachments of the dura to the bones of the skull; the dura splits to form these large venous spaces (review Figure 7.2). A major venous sinus is the **superior sagittal sinus**, which is found along the upper (attached) border of the falx cerebri, in the midline (see Figure 7.1 and also Figure 7.8). Most of the superficial veins of the hemispheres empty into the superior sagittal sinus. This sinus continues posteriorly, and at the back of the interior of the skull it divides to become the laterally placed **transverse venous sinuses**, one on each side, attached to the skull at the lateral edges of the tentorium

cerebelli. As will be explained, venous blood exits the skull via the **sigmoid sinuses**, which continue as the **internal jugular veins** on each side of the neck.

An exception to the rule of the location of the venous sinuses is the **inferior sagittal sinus**, located in the inferior margin of the falx cerebri, where there is no attachment to bone. This sinus drains blood from the medial surface of the brain and is joined by other veins that collect blood from the interior of the brain (see Figure 7.5 and Figure 7.6).

The **straight sinus** is also not located at a bony margin; it is found within the dura itself as the falx "flattens out" to become the tentorium cerebelli (see Figure 7.5 and Figure 7.6). This sinus drains the inferior sagittal sinus and joins with the superior sagittal sinus posteriorly.

This illustration also shows the pituitary fossa (the sella turcica) and the pituitary gland. The basilar artery (to be described subsequently in this section) is seen anterior to the brainstem; other cerebral arteries (not labeled) are also seen.

Note to the Learner: The Magnetic Resonance Venogram (MRV) of a similar view is shown in the upper illustration of Figure 7.7.

CLINICAL ASPECT

Blood flow within the cerebral venous sinuses is low flow, sluggish (as in most veins). Under various clinical conditions, including hyper-coagulability of the blood or marked dehydration, there can be clotting of the blood in the venous sinuses—**venous sinus thrombosis**, most often in the superior sagittal sinus or transverse sinus. This leads to a back-up of blood and resulting back pressure in the area of drainage.

Should this occur in the deep venous system, the clinical consequence will be a venous infarction or hemorrhagic infarction, whereby the brain tissue is compromised because of the lack of venous drainage.

ADDITIONAL DETAIL

The space behind the thalamic area (the diencephalon) is a cistern of the subarachnoid space, outside the brain, not the 3rd ventricle. It is located posterior to the pineal gland and the colliculi and in front of the cerebellum (see also Figure 1.7), hence its name—the quadrigeminal cistern; the four colliculi are also called the quadrigeminal plate (see Figure 3.3 and Figure 7.8). This space also contains some important cerebral veins that drain the interior of the brain including the great cerebral vein of Galen (these have been removed from this specimen).

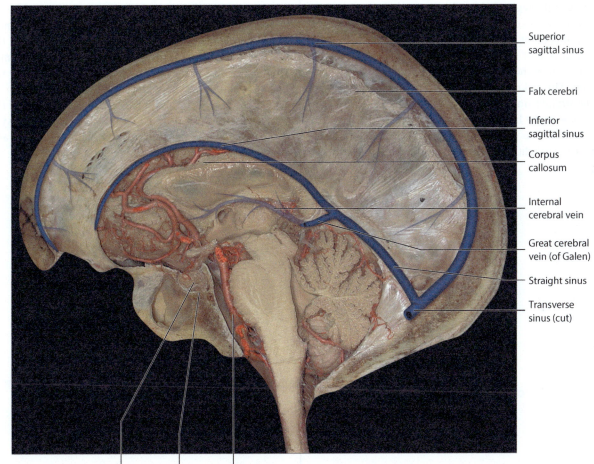

Superior
sagittal sinus

Falx cerebri

Inferior
sagittal sinus

Corpus
callosum

Internal
cerebral vein

Great cerebral
vein (of Galen)

Straight sinus

Transverse
sinus (cut)

Pituitary   Sella   Basilar
              turcica   artery

FIGURE 7.4: Venous Sinuses 1—Venous Circulation Mid-Sagittal (photographic view with overlay)

FIGURE 7.5—VENOUS SINUSES 2

VENOUS CIRCULATION OBLIQUE (PHOTOGRAPHIC VIEW WITH OVERLAY)

This is an illustration of the dural sheaths—the falx and tentorium—with the brain removed. Again, the superior and inferior sagittal sinuses can be seen.

The system of veins that drain the deep structures of the brain emerges medially as the **internal cerebral veins**, one from each hemisphere. These veins join in the midline in the region behind the diencephalon to form the **great cerebral vein** (of Galen).

At this point there is another exception to the rule of the formation of venous sinuses. A sinus is located in the midline where the falx splays out laterally to form the tentorium. This is known as the **straight sinus**. The location of the straight sinus can also be appreciated in a mid-sagittal view of the brain (see Figure 1.7 and Figure 3.2; also Figure 7.8). The great cerebral vein becomes continuous with the straight sinus, in the midline lying above the cerebellum (see Figure 7.6; see also the video in the atlas Web site [www.atlasbrain.com]). At this point it is joined by the *inferior* sagittal sinus.

At the back of skull, the straight sinus joins with the *superior* sagittal sinus. The venous sinuses now divide, and the blood flows into the transverse sinuses, which are seen in Figure 7.6. The venous blood exits the skull via the sigmoid sinus into the internal jugular vein (on both sides, see Figure 7.7).

Note to the Learner: The Magnetic Resonance Venogram (MRV) of a comparable view is shown in the middle illustration of Figure 7.7.

Additional Detail: This oblique view of the meninges, with the brain removed and the brainstem cut at the level of the midbrain, shows the free edge of the tentorium cerebelli (on one side). The "space" created by the dural reflections of the tentorium is called the **tentorial notch** or "**incisura**," and it exists for the passage of the brainstem (discussed as part of Brain Herniation syndromes, known as uncal herniation, with Figure 1.6 and Figure 7.1).

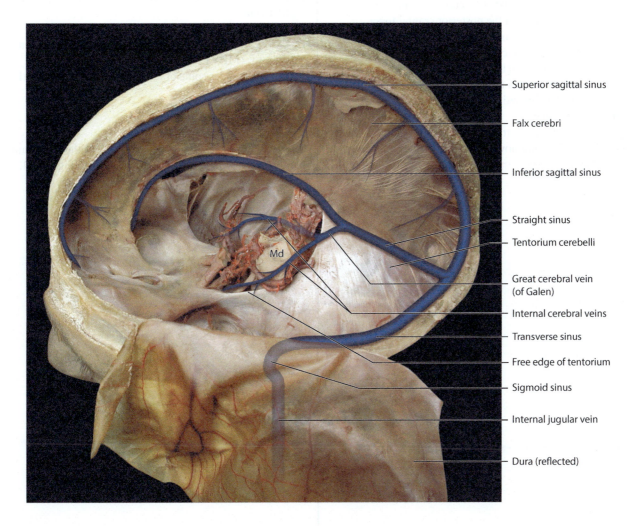

- Superior sagittal sinus
- Falx cerebri
- Inferior sagittal sinus
- Straight sinus
- Tentorium cerebelli
- Great cerebral vein (of Galen)
- Internal cerebral veins
- Transverse sinus
- Free edge of tentorium
- Sigmoid sinus
- Internal jugular vein
- Dura (reflected)

Md

Md = Midbrain (cut)

FIGURE 7.5: Venous Sinuses 2—Venous Circulation Oblique (photographic view with overlay)

FIGURE 7.6—VENOUS SINUSES 3

VENOUS CIRCULATION HORIZONTAL (PHOTOGRAPHIC VIEW WITH OVERLAY)

This is an illustration of the tentorium cerebelli, seen from above in the horizontal (axial) plane. The hemispheres have been removed, and the cerebellum is still present, below the tentorium, in the posterior cranial fossa.

The straight sinus is shown in the midline, where the falx and tentorium are continuous.

The dura to the right side of the straight sinus is created by the falx cerebri (as labeled), which has been rolled up and folded down and held in place with forceps, to produce this specimen.

The sagittal and straight venous sinuses have united (see Figure 7.5) and now divide, sometimes unequally, to form the **transverse sinuses**; these are located where the tentorium is attached to the bone along its lateral margins. The blood then flows (in a forward direction) until it reaches the anterior edge of the posterior cranial fossa (reminder to see the videos of Cerebrospinal Fluid and The Skull on the atlas Web site [www.atlasbrain.com]). At this point the sinuses leave the tentorium and enter the posterior cranial fossa as the sigmoid sinuses (because of their "S" shape), thus making a prominent groove in the skull interior. The venous blood then exits the skull via the jugular foramen (one on each side) to form the internal jugular vein.

Note to the Learner: The Magnetic Resonance Venogram (MRV) shown in the lower illustration of Figure 7.7 includes some of the venous sinuses seen in this illustration.

Additional Detail

This dissection includes an unusual view of the hippocampal formation—to be further discussed with Figure 9.4.

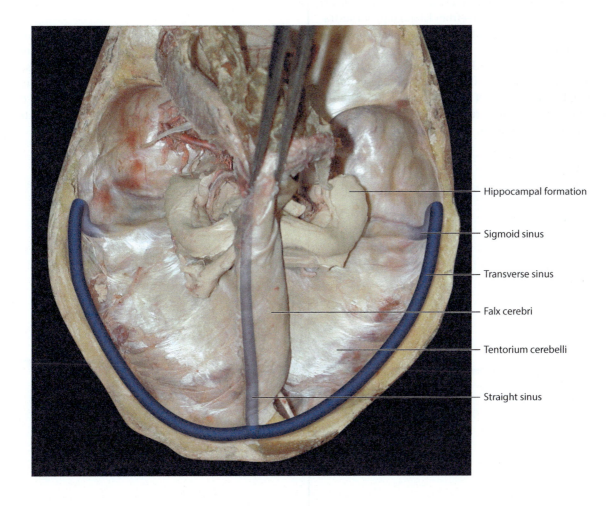

Hippocampal formation

Sigmoid sinus

Transverse sinus

Falx cerebri

Tentorium cerebelli

Straight sinus

FIGURE 7.6: Venous Sinuses 3—Venous Circulation Horizontal (photographic view with overlay)

FIGURE 7.7—VENOUS SINUSES 4

MAGNETIC RESONANCE VENOGRAMS (MRV)

A magnetic resonance venogram (MRV), like a magnetic resonance angiogram (MRA—see Figure 8.2) can be done both with and without contrast (gadolinium-based imaging). In this case, no contrast was used. The sequences are called time-of-flight (TOF), either angiogram or venogram.

The three images correspond to the views shown in the illustrations of the meninges with the venous sinuses (see Figure 7.4, Figure 7.5, and Figure 7.6).

This series depicts the cerebral veins as though the patient is turning their head—from lateral (upper illustration) to oblique (middle illustration) to antero-posterior (lower illustration).

The upper figure, the lateral view, (see Figure 7.4), captures the superior and inferior sagittal sinuses in the upper and lower aspects of the falx cerebri. Some of the collector veins from the cerebral cortex are also seen. The straight sinus is also seen, as well as the transverse and sigmoid sinuses, with the blood flowing into the internal jugular vein in the neck.

The middle image is an oblique view, capturing again the superior and inferior sagittal sinuses. Some of the veins draining the interior of the brain are also seen forming the great cerebral vein (of Galen). Again, the transverse sinus is seen, as well as the sigmoid sinuses, and this time both internal jugular veins are visualized.

The lower image is an anteroposterior view. The superior sagittal sinus, now joined by the straight sinus, is seen dividing into the two transverse sinuses and both internal jugular veins can be seen exiting the skull.

CLINICAL ASPECT

Venograms sometimes reveal an unequal flow of blood in the transverse sinuses that is caused by a narrowing or stenosis of one of the transverse sinuses. If this is severe and thought to predispose the person to clotting in the superior sagittal sinus, procedures are now in place (done by interventional neuroradiologists) to "stent" the transverse sinus involved.

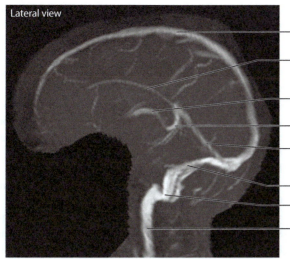

Lateral view

— Superior sagittal sinus

— Inferior sagittal sinus

— Internal cerebral vein

— Great cerebral vein (of Galen)

— Straight sinus

— Transverse sinus

— Sigmoid sinus

— Internal jugular vein

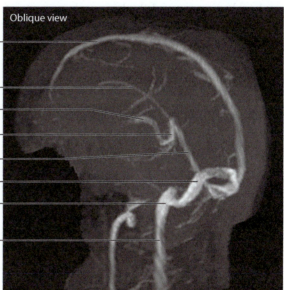

Oblique view

Superior sagittal sinus —

Inferior sagittal sinus —

Internal cerebral vein —

Great cerebral vein (of Galen) —

Straight sinus —

Transverse sinus —

Sigmoid sinus —

Internal jugular vein —

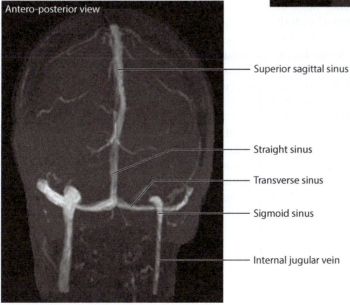

Antero-posterior view

— Superior sagittal sinus

— Straight sinus

— Transverse sinus

— Sigmoid sinus

— Internal jugular vein

FIGURE 7.7: Venous Sinuses 4—Magnetic Resonance Venograms (MRV)

CEREBROSPINAL FLUID (CSF)

FIGURE 7.8—CEREBROSPINAL FLUID CIRCULATION

CEREBROSPINAL FLUID SPACES

This is an illustration of the production, circulation, and reabsorption of cerebrospinal fluid (CSF) slightly modified from the illustration in *The Integrated Nervous System*. The ventricles of the brain are shown, as are the subarachnoid spaces around the brain, enlargements of which are called cisterns including the lumbar cistern, as well as the superior sagittal (and straight) venous sinus. The blood flow in the venous sinuses is also shown.

The cerebral ventricles of the brain (and the central canal of the spinal cord) are found within the brain tissue and are the remnants of the original neural tube from which the nervous system developed. There are four ventricles: one (ventricles I and II) in each of the cerebral hemispheres, also called the lateral ventricles (see Figure 2.1A and Figure 2.1B); the 3rd ventricle in the thalamic (diencephalic) region (see Figure 2.8); and the 4th ventricle in the brainstem region (see Figure 3.1, Figure 3.2, and Figure 3.3).

Choroid plexus is found in the lateral ventricles (see Figure 2.1B and Figure 2.10A and also Figure 9.4; present in the atrium in Figure 6.5 but not labeled), the roof of the 3rd ventricle, and the lower half of the roof of the 4th ventricle. CSF produced in the lateral ventricles flows (as shown with *black arrows*, note the figure legend with the illustration) via the foramen of Monro (from each lateral ventricle—see Figure 2.9B and Figure 2.10A) into the 3rd ventricle, and then through the aqueduct of the midbrain into the 4th ventricle (review Figure 3.2). CSF leaves the ventricular system from the 4th ventricle, as indicated schematically in the diagram. In the intact brain, this occurs via the medially placed foramen of Magendie (see lower illustration in Figure 1.9) and the two laterally placed foramina of Luschka, and CSF enters the enlargement of the subarachnoid space under the cerebellum, the cerebello-medullary cistern, the cisterna magna. This

cistern is located inside the skull, just above the foramen magnum of the skull (see Figure 3.2).

CSF continues to flow through the subarachnoid space, between the pia and the arachnoid (*darker blue* arrows in the figure). The CSF fills the enlargements of the subarachnoid spaces around the brainstem—the various cisterns (each of which has a separate name). The CSF then flows upward around the hemispheres of the brain and is found in all the sulci and fissures. CSF also flows in the subarachnoid space downward around the spinal cord to fill the lumbar cistern (see Figure 1.10 and Figure 7.3).

This slow circulation is completed by the return of CSF to the venous system. The return is through the arachnoid villi, protrusions of arachnoid into the venous sinuses of the brain, particularly along the superior sagittal sinus (see below and also Figure 7.2). These can sometimes be seen on the specimens as collections of villi, called arachnoid granulations, on the surface of the brain lateral to the interhemispheric fissure.

Note to the Learner: These features are well shown in the video on the atlas Web site (www.atlasbrain.com).

The CSF in the lumbar cistern is where lumbar puncture is performed for the sampling of CSF for clinical diagnostic purposes (discussed with Figure 7.3).

CSF CIRCULATION

The normal amount of CSF in the ventricles and cranio-spinal subarachnoid spaces is estimated to be around 150 ml and is replaced roughly every 6 to 8 hours; this indicates a continuous process of production and absorption of CSF, in effect a (slow) CSF circulation. CSF is returned to the venous circulation via the arachnoid granulations (see Figure 7.2) that protrude into the venous sinuses, particularly into the superior sagittal sinus (*light blue* arrows in the illustration). A small pressure differential is thought to account for the transport of CSF across these villi and into the venous sinuses, thus completing the circulation of CSF.

Additional Note: The major arteries of the circle of Willis travel through the subarachnoid space (see Figure 7.2). An aneurysm of these arteries, called a berry aneurysm, that "bursts" (discussed with Figure 8.2) will do so within the CSF space. This is called a subarachnoid hemorrhage (discussed with Figure 7.2).

[TEXT CONTINUED]

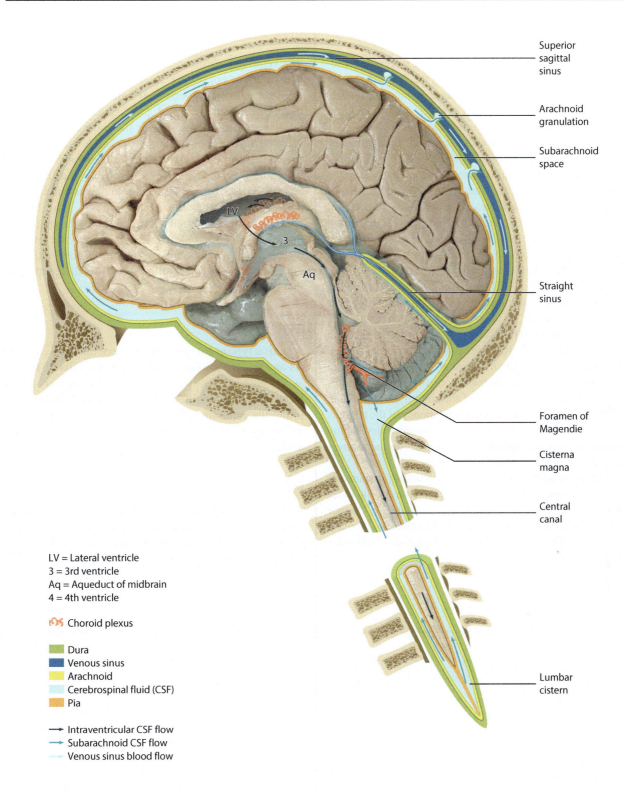

Superior
sagittal
sinus

Arachnoid
granulation

Subarachnoid
space

Straight
sinus

Foramen of
Magendie

Cisterna
magna

Central
canal

LV
3
Aq

LV = Lateral ventricle
3 = 3rd ventricle
Aq = Aqueduct of midbrain
4 = 4th ventricle

Choroid plexus

Dura
Venous sinus
Arachnoid
Cerebrospinal fluid (CSF)
Pia

→ Intraventricular CSF flow
→ Subarachnoid CSF flow
→ Venous sinus blood flow

Lumbar
cistern

FIGURE 7.8: Cerebrospinal Fluid Circulation—Cerebrospinal Fluid Spaces

CLINICAL ASPECT

Blockage of CSF flow, for example, at the level of the midbrain where the aqueduct connecting the 3rd with the 4th ventricle is very narrow, would cause an increase in size, and of pressure, in the cerebral ventricles. This blockage, called obstructive or **non-communicating hydrocephalus**, would be visualized with computed tomography or magnetic resonance imaging as an enlargement of the lateral ventricles of the hemispheres. In an adult, because the sutures of the skull are fused, this process would be accompanied by raised intracranial pressure. In contrast, in a young child (e.g., the first 2 years) with non-fused cranial sutures, the head itself would enlarge, and there would be a separation of the bones of the skull; at a very early age, the anterior fontanelle would bulge.

Blockage of the CSF flow can also occur at the level of the arachnoid granulations, and in fact this does occur following meningitis or other disorder which can cause increased protein in the CSF interfering with the function of the arachnoid villi (further discussed in *The Integrated Nervous System*). If the villi are non-functional or blocked, or if there is a blockage of the venous sinuses (e.g., venous sinus thrombosis) or stenosis of the sinuses, then CSF can no longer return to the venous circulation, and the result is an increase in CSF pressure. Hydrocephalus developing from CSF flow obstruction at a point outside the brain is called **communicating hydrocephalus**.

In young adults with brains of high tissue resistance or low compliance, the CSF pressure increases; in older adults with brains of lower tissue resistance or higher compliance, the CSF spaces such as the ventricles increase in volume. This latter condition can manifest as a disorder called **normal pressure hydrocephalus**.

BLOOD-CSF-BARRIER

The ventricles of the brain are lined with a layer of cells known as the ependyma. In certain loci within each of the ventricles, the ependymal cells and the pia meet, thus forming the **choroid plexus**, which invaginates into the ventricle. Functionally, the choroid plexus has a vascular layer (i.e., the pia) on the inside and the ependymal layer on the ventricular side. The blood vessels of the choroid plexus are freely permeable, but there is a cellular barrier between the choroid plexus and the ventricular space—the **blood-CSF barrier**. The barrier consists of tight junctions between the ependymal cells that line the choroid plexus. CSF is actively secreted by the choroids plexus, and an enzyme is involved. The ionic and protein composition of CSF is different from that of serum.

ADDITIONAL DETAIL

A similar type of barrier exists between the brain capillaries and the extracellular space of the brain tissue, known as the **blood-brain-barrier (BBB)**. This barrier is also formed by tight junctions—between the endothelial cells lining the capillaries. This barrier allows for only small molecules to cross, including glucose and select amino acids.

CLINICAL ASPECT

A breakdown of the BBB occurs in certain diseases, under "toxic" conditions, and within and around tumors of the CNS. The use of a radio-opaque dye during imaging and its escape indicates a breakdown of the BBB.

Chapter 8

Vascular Supply
Cerebral, Brainstem, and Spinal Cord

INTRODUCTION

The central nervous system is totally dependent on a continuous supply of blood; viability of the neurons depends on the immediate and constant availability of both oxygen and glucose. Interruption of this lifeline causes immediate loss of function followed very quickly by death of the nervous tissue (the neurons and axons). Study of the nervous system must include a complete knowledge of the blood supply and the structures (nuclei and tracts) situated in the vascular territory of the various arteries. Failure of the blood supply to a region, because of either occlusion or hemorrhage, causes the expected functional deficits.

Areas of gray matter, where the neurons are located, have a greater requirement for blood supply than white matter. Loss of oxygen and glucose supply to these neurons leads to loss of electrical activity almost immediately (in the adult) and, if continued for several minutes, to neuronal death. Although white matter requires less blood supply, loss of adequate supply leads to destruction of the axons in the area of the infarct and an interruption of pathways. After loss of the cell body or interruption of the axon, the distal portion of the axon (the part on the other side of the lesion separated from the cell body) and the synaptic connections degenerate, resulting in a permanent loss of function.

Every part of the nervous system lies within the vascular territory of an artery, sometimes with an overlap from adjacent arteries. Visualization of the arterial (and venous) branches can be accomplished using the following:

- Arteriogram: By injecting a radiopaque substance into the arteries (this is a procedure which is done by a neuroradiologist), and

following its course through a rapid series of x-ray studies, a detailed view of the vasculature of the brain is obtained (see Figure 8.3). This is an invasive procedure carrying a certain degree of risk.

- Magnetic Resonance Angiogram—**MRA**: Using neuroradiology imaging with MRI, the major blood vessels (such as the circle of Willis) can be visualized (see Figure 8.2). A similar visualization can be obtained with computed tomography (CT) scanning, called a **CTA**.

CLINICAL ASPECT

It is extremely important to know which parts of the brain are located in the territory supplied by each of the major cerebral and brainstem blood vessels and to understand the functional contribution of these parts. This is fundamental for clinical neurology.

A clinical syndrome involving the arteries of the brain is often called a **cerebrovascular accident (CVA)** or a **stroke**. The nature of the process, whether a blood vessel occlusion through infarction or embolus or a hemorrhage, is not specified by the use of this term, nor does the term indicate which blood vessel is involved. The clinical event is a sudden loss of function; the clinical deficit depends on where the occlusion or hemorrhage occurred.

Occlusion is more common than hemorrhage and is often caused by an embolus (e.g., from the heart). Hemorrhage may occur into the brain substance (parenchymal), thereby causing destruction of the brain tissue (discussed with Figure 7.2) and at the same time depriving areas distally of blood.

FIGURE 8.1—CEREBRAL BLOOD SUPPLY 1

ARTERIAL CIRCLE OF WILLIS (PHOTOGRAPHIC VIEW WITH OVERLAY)

The arterial circle (of Willis) is a set of arteries interconnecting the two sources of blood supply to the brain, the vertebral and common carotid arteries. It is located at the base of the brain, surrounding the optic chiasm and the hypothalamus (the mammillary nuclei) (review Figure 1.5A and Figure 1.5B). Within the skull, the circle of Willis is situated above the pituitary fossa (and gland). The major arteries to the cerebral cortex of the hemispheres are branches of this arterial circle. This illustration is a photographic view of the inferior aspect of the brain, including brainstem and cerebral hemispheres, with the blood vessels (as in Figure 1.6). Branches from the major arteries have been added to the photographic image.

The cut end of the **internal carotid** arteries is a starting point. Each artery divides into the **middle cerebral** artery (MCA) and the **anterior cerebral** artery (ACA). The MCA courses within the lateral fissure. The anterior portion of the temporal lobe has been removed on the left side of this illustration to follow the course of the MCA in the lateral fissure. Within the fissure, small arteries are given off to the basal ganglia, called the **striate** arteries (not labeled; see Figure 8.6). The artery emerges at the surface (see Figure 1.3) and courses upward, dividing into branches that are distributed onto the dorsolateral surface of the hemispheres (see Figure 8.4).

By removing (or lifting) the optic chiasm, the ACA can be followed anteriorly. This artery heads into the interhemispheric fissure (see Figure 2.2A) and is followed when viewing the medial surface of the brain (see Figure 1.7, Figure 2.2A, and Figure 8.5). A very short artery connects the ACAs of the two sides, the **anterior communicating** artery (see also Figure 8.2 and Figure 8.3).

The vertebro-basilar system supplies the brainstem, the cerebellum, and the posterior part of the hemispheres. The two **vertebral** arteries unite at the lower border of the pons to form the midline **basilar** artery, which courses in front of the pons. The basilar artery terminates at the midbrain level by dividing into two **posterior** cerebral arteries (PCA). These arteries supply the inferior aspect of the brain and particularly the occipital lobe (see Figure 8.5).

The arterial circle is completed by the **posterior communicating** arteries (normally one on each side, approximately of the same size—note the size difference here), which connects the internal carotid artery (or MCA) artery, often called the *anterior circulation*, with the posterior cerebral artery, the *posterior circulation*.

BRAINSTEM

Small arteries directly from the circle (not shown) provide the blood supply to the diencephalon (thalamus and hypothalamus), some parts of the internal capsule, and part of the basal ganglia. The major blood supply to these regions is from the striate arteries (see Figure 8.6).

The branches from the vertebral artery and basilar artery supply the brainstem. Small branches directly from the vertebral and basilar arteries (not shown), known as **paramedian** arteries, supply the medial structures of the brainstem (further discussed with Appendix Figure A.9). There are three major branches from this arterial tree to the cerebellum—the **posterior inferior cerebellar** artery (PICA), the **anterior inferior cerebellar** artery (AICA), and the **superior cerebellar** artery (SCA). All supply the lateral aspects of the brainstem, including nuclei and tracts, en route to the cerebellum; these are often called the **circumferential** branches.

CLINICAL ASPECT

The vascular territories of the various cerebral blood vessels are shown in color in this diagram. The most common clinical lesion involving the cerebral blood vessels is occlusion, often resulting from an embolus originating from the heart or the carotid bifurcation in the neck. These clinical deficits are described with each of the major branches to the cerebral cortex (with Figure 8.4 and Figure 8.5).

In the eventuality of a slow occlusion of one of the major blood vessels of the circle of Willis, sometimes one of the communicating branches becomes large enough to provide sufficient blood to be shunted to the area deprived (see Figure 8.3).

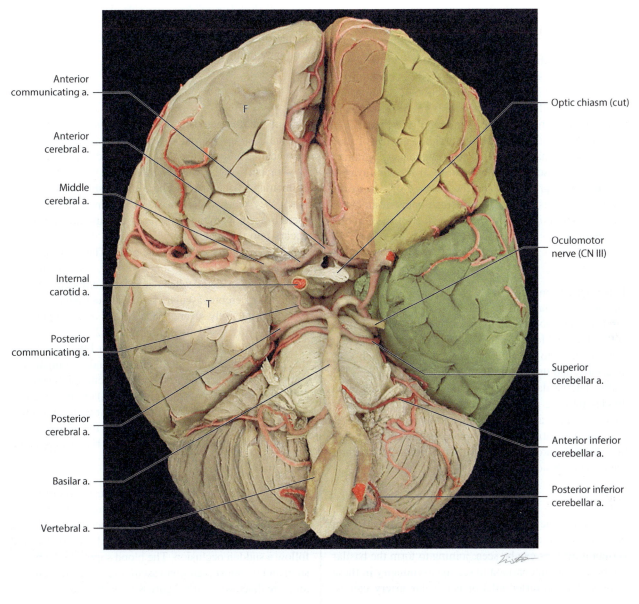

Anterior communicating a.

Anterior cerebral a.

Middle cerebral a.

Internal carotid a.

Posterior communicating a.

Posterior cerebral a.

Basilar a.

Vertebral a.

Optic chiasm (cut)

Oculomotor nerve (CN III)

Superior cerebellar a.

Anterior inferior cerebellar a.

Posterior inferior cerebellar a.

F = Frontal lobe
T = Temporal lobe (cut)

Areas supplied by:
Anterior cerebral a.
Middle cerebral a.
Posterior cerebral a.

FIGURE 8.1: Cerebral Blood Supply 1—Arterial Circle of Willis (photographic view with overlay)

FIGURE 8.2—CEREBRAL BLOOD SUPPLY 2

MAGNETIC RESONANCE ANGIOGRAM (MRA)

Advances in technology have allowed for a visualization of the major blood vessels supplying the brain, notably the arterial circle of Willis, without injection of a radiopaque substance (usually gadolinium based). In this instance, as with the magnetic resonance venogram (see Figure 7.7), no contrast agent was used. Although the quality of such images cannot match the detail seen after an angiogram of select blood vessels (shown in Figure 8.3), the non-invasive nature of this procedure and the absence of risk to the patient clearly establish this investigation as desirable to provide some information about the state of the cerebral vasculature.

UPPER IMAGE

This arteriogram shows the **circle of Willis** as seen as if looking at the brain from below (as in Figure 8.1). The carotid artery goes through the cavernous (venous) sinus of the skull and forms what is called the *carotid siphon*. It then divides into the anterior cerebral artery, which goes anteriorly, and the middle cerebral artery, which goes laterally. The basilar artery is seen at its termination, as it divides into the posterior cerebral arteries. The anterior communicating artery is present, and there are two posterior communicating arteries completing the circle, joining the internal carotid artery with the posterior cerebral artery on each side.

LOWER IMAGE

This is the same angiogram, but displayed at a different orientation—an anterior (somewhat tilted) view. The two vertebral arteries can be seen, joining to form the basilar artery; it is not uncommon to see the asymmetry in these vessels. The posterior inferior cerebellar artery can be seen; this is a branch of the vertebral artery (it is also labeled in the upper image). The basilar artery gives off the superior cerebellar arteries and then ends by dividing into the posterior cerebral arteries. The internal carotid artery can be followed through its curvature in the petrous temporal bone of the skull before it divides into the anterior and middle cerebral arteries.

CLINICAL ASPECT

One of the characteristic vascular lesions in the arteries that make up the arterial circle of Willis is a type of aneurysm, called a **berry aneurysm**. This is caused by a weakness of part of the wall of the artery that causes local ballooning of the artery. These aneurysms often rupture when they reach a certain size (>5 mm), particularly if there is accompanying hypertension. This sudden rupture occurs into the subarachnoid space and may also involve nervous tissue of the base of the brain. The whole event is known as a **subarachnoid hemorrhage**, and this diagnosis must be considered when one is faced clinically with an acute major cerebrovascular event, without trauma, accompanied by intensely severe headache and often loss of consciousness (discussed also with Figure 7.2).

Sometimes, these aneurysms leak a little blood, and this causes irritation of the meninges and accompanying symptoms of headache a so-called "Sentinel Bleed." A computed tomography angiogram (CTA) or magnetic resonance angiogram (MRA) can, at the minimum, visualize whether there is an aneurysm on one of the vessels of the circle of Willis and whether the major blood vessels are patent.

Note to the Learner: One of the best ways of learning the circle of Willis and the arterial supply to the brain is to actually make a sketch drawing, accompanied by a list of the areas supplied and the major losses that would follow a sudden occlusion. The blood supply to the brainstem and the most common vascular lesions affecting this area are discussed with Figure 8.1 and Figure 8.5.

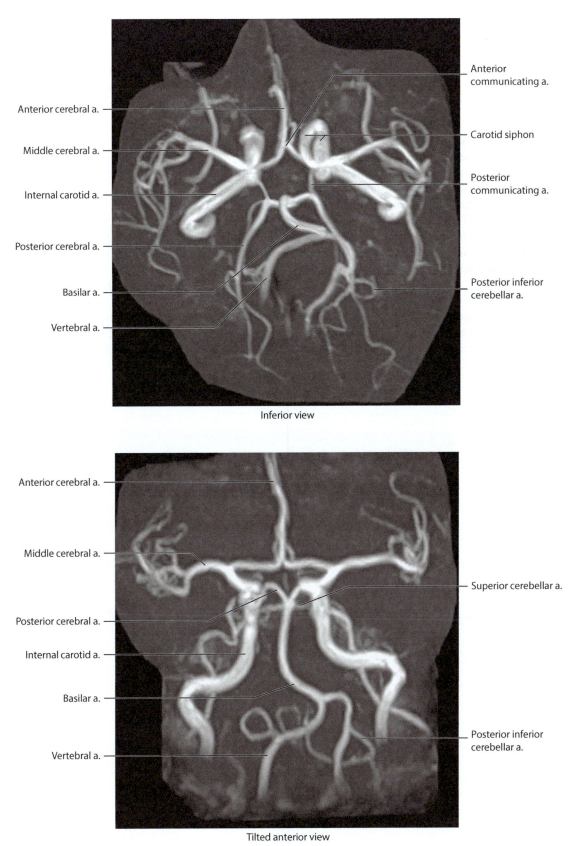

Inferior view

Tilted anterior view

FIGURE 8.2: Cerebral Blood Supply 2—Magnetic Resonance Angiogram (MRA)

FIGURE 8.3—CEREBRAL BLOOD SUPPLY 3

CEREBRAL ANGIOGRAM (RADIOGRAPH)

This radiograph was done by injecting radiopaque dye into the internal carotid artery. Normally, the procedure is to thread a catheter from the groin up the aorta and into the carotid artery, an invasive procedure not without some risk.

In this particular case, a slow occlusion of the right internal carotid artery had allowed time for the anterior communicating artery of the circle of Willis to become widely patent; therefore, blood was shunted into the anterior and middle cerebral arteries on the affected side. This is not usual, and in fact this radiogram was chosen for this reason.

The middle cerebral artery goes through the lateral fissure and breaks up into various branches on the dorsolateral surface of the hemisphere (shown in Figure 8.4). The **striate** arteries—also called the lenticulostriate arteries—are given off en route to supply the interior structures of the hemisphere (discussed with Figure 8.6).

This radiograph shows the profuseness of the blood supply to the brain and the hemispheres, and it is presented to give the student that notion, as well as to show the appearance of an angiogram.

CLINICAL ASPECT

Visualization of the blood supply to the brain is required for the accurate diagnosis of aneurysms and occlusions affecting these blood vessels.

It is important to once again note the radiological convention for radiographs of the brain—the orientation of the laterality (right/left) is as if one is looking *at* the patient.

Had there been a sudden total occlusion of the blood supply to the right hemisphere of the brain, the patient would have experienced motor and sensory deficits affecting the face and the body—on the opposite side—with an extensor plantar response (see Clinical Cases).

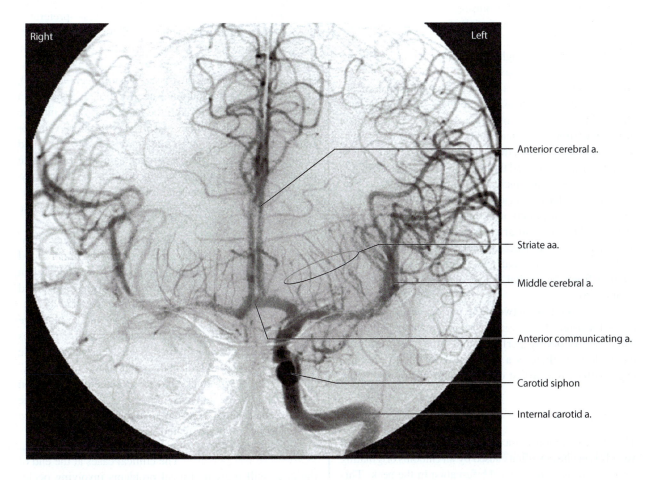

Right

Left

Anterior cerebral a.

Striate aa.

Middle cerebral a.

Anterior communicating a.

Carotid siphon

Internal carotid a.

FIGURE 8.3: Cerebral Blood Supply 3—Cerebral Angiogram (radiograph)

FIGURE 8.4—CEREBRAL BLOOD SUPPLY 4

CORTICAL: DORSOLATERAL (PHOTOGRAPHIC VIEW WITH OVERLAY)

This illustration shows the blood supply to the cortical areas of the dorsolateral aspect of the hemispheres; it has been created by superimposing the blood vessels onto the photographic view of the brain (the same brain as in Figure 4.5).

After coursing through the depths of the lateral fissure (see Figure 8.1 and Figure 8.3), the **middle cerebral artery** emerges and breaks into a number of branches that supply different parts of the dorsolateral cortex—the frontal, parietal, and temporal areas of cortex. Each branch supplies a different territory, as indicated; branches supply the precentral and postcentral gyri, the major motor and sensory areas for the face and head, and the upper limbs (as indicated). On the dominant side, this includes the language areas.

The vascular territories of the various cerebral blood vessels are shown in color in this diagram. The branches of the middle cerebral artery extend toward the midline mid-sagittal fissure, where branches from the other cerebral vessels (anterior and posterior cerebral) are found, coming from the medial aspect of the hemispheres (see Figure 8.5).

A zone remains between the various arterial territories—the arterial **borderzone** region (a watershed area). This area is poorly perfused and prone to infarction, particularly if there is a sudden loss of blood pressure (e.g., with cardiac arrest or after a major hemorrhage).

CLINICAL ASPECT

The most common clinical lesion involving these blood vessels is occlusion, often caused by an embolus originating from the heart or the carotid bifurcation in the neck. This results in infarction of the nervous tissue supplied by that branch—the clinical deficit depends on which branch or branches is or are involved. For example, loss of sensory or motor function, or both, to the arm and face region (of the opposite side) is seen after the blood vessel to the central region is occluded. The type of language loss that occurs depends on the branch affected in the dominant hemisphere: a deficit in expressive language is seen with a lesion affecting Broca's area, whereas a comprehension deficit is found with a lesion affecting Wernicke's area.

Acute strokes are now regarded as an emergency with a narrow therapeutic window. According to current evidence, if the site of the blockage can be identified and the clot (or embolus) removed within 4.5 hours—preferably less—there is a reasonable chance that the individual will have significant if not complete recovery of function. The therapeutic measures include a substance that will dissolve the clot such as tPA (tissue plasminogen activator), combined with interventional neuroradiology whereby a catheter is threaded through the vasculature and into the brain and the clot is either dissolved with the tPA and/or removed. Major hospitals now have a protocol called "Stroke Code" to alert the stroke team and investigate these people immediately when brought to emergency, including a plain CT scan and CT angiogram, so that therapeutic measures can be instituted within the therapeutic window.

A clinical syndrome has been defined in which a temporary loss of blood supply affects one of the major blood vessels. Some would limit this temporary loss to less than 1 hour, whereas others suggest that this period could extend to several hours. This syndrome is called a **transient ischemic attack (TIA)**. Its cause could be blockage of a blood vessel that resolves spontaneously or perhaps an embolus that breaks up on its own. Regardless, people are being educated to look at this event as a "brain attack," much like a heart attack, and to seek medical attention immediately. The statistics indicate that many of these people experiencing TIAs have a high risk of stroke within 30 days if not properly treated.

Note to the Learner: The clinical cases at the end of the atlas will present clinical problems involving occlusion of the various cerebral branches. Students are encouraged to work through these cases—answers will be provided on the Web site.

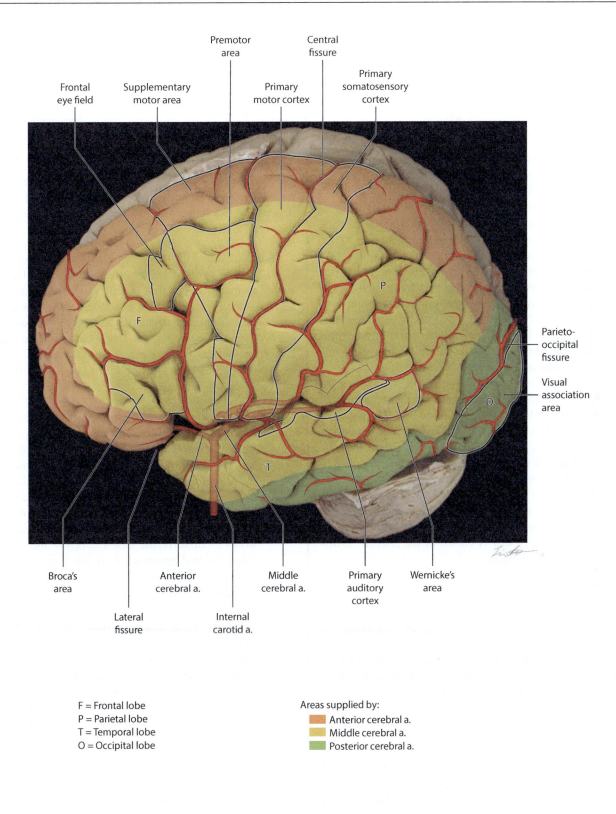

Premotor area

Central fissure

Frontal eye field

Supplementary motor area

Primary motor cortex

Primary somatosensory cortex

Parieto-occipital fissure

Visual association area

Broca's area

Anterior cerebral a.

Middle cerebral a.

Primary auditory cortex

Wernicke's area

Lateral fissure

Internal carotid a.

F = Frontal lobe
P = Parietal lobe
T = Temporal lobe
O = Occipital lobe

Areas supplied by:
Anterior cerebral a.
Middle cerebral a.
Posterior cerebral a.

FIGURE 8.4: Cerebral Blood Supply 4—Cortical: Dorsolateral (photographic view with overlay)

FIGURE 8.5—CEREBRAL BLOOD SUPPLY 5

CORTICAL: MEDIAL (PHOTOGRAPHIC VIEW WITH OVERLAY) AND BRAINSTEM

In this illustration, the blood supply to the medial aspect of the hemispheres has been superimposed onto this view of the brain (see Figure 1.7). Two arteries supply this part—the anterior cerebral artery and the posterior cerebral artery. The vascular territories of the various cerebral blood vessels are shown in color in this illustration.

The **anterior cerebral artery** (ACA) is a branch of the internal carotid artery from the circle of Willis (see Figure 8.1, Figure 8.2, and Figure 8.3). It runs in the interhemispheric fissure, above the corpus callosum (see Figure 2.2A and Figure 9.1B), and it supplies the medial aspects of the frontal lobe and the parietal lobe; this includes the cortical areas responsible for sensory and motor function of the lower limb.

The **posterior cerebral artery** (PCA) supplies the occipital lobe and the visual areas of the cortex, areas 17, 18, and 19 (see Figure 6.5 and Figure 6.6). The posterior cerebral arteries are the terminal branches of the basilar artery from the vertebral or posterior circulation (see Figure 8.1). The demarcation between these arterial territories is the parieto-occipital fissure.

Both sets of arteries have branches that spill over to the dorsolateral surface. As noted in Figure 8.4, there is a gap between these and the territory supplied by the middle cerebral artery that is known as the arterial border-zone or watershed region.

BRAINSTEM

The blood supply to the brainstem and cerebellum is shown from this perspective and should be reviewed with Figure 8.1. The three cerebellar arteries—posterior inferior, anterior inferior, and superior—are branches of the vertebro-basilar artery that supply the lateral aspects of the brainstem en route to the cerebellum.

CLINICAL ASPECT

The deficit most characteristic of an occlusion of the anterior cerebral artery (ACA) is selective weakness and spasticity of the contralateral lower limb. Clinically, the control of micturition seems to be located on this medial area of the brain, perhaps in the supplementary motor area (see Figure 5.8), and symptoms related to voluntary bladder control may also occur with lesions in this area but usually these have to be bilateral.

The clinical deficit found after occlusion of the posterior cerebral artery on one side is a loss of one half of the visual field of both eyes—contralateral homonymous hemianopsia. The blood supply to the calcarine cortex, the visual cortex, area 17, is discussed with Figure 6.6. Depending of the size of the occlusion, there can be deficits of the medial temporal lobe, the thalamus and the parietal lobe.

Note to the Learner: This is an opportune time to review the optic pathway and to review the visual field deficits that are found after a lesion in different parts of the visual system.

Studies indicate that the core of tissue that has lost its blood supply is surrounded by a region where the blood supply is marginal but that is still viable and may be rescued—the **penumbra**, as it is now called. In this area surrounding the infarcted tissue, the blood supply is reduced below the level of nervous tissue functionality and the area is therefore "silent," but the neurons are still viable.

These studies have led to a rethinking of the therapy of strokes:

- In the acute stage, if the patient can be seen quickly (preferably within 3 hours and now up to 4.5 hours) and investigated immediately, the site of the lesion may be identified. Some hospitals have a "stroke code" to expedite the management of such patients. *If* there is an occlusion, then therapeutic measures can be instituted immediately, either with powerful drugs to dissolve the clot and/or the use of interventional neuroradiology (in large centers). If treatment is administered soon enough after the stroke, it may be possible to avert any clinical deficit.

- There may be an additional period beyond this time frame when damaged neurons in the penumbra can be rescued through the use of neuroprotective agents—specific pharmacological agents that protect the neurons from the damaging consequences of loss of blood supply. Such agents are currently an area of active research.

As loss of function and diminished quality of life are the end result of strokes, it is clear that this is a most active area of neuroscience research as to what can be done in the way of preventative measures, for example, the control of high blood pressure (hypertension), and the regulation of the level of blood cholesterol. This takes on added significance in view of the increasing number of people in the "senior" age range.

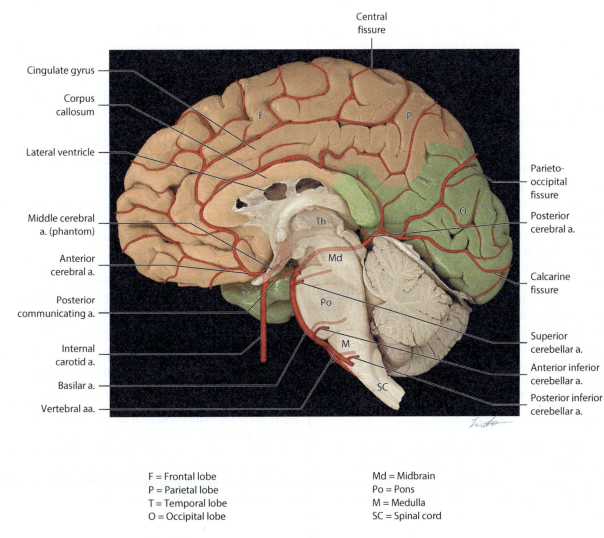

Central fissure

Cingulate gyrus

Corpus callosum

Lateral ventricle

Middle cerebral a. (phantom)

Anterior cerebral a.

Posterior communicating a.

Internal carotid a.

Basilar a.

Vertebral aa.

Parieto-occipital fissure

Posterior cerebral a.

Calcarine fissure

Superior cerebellar a.

Anterior inferior cerebellar a.

Posterior inferior cerebellar a.

F = Frontal lobe
P = Parietal lobe
T = Temporal lobe
O = Occipital lobe

Th = Thalamus

Md = Midbrain
Po = Pons
M = Medulla
SC = Spinal cord

Areas supplied by:
Anterior cerebral a.
Posterior cerebral a.

FIGURE 8.5: Cerebral Blood Supply 5—Cortical: Medial (photographic view with overlay) and Brainstem

FIGURE 8.6—CEREBRAL BLOOD SUPPLY 6

STRIATE ARTERIES (PHOTOGRAPHIC VIEW WITH OVERLAY)

One of the most important sets of branches of the middle cerebral artery is found within the lateral fissure (this artery has been dissected in Figure 8.1). These branches are known as the **striate** arteries, also called lenticulo-striate arteries. They supply most of the internal structures of the hemispheres, including the internal capsule and the basal ganglia (discussed with Figure 4.4).

These branches are seen with the arteriogram in Figure 8.3.

In this illustration, a coronal section of the brain (as shown in the small photographs above, from both the lateral and medial perspectives), the middle cerebral artery is seen traversing the lateral fissure. The artery begins as a branch of the circle of Willis. Several small branches are supplying the area of the lentiform nucleus and the internal capsule, as well as the thalamus. The artery then emerges, after passing through the lateral fissure, to supply the dorsolateral cortex (as previously shown).

These small blood vessels are the major source of blood supply to the internal capsule and the adjacent portions of the basal ganglia (head of caudate nucleus and putamen), as well as the thalamus (see also Figure 3.6A and Figure 3.6B). Additional blood supply to these structures comes directly from small branches of the circle of Willis (discussed with Figure 8.1).

CLINICAL ASPECT

These small-caliber arteries are functionally different from the cortical (cerebral) vessels. First, they are end arteries and do not anastomose. Secondly, they develop degenerative changes due to a chronically elevated blood pressure (hypertension) by a process of degeneration of the muscular wall of the blood vessels—called *fibrinoid necrosis*.

Following this there are two possibilities:

- These blood vessels may occlude, causing small infarcts in the region of the internal capsule that are seen radiographically as small holes—often called lacunes (lakes). Hence, they are known as **lacunar infarcts**, commonly known as a stroke. Lacunes are therefore small areas of white matter infarction and can be seen anywhere in the central nervous system. The extent of the clinical deficit with this type of infarct depends on its location and size in the internal capsule. A relatively small lesion may cause major motor or sensory deficits, or both, on the contralateral side. This may result in devastating incapacity of the person, with contralateral paralysis.

Note to the Learner: At this time, the student should review the major ascending and descending tracts and their course through the internal capsule.

- The other possibility is that these weakened blood vessels can rupture, leading to hemorrhage deep in the hemispheres. (Brain hemorrhage can be visualized by computed tomography; reviewed with Figure 3.6A.)

Although the blood supply to the white matter of the brain is significantly less (because of the lower metabolic demand), this nervous tissue is also dependent on a continuous supply of oxygen and glucose. A loss of blood supply to the white matter results in the loss of the axons (and myelin) and hence interruption of the transmission of information. This type of stroke may lead to a more extensive clinical deficit because the hemorrhage itself causes a loss of brain tissue, as well as a loss of the blood supply to areas distal to the site of the hemorrhage.

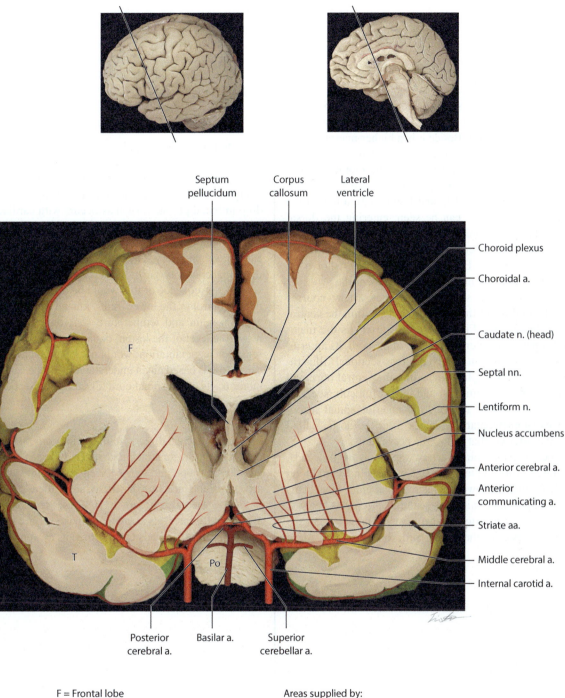

Septum pellucidum   Corpus callosum   Lateral ventricle

Choroid plexus

Choroidal a.

Caudate n. (head)

Septal nn.

Lentiform n.

Nucleus accumbens

Anterior cerebral a.

Anterior communicating a.

Striate aa.

Middle cerebral a.

Internal carotid a.

F

Po

T

Posterior cerebral a.   Basilar a.   Superior cerebellar a.

F = Frontal lobe
T = Temporal lobe
Po = Pons

Areas supplied by:
Anterior cerebral a.
Middle cerebral a.
Posterior cerebral a.

FIGURE 8.6: Cerebral Blood Supply 6—Striate Arteries (photographic view with overlay)

FIGURE 8.7—SPINAL CORD A

BLOOD SUPPLY (PHOTOGRAPHIC VIEW WITH OVERLAY) AND DIAGRAM

The whole spinal cord is shown, from an anterior perspective (see Figure 1.1 and Figure 1.11), with the anterior spinal artery highlighted. Beside it is a higher-magnification photographic image of the cervical region of the spinal cord. Most of the attached roots are the motor (ventral) roots, coming from the ventral horn of the spinal cord (discussed with Figure 1.12 and Figure 7.3); a few of the dorsal (sensory) roots can be seen, entering the dorsal horn of the spinal cord.

The vertebral arteries enter the skull and join to form the basilar artery. Each drops a branch (shown in Figure 8.8), and the branches unite to form the **anterior spinal** artery. This somewhat tortuous artery courses down the midline of the cord anteriorly, within the ventral median fissure (see Figure 3.9). This artery is the major blood supply to the ventral portion of the cord, its anterior two thirds—shown in the axial (cross-sectional) illustration (on the lower right)—including the ventral horn and all the tracts in the anterior and lateral funiculi (see also Figure 6.10). Along its way, the anterior spinal artery receives supplementary branches from the aorta, called radicular arteries, that follow the nerve roots.

The anterior spinal artery tapers as it descends, thereby creating a vulnerable area of the spinal cord blood supply around the lower thoracic (spinal cord) level. This is very important clinically (see Clinical Aspects below).

The vertebral arteries also give off branches on the posterior aspect of the spinal cord, the **posterior spinal** arteries (not shown in Figure 8.8), and these course the full length as separate arteries, supplemented en route by the radicular arteries. These vessels supply the gray matter of the dorsal horn and the region of the dorsal columns (see Figure 5.2 and the cross-sectional view and also Appendix Figure A.11).

CLINICAL ASPECT

As noted, the blood supply to the lower thoracic spinal cord is tenuous; in fact the anterior spinal artery is supplemented in this area by a branch from the thoracic aorta, the artery of Adamkiewicz (shown in Figure 8.8). A dramatic drop in blood pressure, such as occurs with cardiac arrest or excessive blood loss, may lead to an infarction of the lower spinal cord, affecting primarily the territory of the anterior spinal artery. The result can be just as severe as if the spinal cord was (partially) severed by a knife. The most serious consequence of this would be the loss of voluntary motor control of the lower limbs, known as *paraplegia*. In addition, pain and temperature sensation would be lost below the level of the lesion, on both sides; dorsal column sensation (e.g., vibration) would not be affected. The clinical picture is based on an understanding of the sensory and motor tracts of the spinal cord (discussed in Section 2).

ADDITIONAL NOTE

The pia is attached directly to the spinal cord. Sheets of pia are found in the subarachnoid space, between the ventral and dorsal roots, and can be seen attaching to the inner aspect of the arachnoid—these are called **denticulate ligaments** (see also Figure 1.11). These ligaments, which are located at intervals along the spinal cord, are thought to secure the cord, perhaps to minimize its movement.

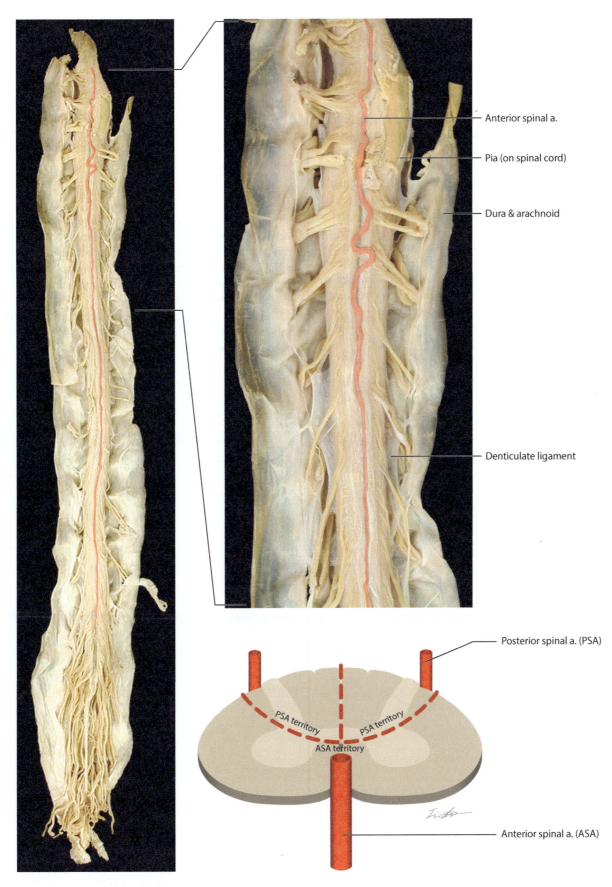

Anterior spinal a.

Pia (on spinal cord)

Dura & arachnoid

Denticulate ligament

Posterior spinal a. (PSA)

PSA territory

PSA territory

ASA territory

Anterior spinal a. (ASA)

FIGURE 8.7: Spinal Cord A—Blood Supply (photographic view with overlay) and diagram

FIGURE 8.8—SPINAL CORD B

BLOOD SUPPLY (DRAWING AND CTA RADIOGRAPH)

This illustration of the brain, brainstem and spinal cord (modified from the *Integrated Nervous System*) shows the blood supply to the spinal cord, notably the anterior spinal artery formed by branches from each vertebral artery. (The posterior spinal arteries are not shown.) A particularly important branch off the aorta supplies the critical region of the spinal cord and supplements the blood supply to the spinal cord. The **great radicular artery of Adamkiewicz**, a branch from the thoracic aorta, joins the anterior spinal artery supplying the spinal cord at a variable level between T8 and L1 and supplementing the flow to the lower spinal cord. Of some interest, this artery enters most frequently on the left side of the person (note orientation).

CTA RADIOGRAPH (COMPOSITE)

This obliquely oriented arteriogram done with computed tomography, called a CTA (discussed in the Introduction to this chapter) is a composite view of the artery of Adamkiewicz entering the vertebral canal and joining with the anterior spinal artery. Careful viewing shows the filling of the anterior spinal artery both upward and downward for a short stretch in both directions. Capturing this view is not routine.

CLINICAL ASPECT

Surgeons who operate on the abdominal aorta (e.g., for aortic aneurysm) must make every effort to preserve the small branches coming off the aorta because these are critical for the vascular supply of the spinal cord. One would not want the end result of an aneurysmal repair to be a paraplegic patient.

Infarction of the cord can also occur from a small embolus occluding the anterior spinal artery via the artery of Adamkiewicz.

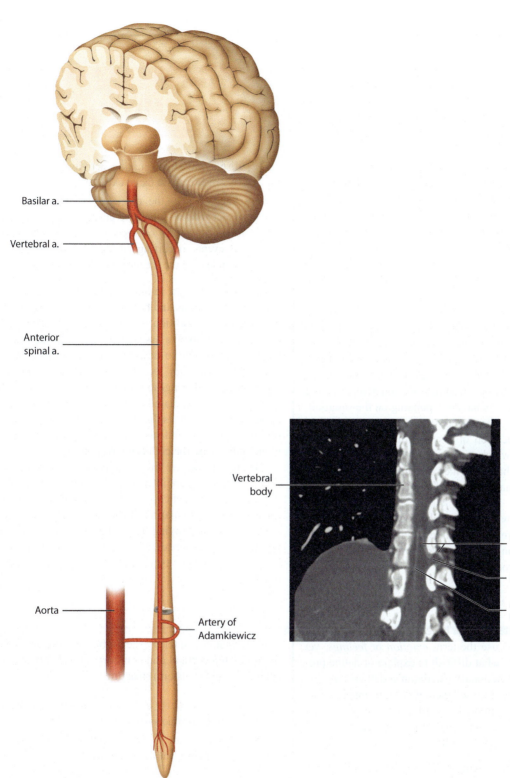

Basilar a.

Vertebral a.

Anterior
spinal a.

Vertebral
body

Artery of
Adamkiewicz

Neural foramen

Anterior
spinal a.

Aorta

Artery of
Adamkiewicz

FIGURE 8.8: Spinal Cord B—Blood Supply (drawing and CTA radiograph)

THE LIMBIC SYSTEM

The term *limbic* is almost synonymous with the term emotional brain—the parts of the brain involved with our emotional state. In 1937, Dr. James Papez initiated the limbic era by proposing that a number of limbic structures in our brain formed the anatomical substratum for emotion (see Figure 10.1A).

EVOLUTIONARY PERSPECTIVE

Dr. Paul MacLean (1913–2007) postulated the *triune model* of brain evolution. The pre-mammalian (reptilian) brain has the capacity to look after the basic life functions, and behaviorally it has organized ritualistic stylized patterns of behavior. In higher species including mammals, neocortical structures evolved that are adaptive, allowing for a modification of behavior depending on the situation.

Dr. MacLean introduced the term **limbic system** (in 1952). He conceived of a set of functional interconnected structures which arose in early mammals that were responsible for behaviors which are thought of as motivational and emotional, including feeding, reproduction, and parenting.

Hence, we now view the limbic system as those parts of the brain that are involved in regulating the "emotional" state (see definition) of the animal in relation to the external and internal worlds.

DEFINITION

Most of us are quite aware or have a general sense of what we mean when we use the term *emotion* or *feelings,* yet the concept is somewhat difficult to explain or define precisely. A medical dictionary (Stedman's) defines emotion as "a strong feeling, aroused mental state, or intense state of drive or unrest directed toward a definite object and evidenced in both behavior and in psychologic changes." Thus, emotions involve the following:

- Physiological changes: These changes include basic drives involving thirst, sexual behavior, and appetite. They are often manifested as involving the autonomic nervous system or endocrine system, or both.
- Behavior: The animal or human does something, that is, performs some type of motor activity (e.g., feeding, fighting, fleeing, displaying anger, mating activity); in humans, this may include facial expression.
- Alterations in the mental state: These can be understood as subjective changes in the way the organism "feels" or reacts to the state of being or to events occurring in the outside world. In humans, we use the term *psychological reaction.*

It is clear, at least in humans, that some of these psychological functions and behaviors must involve the cerebral cortex. In addition, many of these alterations are conscious and involve association areas. In fact, humans are sometimes able to describe and verbalize their reactions or the way they feel. Both cortical and subcortical areas (e.g., basal ganglia) may be involved in the behavioral reactions associated with emotional responses. The hypothalamus controls the autonomic changes, along with brainstem nuclei, and also the activity of the pituitary gland underlying the endocrine responses.

Therefore, we can finally arrive at a definition of the limbic system as an inter-related group of cortical and subcortical (non-cortical) structures and pathways that are involved in the regulation of the mental/emotional state, with the accompanying physiological, behavioral, and psychological responses. This characterization of the limbic system will be revisited in the "synthesis" at the end of this section.

NEURAL STRUCTURES

In neuroanatomical terms, the limbic system now includes the following cortical and non-cortical (subcortical, diencephalic, and brainstem) structures.

- **Core structures** are those definitely associated with the limbic system.
- **Extended structures** are those closely connected with limbic functions.

CORTICAL STRUCTURES

- Core: the hippocampal formation (which consists of three subparts), the parahippocampal gyrus, the cingulate gyrus.
- Extended: parts of the prefrontal and orbitofrontal cortex (the limbic forebrain).

NON-CORTICAL STRUCTURES

- Forebrain:
  - Core: the amygdala, the septal region, ventral portions of the basal ganglia, including the nucleus accumbens.
  - Extended: the basal forebrain.
- Diencephalic and brainstem:
  - Core: certain nuclei of the thalamus, the hypothalamus.
  - Extended: parts of the midbrain (the limbic midbrain).

All these structures are collectively called the **limbic system**. The particular role of the olfactory system and its connections will be discussed in the context of the limbic system (see Figure 10.4).

OVERVIEW OF KEY LIMBIC STRUCTURES

There are key structures of the limbic system that integrate information and relate the external and internal worlds—the hippocampal formation, the parahippocampal gyrus, the amygdala, and the hypothalamus.

- The hippocampal formation is an older (from an evolutionary perspective) cortical region that is involved with integrating information; its role in the formation of memory for facts and events is discussed later.
- The parahippocampal gyrus has widespread connections with many cortical (particularly sensory) areas and is probably the source of the most significant afferents to the hippocampal information.
- The amygdala is in part a subcortical nucleus involved with internal (visceral afferent) information, as well as receiving sensory input about olfaction (our sense of smell).
- The hypothalamus oversees autonomic physiological and endocrine regulation.

Both the amygdala and the hypothalamus are involved with the motor (i.e., behavioral) responses of the organism, the amygdala in part via the hypothalamus, and both are involved along with other structures in generating "emotional" reactions.

LIMBIC CONNECTIONS

The limbic system has internal circuits involving the key structures; these link the hippocampal formation, the parahippocampal gyrus, the amygdala, and the hypothalamus, as well as other structures of the limbic system. There are multiple interconnections within and between these structures, and knowledge of the circuits of the limbic system (which are quite complex) allows one to trace pathways within the limbic system. Only some of these pathways are presented. The best known of these functionally (and for historical reasons) is the Papez circuit (discussed with Figure 10.1A). Also discussed are additional pathways that connect the limbic structures to the remainder of the nervous system and through which the limbic system influences the activity of the nervous system (to be discussed in the limbic "synthesis").

MEMORY

Unfortunately, the definition and description of the limbic system do not include one aspect of brain function that seems to have evolved in conjunction with the limbic system, namely, memory. Memory systems are now divided into two types (further discussed with Figure 9.4):

- Memory for motor skills, called **procedural memory** (also called implicit memory).
- Memory for facts and events, called **declarative or episodic** (also called explicit) memory.

Part of the hippocampal formation is specifically necessary for the initial formation of episodic memories. It is critical to understand that this initial step is an absolute pre-requisite to the formation of any *new* memory trace. Once encoded by the hippocampal formation, the memory trace is then transferred to other parts of the brain for short-term and long-term storage. The limbic system seems not to be involved in the storage and retrieval of long-term memories.

It is interesting to speculate that forgetting may be theoretically more appropriate for this unique aspect of limbic function. This idea proposes that to undo or unlock the fixed behavioral patterns of the old reptilian brain, some part of the brain must be assigned the function of "recording" that something has happened. To change a response, the organism needs to "remember" what happened the last time when faced with a similar situation. Hence, the development of memory functions of the brain may have occurred in association with the evolution of the limbic system. The availability of stored memories makes it possible for mammals to over-ride or over-rule the stereotypical behaviors of the reptile, thus allowing for more flexibility and adaptiveness when faced with a changing environment or altered circumstances.

In summary, the limbic system—both cortical and non-cortical components—includes a set of "F" functions: feeding (and other basic drives), fornication (reproduction), fighting and fleeing (behavioral), feeling (psychological), "forgetting" (memory), and family, as well as perhaps a consciousness of self.

Chapter 9

Major Limbic Structures

FIGURE 9.1A—LIMBIC LOBE 1

CORTICAL STRUCTURES

The limbic lobe refers to cortical areas of the limbic system. These cortical areas, which were given the name "limbic," form a border (limbus) around the inner structures of the diencephalon and midbrain (see Figure 1.7 and Figure 9.1B). The core cortical areas include the hippocampal formation, the parahippocampal gyrus, and the cingulate gyrus.

There are a number of cortical areas located in the most medial (also called mesial) aspects of the temporal lobe in humans that form part of this limbus. These areas are collectively called the **hippocampal formation**; there are three portions—the hippocampus proper, the dentate gyrus, and the subicular region (see Figure 9.3A and Figure 9.3B). The **hippocampus proper** is, in fact, no longer found at the surface of the brain, as would be expected for any cortical area. The **dentate gyrus** is a very small band of cortex, part of which can be found at the surface, and the **subicular** region is located at the surface but far within the temporal area. These structures are the central structures of the limbic lobe.

The typical cortex of the various lobes of the human brain consists of six layers (and sometimes sublayers), called neocortex. One of the distinguishing features of the limbic cortical areas is that, for the most part, these are older (from an evolutionary perspective) cortical areas consisting of three to five layers, termed allocortex. The hippocampus proper and the dentate cortex are three-layered cortical areas (archicortex), whereas the subicular region has four to five layers (paleocortex).

Note to the Learner: At this stage, it is very challenging to understand where these structures are located. The component parts of the hippocampal formation are "buried" in the temporal lobe and remain somewhat obscure. It is suggested that the student preview some of the illustrations of the "hippocampus" (see Figure 9.4), as well as sections through the hippocampal formation (see Figure 9.3B) to understand the configuration of the three component parts and the relationship with the parahippocampal gyrus more clearly. The details of these various limbic structures, including their important connections and the functional aspects, are discussed with the appropriate diagram.

The **parahippocampal gyrus**, which is situated on the inferior aspect of the brain (see Figure 1.5 and Figure 1.6), is a foremost structure of the limbic lobe; it is mostly a five-layered cortical area. It is heavily connected (reciprocally) with the hippocampal formation. The anterior portion of this gyrus is called the *entorhinal cortex.* This gyrus also has widespread connections with many areas of the cerebral cortex, including all the sensory cortical regions, as well as the cingulate gyrus. It is thought to play a key role in memory function.

The **cingulate gyrus**, which is situated above the corpus callosum (see Figure 1.7), consists of five-layered cortex, as well as neocortex. The cingulate gyrus is connected reciprocally with the parahippocampal gyrus via by a bundle of fibers in the white matter, known as the cingulum bundle (see Figure 9.1B). This connection unites the various portions of the limbic "lobe." It also has widespread connections with the frontal lobe.

Of the many tracts of the limbic system, two major tracts have been included in this diagram—the fornix and the anterior commissure.

- The **fornix** is one of the more visible tracts and is often encountered during dissections of the brain (e.g., see Figure 1.7). This fiber bundle connects the hippocampal formation with other areas (discussed with Figure 9.3A and Figure 9.3B).

- The **anterior commissure** is an older commissure than the corpus callosum and connects several structures of the limbic system on the two sides of the brain; these include the amygdala, the hippocampal formation, and parts of the parahippocampal gyrus, as well as the anterior portions of the temporal lobe; the anterior commissure is seen on many of the limbic diagrams and can also be a useful reference point for orientation (e.g., see Figure 9.6B).

The other structures shown in this diagram include the diencephalon (the thalamus) and the brainstem. The corpus callosum "area" is indicated as a reference point in these illustrations (see Figure 9.1B).

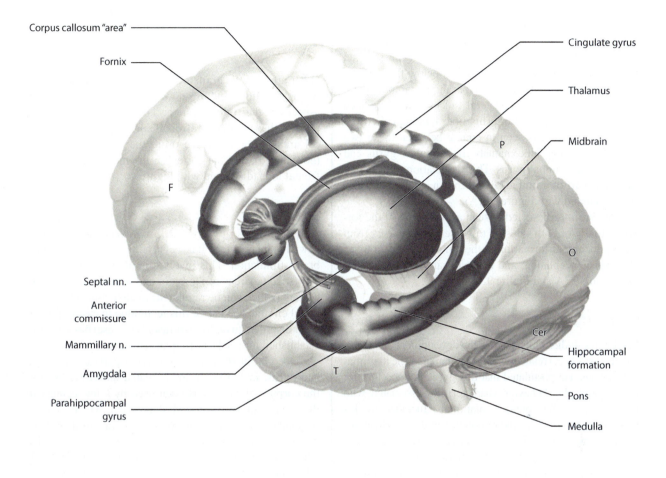

Corpus callosum "area"

Fornix

Cingulate gyrus

Thalamus

Midbrain

P

F

Septal nn.

Anterior
commissure

Mammillary n.

Amygdala

Parahippocampal
gyrus

Cer

T

O

Hippocampal
formation

Pons

Medulla

F = Frontal lobe
P = Parietal lobe
T = Temporal lobe
O = Occipital lobe

Cer = Cerebellum

FIGURE 9.1A: Limbic Lobe 1—Cortical Structures

FIGURE 9.1B—LIMBIC LOBE 2

CINGULUM BUNDLE (PHOTOGRAPH)

This is a dissection of the brain, from the medial perspective (see also Figure 1.7). The thalamus (diencephalon) has been removed, revealing the fibers of the internal capsule (see Figure 4.4). The specimen has been tilted slightly to show more of the inferior aspect of the temporal lobe.

The cortex of the cingulate gyrus has been scraped away, revealing a bundle of fibers just below the surface. The dissection is continued to the parahippocampal gyrus, as demarcated by the collateral sulcus or fissure (see Figure 1.5 and Figure 1.6). This fiber bundle, called the **cingulum bundle**, is seen to course between these two gyri of the limbic lobe, namely the cingulate gyrus and the parahippocampal gyrus. This association tract is discussed as part of a limbic circuit known as the Papez circuit (discussed with Figure 10.1A).

The brain is dissected in such a way to reveal the fornix (actually of both sides) as this fiber tract courses from the hippocampal formation in the temporal lobe, courses over the diencephalon, and heads toward its connections (see Figure 9.3A and Figure 9.3B).

CINGULATE GYRUS

MacLean's studies indicated that the development of the cingulate gyrus is correlated with the evolution of the mammalian species. He postulated that this gyrus is important for nursing and play behavior, characteristics associated with the rearing of the young in mammals. Included in this behavior is recognizing and responding to the vocalizations of the young (studied in rodents, in cats, and in other animals); a mother responds to the unique tone of her own baby's crying. This cluster of behavioral patterns forms the basis of the other "F" in the list of functions of the limbic system, namely, family.

The cingulate gyrus also seems to have an important role in attention, a critical aspect of behavior.

A small cortical region under the anterior part (the rostrum) of the corpus callosum is also included with the limbic system. These small gyri (not labeled; see Figure 1.7) are part of the septal region (see Figure 9.1A) and are considered along with the septal nuclei (see Figure 10.3).

EXTENDED LIMBIC LOBE

Other areas of the brain are now known to be involved in limbic functions and are now included in the functional aspects of the limbic system. These areas include large parts of the so-called prefrontal cortex, particularly cortical areas lying above the orbit, the orbitofrontal cortex (not labeled), and the cortex on the medial aspect of the frontal lobe (discussed with Figure 10.1B).

CLINICAL ASPECT

Deep Brain Stimulation (DBS): In very recent times, a new therapeutic tool has been added—the use of deep brain stimulation. This involves implanting electrodes in very specific areas of the brain and using a small electric current (which can be regulated) to alter some brain functional circuit. The use of this technique in cases of movement disorders (usually Parkinson's disease) has been well established for a select group of patients.

Deep brain stimulation involving parts of the cingulate gyrus, including an area located anteriorly "below" the corpus callosum, has been used by a few centers for the treatment of profound depression which has not responded to other forms of therapy, as well as some behavioral maladaptive conditions such as obsessive-compulsive disorder. Time will tell if DBS is useful for the alleviation of these conditions, and perhaps others.

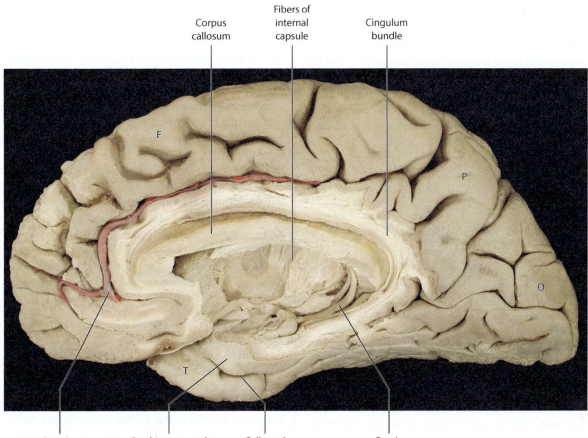

Corpus callosum

Fibers of internal capsule

Cingulum bundle

F

P

O

T

Anterior cerebral artery

Parahippocampal gyrus

Collateral sulcus

Fornix

F = Frontal lobe
P = Parietal lobe
T = Temporal lobe
O = Occipital lobe

FIGURE 9.1B: Limbic Lobe 2—Cingulum Bundle (photograph)

FIGURE 9.2—LIMBIC SYSTEM

NON-CORTICAL STRUCTURES

The term *limbic system* is the concept now used to include those parts of the brain that are associated with the functional definition of the limbic system as discussed in the Introduction to this section.

This is an overall diagram focusing on the non-cortical components of the limbic system, both core and extended (see Introduction to this section). These structures are found in the forebrain, the diencephalon, and the midbrain. Each of the structures, including the connections, is discussed in greater detail in subsequent illustrations while this diagram, indicated appropriately, is used to show only the structures of the limbic system that are being described.

The non-cortical structures shown in this diagram include:

- The amygdala.
- The septal region.
- The nucleus accumbens.
- The thalamus.
- The hypothalamus.
- The limbic midbrain.
- The olfactory system.

FOREBRAIN

The **amygdala**, also called the amygdaloid nucleus, a core limbic structure, is anatomically one of the basal ganglia (as discussed with Figure 2.5A and Figure 2.6). Functionally, and through its connections, it is part of the limbic system. Therefore, it is considered in this section of the atlas (see Figure 9.6A and Figure 9.6B).

The **septal region** includes two components—the cortical gyri below the rostrum of the corpus callosum (see Figure 9.1B) and some nuclei deep to them. These nuclei are not located within the septum pellucidum in the human (see Figure 8.6). The term *septal region* includes both the cortical gyri and the nuclei (see Figure 9.1A and Figure 10.3).

BASAL GANGLIA

The ventral portions of the striatum and globus pallidus are now known to be connected with limbic functions and are part of the extended limbic system. The **nucleus accumbens** is located inferior to the neostriatum (specifically, the head of the caudate nucleus; see Figure 2.5B and Figure 2.7). It has been found to have a critically important function in activities where there is an aspect of reward; this is now thought to be the critical area of

the brain involved in addiction (discussed with Figure 10.7).

Not represented in this diagram is the region known as the **basal forebrain**. This subcortical region is composed of several cell groups located beside the hypothalamus and below the anterior commissure (see Figure 10.5A and Figure 10.5B). This somewhat obscure region has connections with several limbic areas and the prefrontal cortex.

DIENCEPHALON

Two of the nuclei of the **thalamus**, the anterior group of nuclei and the dorsomedial nucleus (see Figure 4.3 and Figure 6.13), are part of the pathways of the limbic system, relaying information from subcortical nuclei to limbic parts of the cortex (the cingulate gyrus and areas of the prefrontal cortex—discussed with Figure 10.1A and Figure 10.1B).

The **hypothalamus** lies below and somewhat anterior to the thalamus (see Figure 1.7). Many nuclei of the hypothalamus function as part of the core limbic system. Only a few of these nuclei are shown, and among these is the prominent **mammillary nucleus**, which is visible on the inferior view of the brain (see Figure 1.5). The connection of the hypothalamus to the pituitary gland is not shown.

MIDBRAIN

The extended limbic system also includes nuclei of the midbrain, the so-called **limbic midbrain**. Some of the descending limbic pathways terminate in this region, and it is important to consider the role of this area in limbic functions. An important limbic pathway, the medial forebrain bundle, interconnects the septal region, the hypothalamus, and the limbic midbrain (see Figure 10.3). The ventral tegmental area (VTA), now part of the limbic circuitry, is located in the upper midbrain (see Figure 10.6).

OLFACTORY SYSTEM

The **olfactory system** is described with the limbic system because many of its connections are directly with limbic areas. Years ago, it was commonplace to think of various limbic structures as part of the "smell brain," the rhinencephalon. We now know that this is only partially correct. The olfactory input connects directly into the limbic system (and not via the thalamus; see Figure 10.4), but the limbic system is now known to have many other functional capabilities.

TRACTS

The various tracts which interconnect the limbic structures—fornix, stria terminalis, ventral amygdalofugal pathway—are discussed at the appropriate time with the relevant structure or structures.

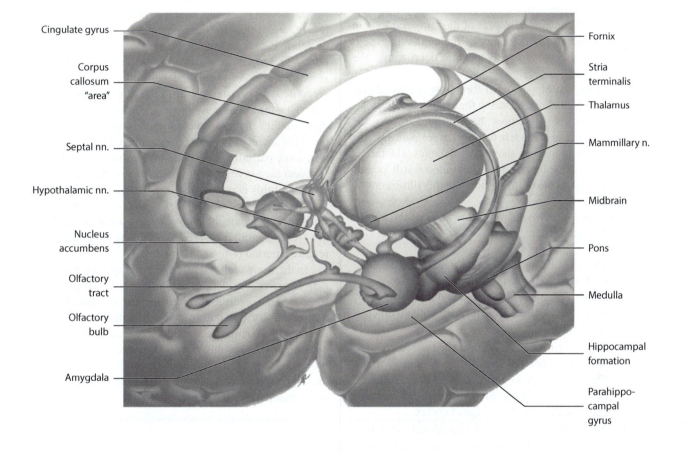

Cingulate gyrus

Corpus
callosum
"area"

Septal nn.

Hypothalamic nn.

Nucleus
accumbens

Olfactory
tract

Olfactory
bulb

Amygdala

Fornix

Stria
terminalis

Thalamus

Mammillary n.

Midbrain

Pons

Medulla

Hippocampal
formation

Parahippo-
campal
gyrus

FIGURE 9.2: Limbic System—Non-Cortical Structures

FIGURE 9.3A— "HIPPOCAMPUS" 1

HIPPOCAMPAL FORMATION

This diagram, which is the same as Figure 9.2, highlights the functional portion of the limbic lobe to be discussed— the "hippocampus" (i.e., the hippocampal formation and the pathway known as the fornix).

The **hippocampal formation** includes older cortical regions, all consisting of fewer than six layers, which are located in the medial aspect of the temporal lobe in humans. Much of the difficulty of understanding these structures is their anatomical location deep within the temporal lobe.

In the rat, the hippocampal formation is located dorsally, above the thalamus. During the evolution of the temporal lobe, these structures have migrated into the temporal lobe and have left behind a fiber pathway, the fornix, which is located above the thalamus. (In fact, a vestigial part of the hippocampal formation is still located above the corpus callosum, as shown in this illustration— not labeled.)

The term *hippocampal formation* includes the following (see Figure 9.3B):

- The **hippocampus proper**, a three-layered cortical area that during development becomes "rolled up" and is no longer found at the surface of the hemispheres (as is the case with all other cortical regions).
- The **dentate gyrus**, a three-layered cortical area that is partly found on the surface of the brain, although its location is so deep that it presents a challenge to non-experts to locate and visualize this thin ridge of cortex.
- The **subicular region**, a transitional cortical area of three to five layers that becomes continuous with the parahippocampal gyrus (located on the inferior aspect of the brain; see Figure 1.5).

The **fornix** is a fiber bundle that is visible on medial views of the brain (at the lower edge of the septum pellucidum) (see Figure 1.7). These fibers emerge from the hippocampal formation (shown in Figure 9.4; see also Figure 9.1B) and course over the thalamus, where they are found just below the corpus callosum (see coronal sections; Figure 2.9A and Figure 9.5A). The fibers end in the septal region and in the mammillary nucleus of the hypothalamus (shown in Figure 9.3B). Some fibers in the fornix are conveying information from these regions to the hippocampal formation. It is perhaps best to regard the fornix as an association bundle, part of the limbic pathways. It has attracted much attention because of its connections and because of its visibility and accessibility for research into the function of the hippocampal formation, particularly with regard to memory.

MEMORY

Studies in animals have indicated that the neurons located in one portion of the hippocampus proper, called the CA3 region, are critical for the formation of new memories— declarative or episodic types of memories (not procedural). This means that in order for the brain to "remember" some new fact or event, the new information must be "recorded" or registered within the hippocampal formation. This information is "processed" through some complex circuitry in these structures and is retained for a brief period of seconds. For it to be remembered for longer periods, some partially understood process occurs so that the transient memory trace is transferred to other parts of the brain, and this is now stored in working memory or as a long-term memory. (An analogy to computers may be useful here: if a document is not saved, it is not available for use—ever again.) The process of memory storage may require a period of hours, if not days.

In the study of the function of the hippocampus in animals, there is considerable evidence that the hippocampal formation is involved in constructing a "spatial map." According to this literature, this part of the brain is needed to orient in a complex environment (e.g., a maze). It is not quite clear whether this is a memory function or whether this spatial representation depends on the connections of the hippocampal formation and parahippocampal gyrus with other parts of the brain.

CLINICAL ASPECT

The clinical implications of the functional involvement of the hippocampal formation in memory are further elaborated with Figure 9.4.

It is now possible to view the hippocampal area in detail on magnetic resonance imaging scans (MRI) and to assess the volume of tissue (see Figure 9.5B). Bilateral damage here apparently correlates with the loss of memory function in humans with Alzheimer's dementia, particularly for the formation of memories for new events or for new information (further discussed with Figure 9.4).

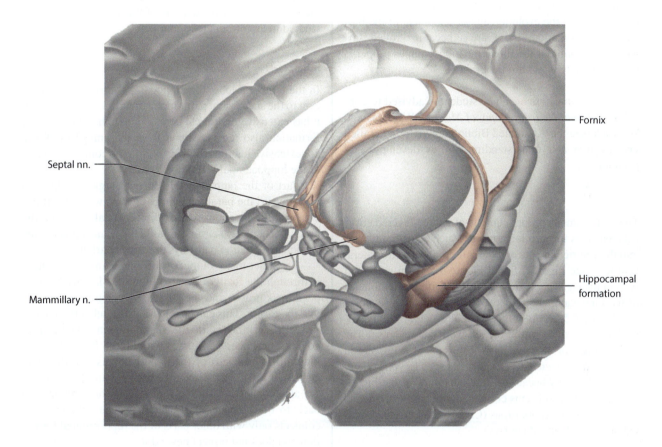

Septal nn.

Fornix

Mammillary n.

Hippocampal
formation

FIGURE 9.3A: "Hippocampus" 1—Hippocampal Formation

FIGURE 9.3B— "HIPPOCAMPUS" 2

HIPPOCAMPAL FORMATION: THREE PARTS

The hippocampal formation is one of the most important components of the limbic system in humans. It is certainly the most complex. This diagram isolates the component parts of the hippocampal formation, on both sides.

One expects a cortical area to be found at the surface of the brain, even if this surface is located deep within a fissure. During the evolution and development of the hippocampal formation, these areas became "rolled up" within the brain. Of the three, the hippocampus proper is found completely "within the brain."

Note to the Learner: The student is advised to consult *Functional Neuroanatomy of Man,* by Williams and Warwick (see the Annotated Bibliography), for a detailed visualization and understanding of this developmental phenomenon.

HIPPOCAMPUS PROPER

The hippocampus proper consists of a three-layered cortical area. This forms a large mass, which actually intrudes into the ventricular space of the inferior horn of the lateral ventricle (see Figure 7.5 and Figure 9.5A). In a coronal section through this region, there is a certain resemblance of the hippocampal structures to the shape of a seahorse (see Figure 9.5A). From this shape, the name "hippocampus" is derived, from the French word for seahorse. The other name for this area is **Ammon's horn** or cornu ammonis (abbreviated CA), named after an Egyptian deity with ram's horns because of the curvature of the hippocampus in the brain. This cortical region has been divided into a number of subportions (CA1 to CA4, usually studied in more advanced courses).

DENTATE GYRUS

The dentate gyrus is also a phylogenetically older cortical area consisting of only three layers. During the formation discussed earlier, the leading edge of the cortex detaches itself and becomes the dentate gyrus. Parts of it remain visible at the surface of the brain.

Because this small surface is buried on the most medial aspect of the temporal lobe and is located deep within a fissure, it is rarely located in studies of the gross brain. Its cortical surface has serrations, which led to its name dentate (referring to teeth; (see also Figure 9.7)).

The appearance of the dentate gyrus is shown on the view of the medial aspect of the temporal lobe (on the far side of the illustration). A "cut" section through the temporal lobe (as seen in the lower portion of this illustration) indicates that the dentate gyrus is more extensive than its exposed surface portion.

SUBICULAR REGION

The next part of the cortically rolled-in structures that make up the hippocampal formation is the subicular region. The cortical thickness is transitional, starting from the three-layered hippocampal formation to the more layered parahippocampal gyrus (see also Figure 1.5). (Again, there are a number of subparts of this area that are rarely studied in an introductory course.)

CONNECTIONS AND FUNCTION

In the temporal lobe, the hippocampal formation is adjacent to the six-layered parahippocampal gyrus, with which it has extensive reciprocal connections. The hippocampal formation also receives input from the amygdala. There are extensive interconnections within the component parts of the hippocampal formation itself.

Part of the output of this cortical region is directed back toward the parahippocampal gyrus, which itself has extensive connections with other cortical areas of the brain, particularly sensory areas. This is analogous to the cortical association pathways described earlier.

The other major output of the hippocampal formation is through the **fornix** (see also Figure 9.4). Only the hippocampus proper and the subicular region project fibers into the fornix. This can be regarded as a subcortical pathway that terminates in the septal region (via the precommissural fibers; discussed with Figure 10.3) and in the mammillary nucleus of the hypothalamus (via the postcommissural fibers; discussed with Figure 10.2). There are also connections in the fornix from the septal region back to the hippocampal formation. The dentate gyrus connects only with other parts of the hippocampal formation and does not project beyond it.

CLINICAL ASPECT

The term *medial* or *mesial temporal sclerosis* is a general term for damage to the hippocampal region located in this part of the brain. Lesions in this area are known to be associated with epilepsy, particularly psychomotor seizures, classified as a partial complex seizure disorder.

MRI imaging is very useful for the diagnosis of mesial temporal sclerosis. Provided that the hippocampus on the other side is functioning (discussed with Figure 9.4), surgical removal of the anterior temporal region may be recommended, in select cases, to alleviate the epileptic condition when anti-epileptic medication has not been successful.

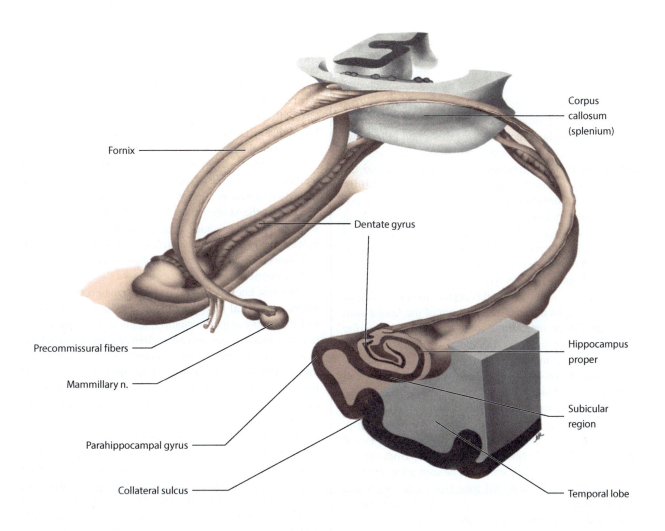

Fornix

Corpus callosum (splenium)

Dentate gyrus

Precommissural fibers

Mammillary n.

Hippocampus proper

Parahippocampal gyrus

Subicular region

Collateral sulcus

Temporal lobe

FIGURE 9.3B: "Hippocampus" 2—Hippocampal Formation: Three Parts

FIGURE 9.4—
"HIPPOCAMPUS" 3

HIPPOCAMPAL FORMATION (PHOTOGRAPH AND T1 MAGNETIC RESONANCE IMAGING SCAN)

The brain is being shown from the lateral aspect (as in Figure 1.3 and Figure 1.4). The left hemisphere has been dissected by removing cortex and white matter above the corpus callosum: the lateral ventricle has been exposed from this perspective (see Figure 2.1A). The choroid plexus tissue has been removed from the ventricle to improve visualization of the structures. This dissection also shows the lateral aspect of the lentiform nucleus, the putamen, and the fibers of the internal capsule between it and the thalamus (see Figure 1.7, Figure 2.10A, and Figure 2.10B).

A similar dissection has been performed in the temporal lobe, thereby exposing the inferior horn of the lateral ventricle (see Figure 2.8). A large mass of tissue is found protruding into the inferior horn of this ventricle—named the hippocampus, a visible gross brain structure. In fact, the correct term now used is the **hippocampal formation**. In a coronal section through this region, the protrusion of the hippocampus into the inferior horn of the lateral ventricle can also be seen, almost obliterating the ventricular space (shown in Figure 9.5A; see also Figure 2.9A and Figure 2.9B). An unusual view of the hippocampal formation—from above—is seen in Figure 7.6 (particularly well seen on the right side of the illustration).

The hippocampal formation is composed of three distinct regions—the hippocampus proper (Ammon's horn), the dentate gyrus, and the subicular region, as explained with Figure 9.3B. The fiber bundle that arises from the visible "hippocampus," the fornix, can be seen adjacent to the hippocampus in the temporal lobe (see Figure 9.1A and Figure 9.1B), and it continues over the top of the thalamus (discussed with Figure 9.3B; see also Figure 1.7).

T1 MAGNETIC RESONANCE IMAGING SCAN

The T1 MRI (magnetic resonance imaging scan)—from the same perspective—captures the whole lateral ventricle and the hippocampal formation in the inferior horn in the temporal lobe, as well as structures in the "core" (the putamen of the lentiform nucleus and fibers of the internal capsule). Note the choroid plexus in the atrium of the lateral ventricle (also seen in Figure 6.5).

CLINICAL ASPECT—MEMORY

It is now known that the hippocampal formation is the critical structure for memory. This function of the hippocampal formation became understood because of an individual known in the literature as H.M.—Henry Molaison (1926–2008). H.M. suffered from intractable epilepsy (that may have been caused by a bicycle accident at the age of 7). In 1953, at the age of 27, he had neurosurgery to alleviate his condition (which it did) consisting of bilateral medial temporal lobe resection, including most of the hippocampal formation and adjacent structures.

After the surgery it was found that H.M. could not form new memories for facts or events (episodic memory, also called explicit memory), although subsequent studies found that he could learn new motor skills (called implicit or procedural memory).

Our understanding of memory, the different types of memory, and the role of the hippocampal formation in the formation of new memories for facts or events was formulated from the studies of H.M. by Dr. Brenda Milner, a well known Canadian neuropsychologist (who later moved to the Montreal Neurological Institute, the MNI, and worked with the world-famous neurosurgeon Dr. Wilder Penfield).

We now know that **bilateral** damage or removal of the anterior temporal lobe structures, including the amygdala and the hippocampal formation, leads to a unique condition in which the person can no longer form new declarative or episodic memories, although older memories are intact. The individual cannot remember what occurred moments before. Therefore, the individual is unable to learn (i.e., to acquire new information) and is not able to function independently.

In order to determine the functionality of the hippocampus, a test called the Wada Test was devised. This involves injecting a short acting barbiturate into each carotid artery and observing its effect on speech and short term memory on each half of the brain. This test has been supplemented with the availability of functional MRI (fMRI).

If a surgical procedure is to be performed in this region nowadays, additional special neuropsychological testing must be done to ascertain that the hippocampus contralateral to the operation is intact and functioning.

Note to the Learner: It is suggested that the learner read further about memory, H.M. and other similar cases in the neuropsychology literature (see Annotated Bibliography).

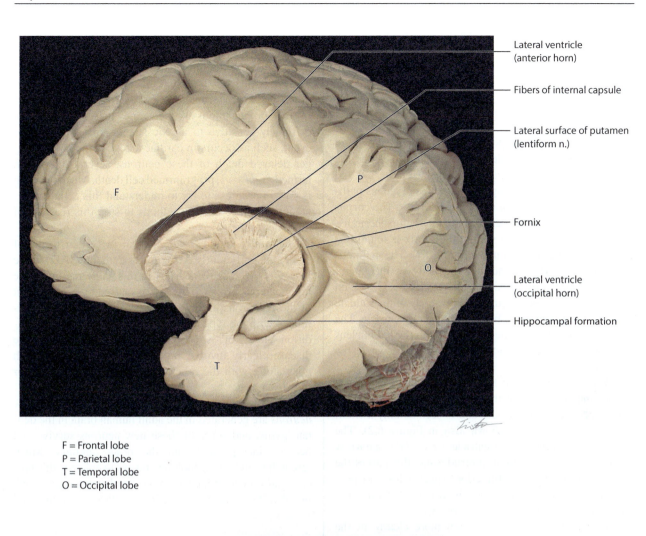

Lateral ventricle
(anterior horn)

Fibers of internal capsule

Lateral surface of putamen
(lentiform n.)

Fornix

Lateral ventricle
(occipital horn)

Hippocampal formation

F = Frontal lobe
P = Parietal lobe
T = Temporal lobe
O = Occipital lobe

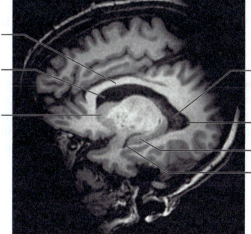

Corpus callosum

Lateral ventricle (anterior horn)

Caudate n. (head)

Choroid plexus

Lateral ventricle (occipital horn)

Hippocampal formation

Lateral ventricle (inferior horn)

FIGURE 9.4: "Hippocampus" 3—Hippocampal Formation (photograph and T1 magnetic resonance imaging scan)

FIGURE 9.5A— "HIPPOCAMPUS" 4

CORONAL BRAIN SECTION (PHOTOGRAPH)

This section is taken posterior to the one shown in Figure 2.9A (see upper left image), and it includes the inferior horn of the lateral ventricle (see Figure 2.8 and Figure 9.4). The basal ganglia, putamen, and globus pallidus are no longer present (see Figure 2.10A). The corpus callosum is seen in the depth of the interhemispheric fissure, and at this plane of section the fornix is found just below the corpus callosum (see Figure 1.7). The thalamus is below. The section includes the body of the lateral ventricle (see Figure 2.1A), and choroid plexus is seen on its medial corner (see Figure 7.8). The body of the caudate nucleus is adjacent to the body of the lateral ventricle. The section passes through the midbrain (with the red nucleus and the substantia nigra) and the pons, as shown in the upper right image.

The inferior horn of the lateral ventricles (shown at higher magnification in the lower inset) is found in the temporal lobes on both sides and is seen as only a small crescent-shaped cavity (shown also in Figure 6.2). The inferior horn of the lateral ventricle is reduced to a narrow slit because a mass of tissue protrudes into this part of the ventricle from its medial-inferior aspect. Closer inspection of this tissue reveals that it is gray matter; this gray matter is in fact the hippocampus proper.

This plane of viewing (shown more clearly in the enlarged view below) allows one to follow the gray matter from the hippocampus proper medially and through an intermediate zone, known as the subiculum or subicular region (see Figure 9.3B), until it becomes continuous with the gray matter of the parahippocampal gyrus. The hippocampus proper has only three cortical layers. The subicular region consists of four to five layers; the parahippocampal gyrus is to a large part a five-layered cortex. This view also allows us to understand that the parahippocampal gyrus is so named because it lies beside the "hippocampus."

CLINICAL ASPECT

The neurons of the hippocampal area are prone to damage for a variety of reasons, including vascular conditions. The key neurons for memory function are located in the hippocampus proper, and these neurons are extremely sensitive to anoxic states. An acute hypoxic event, such as occurs in cardiac arrest, is thought to trigger delayed death of these neurons, several days later, termed **apoptosis**, programmed cell death. Much research is now in progress to try to understand this cellular phenomenon and to devise methods to stop this reaction of these neurons.

Currently, studies indicate that in certain forms of dementia, particularly Alzheimer's dementia (AD), there is a *selective* loss of neurons in this same region of the hippocampus proper. This loss is caused by involvement of these neurons in the disease process. Again, this correlates with the type of memory deficit seen in this condition—loss of short-term memory—although the disease clearly involves other neocortical areas, which goes along with the other cognitive deficits typical for this disease.

We now know, based on extensive research, that *new neurons* are generated in the adult human brain in the dentate gyrus, and some of these new neurons survive and become incorporated into the brain circuitry. Much research is under way looking at ways to enhance this process and how to induce a similar "replacement program" in other brain areas, particularly following injury and strokes.

Note to the Learner: The only other area where new neurons are generated in the adult human brain is the olfactory system. Adult human neurogenesis is further discussed with Figure 9.5B.

ADDITIONAL DETAIL

The relationship of the caudate nucleus with the lateral ventricle is shown in two locations—with the body of the ventricle and with the tail in the "roof" of the inferior horn (see Figure 2.5B and Figure 2.7).

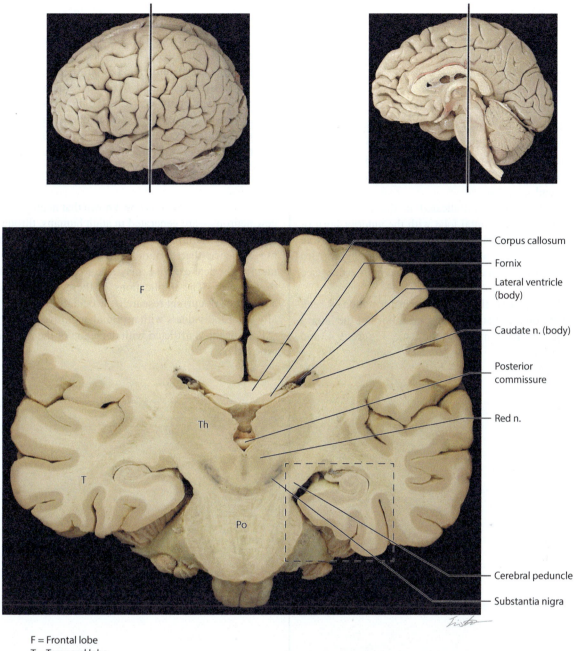

F = Frontal lobe
T = Temporal lobe

Th = Thalamus

Po = Pons

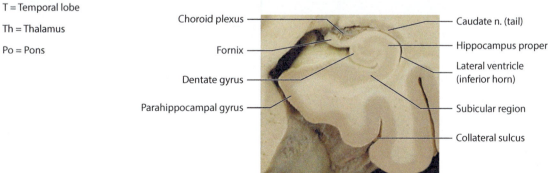

FIGURE 9.5A: "Hippocampus" 4—Coronal Brain Section (photograph)

FIGURE 9.5B—"HIPPOCAMPUS" 5

CORONAL VIEW (T1 MAGNETIC RESONANCE IMAGING SCAN)

This coronal image of the brain is taken particularly to view the medial aspect of the temporal lobe as part of a "seizure protocol" (because of the known association between lesions of the medial, also called mesial, temporal lobe and seizures).

The plane of view indicated in the locator image above includes the frontal lobe with the anterior horn of the lateral ventricles separated by the septum. Note the connection between the lateral ventricles and the 3rd ventricle—the foramen of Monro, on both sides (see Figure 7.8).

With this view, the hippocampal formation is seen infringing on the space of the inferior horn of the lateral ventricle in the temporal lobe (see Figure 9.4 and Figure 9.5A).

This image can be used to "quantify" the volume of the hippocampal formation, by noting especially whether there is a reduction in the size of the hippocampus in patients being investigated for memory loss and dementia. This volumetric quantification is still not—as yet—used as a diagnostic criterion for Alzheimer's dementia.

In this view, it appears as if the fibers of the internal capsule are continuing into the midbrain which in fact they do, forming the cerebral peduncle (see Figure 4.4 and Figure 5.15 and also Figure 9.5A). The outline of the pons is recognizable, as is the narrower medulla.

Additional note: It is now known that neurogenesis—new neurons—are generated in adult humans, throughout life, although diminishing with age. There is accumulating evidence that certain adult lifestyle modifications, mainly physical exercise, may lead to an enhancement of this process. The implications of these findings are an active area of current research, particularly for those usually older individuals with minimal cognitive impairment (MCI), or those afflicted with dementia of the Alzheimer type (AD).

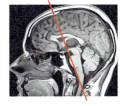

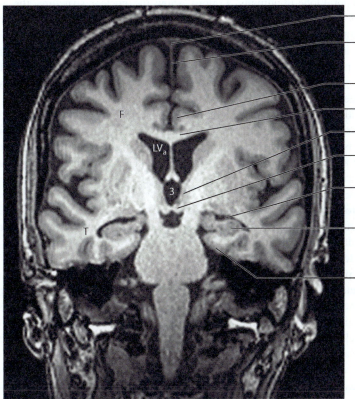

F = Frontal lobe
T = Temporal lobe

LV$_a$ = Lateral ventricle (anterior horn)
3 = 3rd ventricle

FIGURE 9.5B: "Hippocampus" 5—Coronal View (T1 magnetic resonance imaging scan)

FIGURE 9.6A—AMYGDALA 1

AMYGDALA: LOCATION AND FUNCTION (WITH T2 AND FLAIR MAGNETIC RESONANCE IMAGING SCANS)

This diagram, which is the same one as Figure 9.2, highlights a functional portion of the limbic system—the amygdala and its pathways, the stria terminalis and the ventral amygdalofugal pathway. The septal region (nuclei) is also included.

The amygdala (amygdaloid nucleus) is a subcortical nuclear structure located in the temporal lobe in humans (see Figure 2.5A and Figure 2.5). As a subcortical nucleus of the forebrain, it belongs by definition with the basal ganglia, but because its connections are with limbic structures, it is now almost always described with the limbic system.

The amygdala is located between the temporal pole (the most anterior tip of the temporal lobe) and the end of the inferior horn of the lateral ventricle (in the temporal lobe; see Figure 2.1A). The nucleus is located "inside" the uncus, which is seen on the inferior aspect of the brain as a large medial protrusion of the anterior aspect of the temporal lobe (see Figure 1.5 and Figure 1.6).

The amygdala receives input from the olfactory system, as well as from visceral structures. Two fiber tracts are shown connecting the amygdala to other limbic structures—a dorsal tract (the stria terminalis) and a ventral tract (the ventral amygdalofugal pathway, consisting of two parts). These are described in detail with Figure 9.6B.

The amygdala in humans is now being shown, using functional magnetic resonance imaging (fMRI), to be the area of the brain that is best correlated with emotional reactions, often associated with fearful situations. The emotional aspect of the response of the individual is passed onto the frontal cortex (discussed with the connections in Figure 9.6B), where "decisions" are made regarding possible responses. In this way, the response of the individual incorporates the emotional aspect of the situation.

Stimulation of the amygdaloid nucleus produces a variety of vegetative responses, including licking and chewing movements. Functionally, in animal experimentation, stimulation of the amygdala may produce a rage response, whereas removal of the amygdala (bilaterally) results in docility. Similar responses are also seen with stimulation or lesions in the hypothalamus. Some of these responses may occur through nuclei in the midbrain and medulla.

In monkeys, bilateral removal of the anterior parts of the temporal lobe (including the amygdala) produces some behavioral effects that are collectively called the **Kluver-Bucy syndrome**. The monkeys evidently become tamer after the surgical procedure, put everything into their mouths, and display inappropriate sexual behavior.

The amygdala is also known to contain a high amount of enkephalins. It is not clear why this is so and what may be the functional significance.

MAGNETIC RESONANCE IMAGING SCANS (T2 AND FLAIR)

The MRI scans in the lower part of the illustration, taken from two perspectives as indicated in the small illustrations, show the nuclear structure, the amygdala, in the temporal lobe. The image on the left in the horizontal (axial) plane is a T2-weighted image, and the amygdala is seen occupying the temporal "pole." The image on the right in the coronal plane is a FLAIR (fluid attenuation inversion recovery) image, and the amygdala is seen just as the inferior horn of the lateral ventricle is disappearing, at its very anterior end; on the right side of the image, the amygdala is located under the protruding tissue mass called the uncus (see Figure 1.5 and Figure 1.6). In fact, it is the amygdala nucleus that forms the uncus.

CLINICAL ASPECT

The amygdala is known to have a low threshold for electrical discharges, and this may make it prone to be the focus for the development of seizures. This has been found to occur in *kindling,* an experimental model of epilepsy. In humans, epilepsy from this part of the brain (anterior and medial temporal regions) usually gives rise to complex partial seizures, sometimes called temporal lobe seizures, in which automatisms such as lip smacking movements and fumbling of the hands are often seen, along with transient cognitive impairment (see also Figure 9.3B).

In very rare circumstances, bilateral destruction of the amygdala is recommended in humans for individuals whose violent behavior cannot be controlled by other means. This type of treatment is called psychosurgery.

The role of the amygdala in the formation of memory is not clear. Bilateral removal of the anterior portions of the temporal lobe in humans for the treatment of severe cases of epilepsy results in a memory disorder, which has been described with the hippocampal formation (discussed with Figure 9.4 and Figure 9.5A). It is possible that the role of the amygdala in the formation of memories is mediated either through the connections of this nuclear complex with the hippocampus or with the dorsomedial nucleus of the thalamus (see Figure 10.1B).

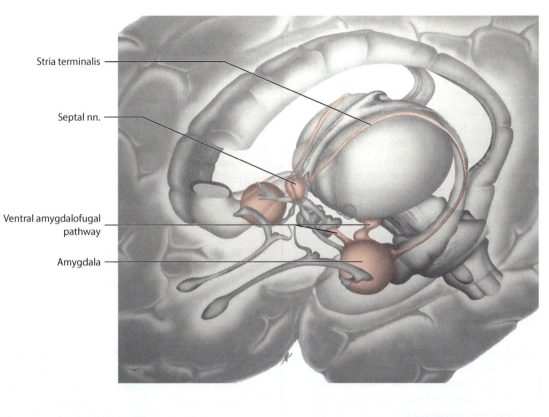

Stria terminalis

Septal nn.

Ventral amygdalofugal pathway

Amygdala

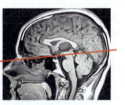

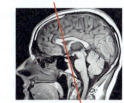

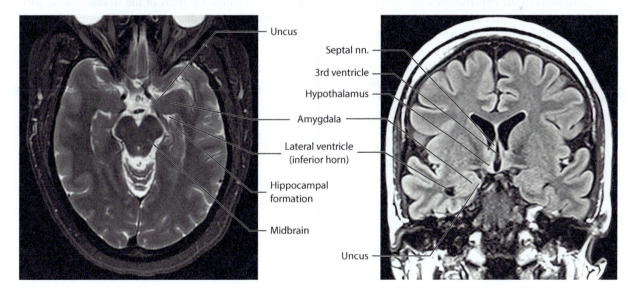

Uncus

Septal nn.

3rd ventricle

Hypothalamus

Amygdala

Lateral ventricle (inferior horn)

Hippocampal formation

Midbrain

Uncus

FIGURE 9.6A: Amygdala 1—Amygdala: Location and Function (with T2 and FLAIR magnetic resonance imaging scans)

FIGURE 9.6B—AMYGDALA 2

AMYGDALA: CONNECTIONS

One of the major differences between the amygdala and the other parts of the basal ganglia is that the amygdala is not a homogeneous nuclear structure but is composed in fact of different component parts. These parts are not usually studied in an introductory course.

The amygdala receives a variety of inputs from other parts of the brain, including the adjacent parahippocampal gyrus (not illustrated). It receives olfactory input directly (via the lateral olfactory stria; see Figure 10.4) and indirectly (from the cortex of the uncal region, as shown on the left side of the diagram).

The amygdaloid nuclei are connected to the hypothalamus, the thalamus (mainly the dorsomedial nucleus), and the septal region. The connections, which are reciprocal, travel through two routes:

- A dorsal route, known as the **stria terminalis**, which follows the ventricular curve and is found on the upper aspect of the thalamus (see Figure 9.6A). The stria terminalis lies adjacent to the body of the caudate nucleus in this location. This connects the amygdala with the hypothalamus and the septal region.

- A ventral route, known as the ventral pathway or the **ventral amygdalofugal** pathway. This pathway, which goes through the basal forebrain region (see Figure 10.5A), connects the amygdala to the hypothalamus (as shown) and to the thalamus (the fibers are shown en route), particularly the dorsomedial nucleus (see Figure 6.13 and Figure 10.1B).

The connection with the hypothalamus is likely the basis for the similarity of responses seen in animals with stimulation of the amygdala and the hypothalamus (see Figure 9.6A and Figure 10.2). This pathway to the hypothalamus may result in endocrine responses, and the connections with the midbrain and medulla may lead to autonomic response (see Figure 10.2).

Further possible connections of the amygdala with other limbic structures and other parts of the brain can occur via the septal region (see Figure 10.3), as well as via the dorsomedial nucleus of the thalamus to the prefrontal cortex (see Figure 10.1B).

The anterior commissure conveys connections between the nuclei of the two sides.

CLINICAL ASPECT

Seizure activity in the anterior temporal region may spread to the orbitofrontal region, via a particular group of fibers called the uncinate bundle (see Additional Detail below).

ADDITIONAL DETAIL

There is a white matter association pathway between the anterior pole of the temporal lobe and the inferior (orbital) part of the frontal lobe called the **uncinate** fasciculus, a "U-shaped" bundle of fibers. It is suggested that the student consult other texts for an illustration of this structure (see the Annotated Bibliography—for example, Figure 22-9 in *The Human Brain,* by Nolte 6th edition; a clearer image of a dissection of the uncinate fasciculus is seen in Nolte's 5th edition, Figure 22-11). The role of this pathway in the epileptic condition unfortunately named "uncinate fits" is discussed with Figure 10.4. The role of the orbital and medial parts of the frontal lobe as part of the limbic system is discussed with Figure 10.8 (see also Figure 6.13) and under "synthesis" at the end of the Limbic section.

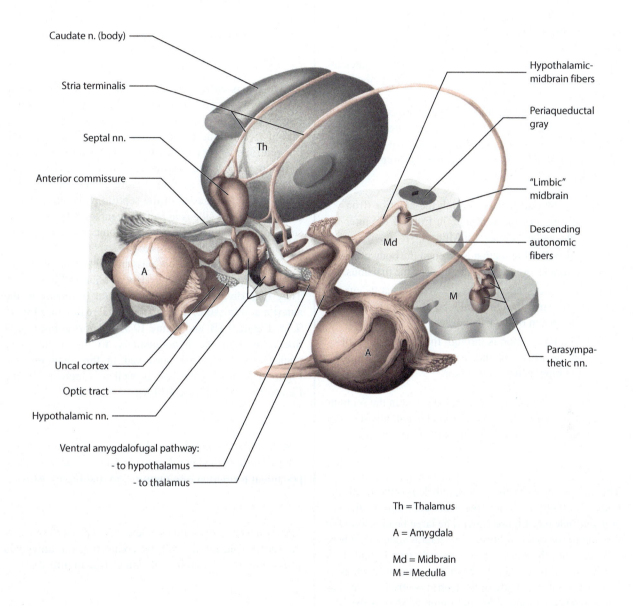

Caudate n. (body)

Stria terminalis

Septal nn.

Anterior commissure

Th

A

Md

M

A

Hypothalamic-
midbrain fibers

Periaqueductal
gray

"Limbic"
midbrain

Descending
autonomic
fibers

Parasympa-
thetic nn.

Uncal cortex

Optic tract

Hypothalamic nn.

Ventral amygdalofugal pathway:
- to hypothalamus
- to thalamus

Th = Thalamus

A = Amygdala

Md = Midbrain
M = Medulla

FIGURE 9.6B: Amygdala 2—Amygdala: Connections

FIGURE 9.7—LIMBIC "CRESCENT"

LIMBIC STRUCTURES AND THE LATERAL VENTRICLE

The temporal lobe is a more recent addition in the evolution of the hemispheres, and it develops later in the formation of the brain. During the development of the temporal lobe, certain structures migrate into it—the lateral ventricle, the hippocampal formation, and the caudate nucleus, as well as various tracts, the fornix, and the stria terminalis.

The lateral ventricle and associated structures form a crescent in the shape of a reverse letter "C" (see Figure 2.1A). These relationships are shown in this diagram by detailed "cuts" at various points along the lateral ventricle:

- The first section is through the anterior (frontal) horn of the ventricle, in front of the interventricular foramen (of Monro).
- The following section is through the body of the ventricle, over the dorsal aspect of the thalamus.
- The next section shows the ventricle at its curvature into the temporal lobe (this area is called the atrium or the trigone).
- The last section is through the inferior horn of the ventricle, in the temporal lobe, including the hippocampal formation.

Note to the Learner: The initials used in the sections (insets) to identify structures are found in parentheses after the labeled structure in the main part of the diagram.

CAUDATE NUCLEUS

The various parts of the caudate nucleus—the head, the body, and the tail—follow the inner curvature of the lateral ventricle (see Figure 2.5A). The large head is found in relation to the anterior horn of the lateral ventricle, where it bulges into the space of the ventricle (see Figure 2.9A and Figure 2.10A). The body of the caudate nucleus is coincident with the body of the lateral ventricle, on its lateral aspect (see Figure 2.9A and Figure 9.5A). As the caudate follows the ventricle into the temporal lobe, it becomes the tail of the caudate nucleus, where it is found on the upper aspect of the inferior horn, its roof (see Figure 6.2 and Figure 9.5A).

HIPPOCAMPAL FORMATION

The hippocampal formation is found in the temporal lobe situated medial and inferior to the ventricle (see Figure 9.3A, Figure 9.3B, and Figure 9.4). It bulges into the ventricle, almost obliterating the space; it is often difficult to visualize the small crevice of the ventricle in specimens and radiograms. The dentate gyrus is again seen (on the far side) with its indented surface (see also Figure 9.3B). The configuration of the three parts of the hippocampal formation is shown in the lower section.

FORNIX

The fornix is easily found in studies of the gross brain (e.g., see Figure 1.7). Its fibers can be seen as a continuation of the hippocampal formation (see Figure 9.3B and Figure 9.4), and these fibers course on the inner aspect of the ventricle as they sweep forward above the thalamus. In the area above the thalamus and below the corpus callosum (see coronal section; Figure 2.9A), the fornix is found at the lower edge of the septum pellucidum. Here, the fornix of one side is in fact adjacent to that of the other side (see also Figure 9.2); there are some interconnections between the two sides in this area.

The fibers of the fornix pass in front of the interventricular foramen (see medial view of brain in Figure 1.7). They then divide into pre-commissural (referring to the anterior commissure) fibers to the septal region (see Figure 9.3A, Figure 9.3B, and Figure 10.3) and post-commissural fibers, through the hypothalamus, to the mammillary nucleus (which is not portrayed in this diagram; see Figure 1.5 and Figure 1.6, as well as Figure 9.3B and Figure 10.3; also Figure 10.1A).

AMYGDALA

The amygdala is clearly seen to be situated anterior to the inferior horn of the lateral ventricle and in front of the hippocampal formation (see Figure 2.5A and Figure 9.6A).

STRIA TERMINALIS

The stria terminalis follows essentially the same course as the fornix (see Figure 9.2), by connecting the amygdala with septal region and hypothalamus (see Figure 10.3).

ADDITIONAL DETAIL

The stria terminalis is found slightly more medially on the dorsal aspect of the thalamus, in the floor of the body of the lateral ventricle. In the temporal lobe, the stria is found in the roof of the inferior horn of the lateral ventricle.

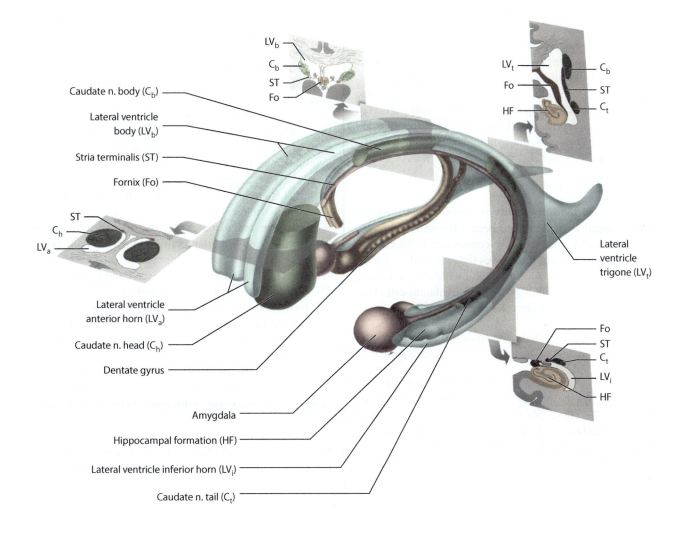

FIGURE 9.7: Limbic "Crescent"—Limbic Structures and the Lateral Ventricle

Chapter 10

Limbic Non-Cortical Structures

FIGURE 10.1A—LIMBIC DIENCEPHALON 1

THALAMUS: ANTERIOR NUCLEUS

This detailed diagram shows one of the major connections of the limbic system via the thalamus. This diagram shows an enlarged view of the thalamus of one side and the internal capsule (see Figure 4.3 and Figure 4.4), the head of the caudate nucleus (see Figure 2.5A), and a small portion of the cingulate gyrus (see Figure 1.7 and Figure 6.13). Immediately below is the hypothalamus, with only the two mammillary nuclei shown (see Figure 9.3A).

ANTERIOR NUCLEUS—CINGULATE GYRUS

The fibers of the fornix (carrying information from the hippocampal formation) have been followed to the mammillary nuclei (as the post-commissural fibers; see Figure 9.3B). A major tract leaves the mammillary nuclei, the **mammillo-thalamic tract**, and its fibers are headed for a group of association nuclei of the thalamus called the **anterior nuclei** (see Figure 4.3 and Figure 6.13).

   Note to the Learner: The student is advised to refer to the classification of the thalamic nuclei (see Figure 4.3).

   Axons leave the anterior nuclei of the thalamus and course through the anterior limb of the internal capsule (see Figure 4.4). These fibers course between the caudate nucleus (head and body) and the lentiform nucleus (which is just visible in the background). The axons terminate in the cortex of the cingulate gyrus after passing through the corpus callosum (see Figure 1.7 and Figure 6.13). The continuation of this circuit—the Papez circuit—is discussed below.

PAPEZ CIRCUIT

James Papez, in 1937, described a pathway involving some limbic and cortical structures and associated pathways. These, he postulated, formed the anatomical substrate for emotional experiences. The pathway forms a series of connections that has since been called the Papez circuit. We have continued to learn about many other pathways and structures involved in processing "emotion," but this marked the beginning of the unfolding of our understanding.

   To review, fibers leave the hippocampal formation and proceed through the fornix, and some of these fibers have been shown to terminate in the mammillary nuclei of the hypothalamus. From here, a new pathway, the mammillo-thalamic tract, ascends to the anterior group of thalamic nuclei. This group of nuclei projects to the cingulate gyrus (see also Figure 6.13).

   From the cingulate gyrus, there is an association bundle, the cingulum, that connects the cingulate gyrus (reciprocally) with the parahippocampal gyrus as part of the limbic lobe (refer to Figure 9.1A and Figure 9.1B). The parahippocampal gyrus projects to the hippocampal formation, which processes the information and sends it via the fornix to the mammillary nuclei of the hypothalamus (and the septal region). Hence the circuit is formed.

   We now have a broader view of the limbic system, and the precise functional role of the Papez circuit is not completely understood. Although there is a circuitry that forms a loop, the various structures have connections with other parts of the limbic system and other areas of the brain and thus can influence other neuronal functions (to be discussed in the "synthesis" at the end of this section).

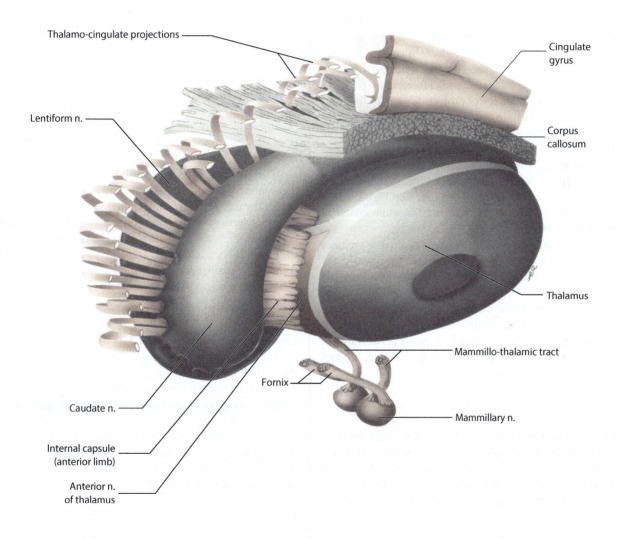

FIGURE 10.1A: Limbic Diencephalon 1—Thalamus: Anterior Nucleus

FIGURE 10.1B—LIMBIC DIENCEPHALON 2

THALAMUS: DORSOMEDIAL NUCLEUS

The thalamus of both sides is shown in this diagram, focusing on the medial nuclear mass of the thalamus, the dorsomedial nucleus, one of the most important of the association nuclei of the thalamus (see Figure 4.3). Sometimes this is called the mediodorsal nucleus.

Shown below is the amygdala with one of its pathways, the ventral amygdalofugal fibers, projecting to the dorsomedial nucleus (see Figure 9.6A and Figure 9.6B). This pathway brings "emotional" information to the thalamus. The dorsomedial nucleus collects information from a variety of sources, including other thalamic nuclei, as well as from various hypothalamic nuclei (see Figure 6.13).

The dorsomedial nucleus projects heavily to the frontal lobe, particularly to the cortical area which has been called the prefrontal cortex (see Figure 1.3 and Figure 6.13). The projection thus includes the emotional component of the experience. This pathway passes through the anterior limb of the internal capsule, between the head of the caudate nucleus and the lentiform nucleus (see Figure 2.10A and Figure 4.4). The fibers course in the white matter of the frontal lobes.

Our expanded view of the limbic system now includes its extension to this prefrontal cortex, specifically the orbital and medial portions of the frontal lobe; this has been termed the **limbic forebrain**. Widespread areas of the limbic system and association cortex of the frontal lobe, particularly the medial and orbital portions, are involved with human reactions to pain, particularly to chronic pain, as well as the human experiences of grief and reactions to the tragedies of life.

CLINICAL ASPECT: PSYCHOSURGERY

The projection of the dorsomedial nucleus to the prefrontal cortex has been implicated as the key pathway that is interrupted in a now banned surgical procedure. Before the era of medication for psychiatric disorders, when up to one half of state institutions were filled with patients with mental illness, a psychosurgical procedure was attempted to help alleviate the distressing symptoms of these diseases.

The procedure involved the introduction of a blunt instrument into the frontal lobes by inserting an instrument (bilaterally) through the orbital bone above the eye (which is a very thin plate of bone). This procedure interrupts the fibers projecting through the white matter, presumably including the projection from the dorsomedial nucleus. This operation became known as a *frontal lobotomy*.

Long-term studies of individuals who have had frontal lobotomies have shown profound personality changes in these individuals. These people become emotionally "flat" and lose some human quality in their interpersonal interactions that is difficult to define. In addition, such an individual may perform socially inappropriate acts that are not in keeping with the personality of that individual before the surgical procedure.

Once the long-term effects of this surgical procedure became clear, and because powerful and selective drugs became widely available for various psychiatric conditions, this operation was abandoned in the 1960s and is never performed nowadays.

This same procedure had also been recommended for the treatment of pain in patients with terminal cancer, as part of the palliative care of an individual. After the surgical procedure, the individual is said to still have the pain but no longer "suffers" from it (i.e., the psychic aspect of the pain has been removed). There may even be a reduced demand for pain medication such as morphine. Again, other approaches to pain management are now used.

Phineas Gage

This person has become a legendary figure in the annals of the history of the brain. In brief, Phineas was working on the construction of a railway in the 1800s when an untimely explosion drove a steel peg through his brain. The steel peg is said to have penetrated the orbit and the frontal lobes, much like the surgical procedure described earlier, and emerged through the skull. He did survive and lived on; the personality changes that were well documented subsequent to this accident concur with those described following a frontal lobotomy. The brain injury sustained by this man has been reconstructed and published (in the *New England Journal of Medicine*, Dec 2, 2004).

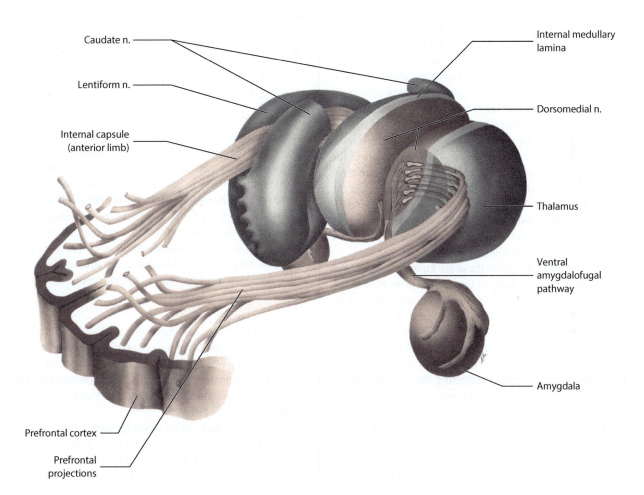

Caudate n.

Lentiform n.

Internal capsule
(anterior limb)

Internal medullary
lamina

Dorsomedial n.

Thalamus

Ventral
amygdalofugal
pathway

Amygdala

Prefrontal cortex

Prefrontal
projections

FIGURE 10.1B: Limbic Diencephalon 2—Thalamus: Dorsomedial Nucleus

FIGURE 10.2— LIMBIC DIENCEPHALON 3

NEURAL HYPOTHALAMUS

This diagram, which is the same as Figure 9.2, highlights the hypothalamus, one of the core structures of the limbic system, with the prominent mammillary nuclei as part of the hypothalamus. The 3rd ventricle is situated between the two diencephalic parts of the brain, and the hypothalamic tissue of both sides joins together at its inferior portion as the median eminence (see Figure 10.3, as well as Figure 1.5 and Figure 1.6).

The hypothalamus is usually divided into a medial group and a lateral group of nuclei (see Figure 10.3). Certain nuclei that control the anterior pituitary gland are located in the medial group. This occurs via the median eminence and the portal system of veins along the pituitary stalk; other nuclei in the supraoptic region (above the optic chiasm) connect directly with the posterior pituitary via the pituitary stalk (see Figure 1.5 and Figure 1.6).

Some of the major inputs to the hypothalamus come from limbic structures, including the amygdala (via the stria terminalis and the ventral pathway; see Figure 9.6A and Figure 9.6B) and the hippocampal formation (via the fornix; see Figure 9.3B). Stimulation of particular small areas of the hypothalamus can lead to a variety of behaviors (e.g., sham rage), similar to those occurring after stimulation of the amygdala.

Certain basic drives (as these are known in the field of psychology) such as hunger (feeding), thirst (drinking), sex (fornication), and body temperature are regulated by the hypothalamus through limbic connections. Many of the receptor mechanisms for these functions are now known to be located in highly specialized hypothalamic neurons. The hypothalamus responds in two ways—as a neuroendocrine structure controlling the activities of the pituitary gland and as a neural structure linked into the limbic system.

In its neural role, the hypothalamus acts as the "head ganglion" of the autonomic nervous system, by influencing both sympathetic and parasympathetic activities. The response to hunger or thirst or a cold environment usually leads to a complex series of motor activities that are almost automatic, as well as the autonomic adjustments and endocrine changes. In addition, in humans, there is an internal state of discomfort to being cold, or hungry, or thirsty that we call an emotional response. Additional connections are required for the behavioral (motor) activities, and the accompanying psychological reaction requires the forebrain, as well as the limbic cortical areas (further discussed in the "synthesis" at the end of this section).

The **mammillary nuclei** are of special importance as part of the limbic system. They receive direct input from the hippocampal formation via the fornix (see Figure 9.3B) and give rise to the mammillo-thalamic tract to the thalamic anterior group of nuclei as part of the Papez circuit (discussed with Figure 10.1A). In addition, there are fibers that connect directly to the limbic midbrain (shown in Figure 10.3).

Running through the lateral mass of the hypothalamus is a prominent fiber tract, the medial forebrain bundle, which interconnects the hypothalamus with two areas—the septal region of the forebrain and certain midbrain nuclei associated with the limbic system, the "limbic midbrain" (both discussed with Figure 10.3). Other fiber bundles connect the hypothalamus with the limbic midbrain. There are also some indirect connections to nuclei of the medulla via descending autonomic fibers (shown in Figure 10.3).

CLINICAL ASPECT

Korsakoff's syndrome is a disease associated primarily with chronic alcoholism. In addition to neurological problems (affecting vision and gait), there is a deficit in memory for events, recent and past. This leads to confabulation which is the primary behavioral symptom. Often the syndrome is called Korsakoff's psychosis (and also Wernicke-Korsakoff syndrome). The disease is associated with degeneration of the mammillary nuclei and the dorsomedial nucleus of the thalamus, and perhaps other areas (e.g., the hippocampus). It is considered a neurometabolic disorder due to thiamine (vitamin B_1) deficiency.

ADDITIONAL DETAIL: HABENULA (NOT ILLUSTRATED)

In many texts, the habenular region of the diencephalon is labeled, and discussed, in the context of the limbic system. The habenular nuclei comprise a group of small nuclei situated at the posterior end of thalamus, on its upper surface. The pineal gland is attached to the brain in this region (see Figure 1.9).

There is another circuit whereby septal influences are conveyed to the midbrain. The first part of the pathway is the stria medullaris (note the possible confusion of terminology), which connects the septal nuclei (region) with the habenular nuclei. The stria medullaris is found on the medial surface of the thalamus. From the habenular nuclei, the habenulo-interpeduncular tract descends to the midbrain reticular formation, mainly to the interpeduncular nucleus located between the cerebral peduncles (see midbrain cross-section; Appendix Figure A.4). (This tract is also called the fasciculus retroflexus.)

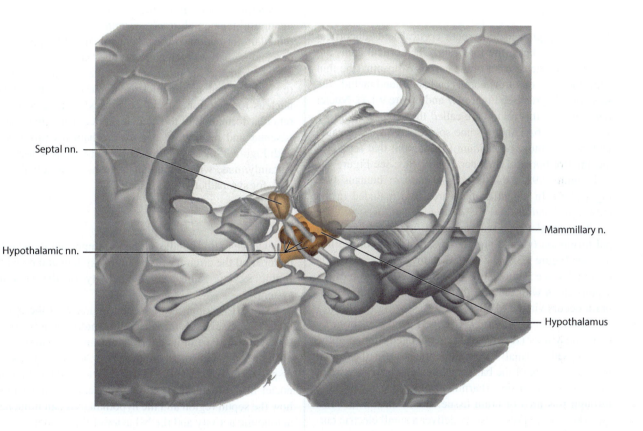

Septal nn.

Hypothalamic nn.

Mammillary n.

Hypothalamus

FIGURE 10.2: Limbic Diencephalon 3—Neural Hypothalamus

FIGURE 10.3—MEDIAL FOREBRAIN BUNDLE

SEPTAL REGION AND LIMBIC MIDBRAIN

This illustration provides detailed information about other important parts of the limbic system—the septal region and the limbic midbrain. The pathway that interconnects the hypothalamus and these areas is the **medial forebrain bundle (MFB)**.

SEPTAL REGION

The septal region includes both cortical and subcortical areas that belong to the forebrain. The cortical areas are found under the rostrum of the corpus callosum, (the thin "inferior" portion of the corpus callosum) and include the subcallosal gyrus (see Figure 1.7 and Figure 9.1A). Nuclei lying deep to this region are called the septal nuclei (see also Figure 9.6A) and in some species are in fact located within the septum pellucidum (the septum that separates the anterior horns of the lateral ventricles; see Figure 1.7 and Figure 2.9A); this is not the case in humans (see Figure 8.6). In this atlas, both areas are included in the term *septal region*.

The septal region receives input from the hippocampal formation (via the pre-commissural fibers of the fornix; see Figure 9.3B) and from the amygdala (via the stria terminalis; see Figure 9.6B). The major connection of the septal region with the hypothalamus and the limbic midbrain occurs via the MFB.

Several decades ago (1954), experiments were done by Olds and Milner (Reference: see Annotated Bibliography) in rats with a small electrode implanted in the septal region; pressing of the bar completed an electrical circuit that resulted in a tiny (harmless) electric current going through this area of brain tissue. It was shown that rats quickly learn to press a bar to deliver a small electric current to the septal region. In fact, the animals continue pressing the bar virtually non-stop. From this result, it has been inferred that the animals derive some type of a "pleasant sensation" from stimulation of this region, and it was named the "pleasure center." It has since been shown that there are other areas where similar behavior can be produced. However, this type of positive effect is not seen in all parts of the brain, and in fact in some areas an opposite (negative) reaction may be seen.

LIMBIC MIDBRAIN

Certain limbic pathways terminate within the reticular formation of the midbrain, including the periaqueductal gray, thus leading to the notion that these areas are to be incorporated in the structures that comprise the extended limbic system (discussed in the introduction to Section 4). This concept has led to the use of the term *limbic midbrain*.

The two major limbic pathways, the MFB and a descending tract from the mammillary nuclei (the mammillo-tegmental tract), terminate in the midbrain reticular formation. From here, there are apparently descending pathways that convey the "commands" to the parasympathetic and other nuclei of the pons and medulla (e.g., the dorsal motor nucleus of the vagus, the facial nucleus for emotional facial responses) and areas of the reticular formation of the medulla concerned with cardiovascular and respiratory control mechanisms (discussed with Figure 3.6A and Figure 3.6B). Other connections are certainly made with autonomic neurons in the spinal cord (i.e., for sympathetic-type responses).

MEDIAL FOREBRAIN BUNDLE

Knowledge of this bundle of fibers is important if one is to understand the circuitry of the limbic system and how the limbic system influences the activity of the nervous system.

The MFB connects the septal region with the hypothalamus and extends into the limbic midbrain; it is a two-way pathway. Part of its course is through the lateral part of the hypothalamus, where the fibers become somewhat dispersed (as illustrated). There are further connections to nuclei in the medulla. It is relatively easy to understand how the septal region and the hypothalamus can influence autonomic activity and the behavior of the animal.

ADDITIONAL DETAIL

There are other pathways from the hypothalamus to the limbic midbrain, such as the dorsal longitudinal bundle (as shown in the illustration).

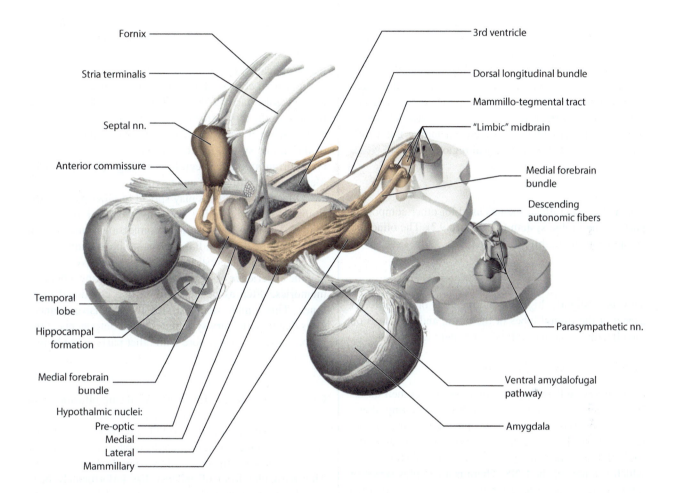

Fornix

Stria terminalis

Septal nn.

Anterior commissure

3rd ventricle

Dorsal longitudinal bundle

Mammillo-tegmental tract

"Limbic" midbrain

Medial forebrain bundle

Descending autonomic fibers

Parasympathetic nn.

Temporal lobe

Hippocampal formation

Medial forebrain bundle

Hypothalmic nuclei:
Pre-optic
Medial
Lateral
Mammillary

Ventral amydalofugal pathway

Amygdala

FIGURE 10.3: Medial Forebrain Bundle—Septal Region and Limbic Midbrain

FIGURE 10.4—OLFACTORY SYSTEM

SENSE OF SMELL (ILLUSTRATION AND PHOTOGRAPH)

The olfactory system, our sense of smell, is a sensory system that inputs directly into the limbic system and does not have a thalamic nucleus (see Figure 4.3).

The olfactory system is a phylogenetically older sensory system. Its size depends somewhat on the species and is larger in animals that have a highly developed sense of smell; this is not the case in humans, in whom the olfactory system is small. Its component parts are the olfactory nerve, bulb, and tract, as well as various areas where the primary olfactory fibers terminate, including the amygdala and the cortex over the uncal region.

ILLUSTRATION

This is the same illustration as used for other component parts of the limbic system (see Figure 9.2). The olfactory components are highlighted.

PHOTOGRAPH

This is a slightly higher magnification of the inferior aspect of the frontal lobes of the brain, and it focuses on the olfactory system (see Figure 1.5 and Figure 1.6).

OLFACTORY NERVE, BULB, AND TRACT

The sensory cells in the nasal mucosa project their axons into the central nervous system (CNS). These tiny fibers, which constitute the actual peripheral olfactory nerve (cranial nerve I), pierce the bony (cribriform) plate in the roof of the nose and terminate in the olfactory bulb, which is a part of the CNS. There is a complex series of interactions in the olfactory bulb, and one cell type then projects its axon into the olfactory tract, a CNS pathway.

The olfactory tract runs posteriorly along the inferior surface of the frontal lobe (see Figure 1.5 and Figure 1.6) and divides into lateral and medial tracts, called stria. At this dividing point, there are a number of small holes for the entry of several blood vessels to the interior of the brain, the striate arteries (see Figure 8.5 and Figure 10.5B); this triangular area is known as the anterior perforated space or substance.

It is best to remember only the lateral tract as the principal tract of the olfactory system. This tract is said to have cortical tissue along its course for the termination of some olfactory fibers. The lateral tract ends in the cortex of the uncal area (see Figure 1.5 and Figure 1.6), with some of the fibers terminating in an adjacent part of the amygdaloid nucleus (see Figure 9.6A and Figure 9.6B). The olfactory system terminates directly in primary olfactory areas of the cortex without a thalamic relay.

OLFACTORY CONNECTIONS

The connections of the olfactory system involve the limbic cortex. These are called secondary olfactory areas, and they include the cortex in the anterior portion of the parahippocampal gyrus, an area that has been referred to as entorhinal cortex. (The term *rhinencephalon* refers to the olfactory parts of the CNS, the "smell brain.") This input of olfactory information into the limbic system makes sense if one remembers that one of the functions of the limbic system is procreation of the species. Smell is important in many species for mating behavior and for identification of the nest and territory.

Olfactory influences may spread to other parts of the limbic system, including the amygdala and the septal region. Through these various connections, information may reach the dorsomedial nucleus of the thalamus.

Smell is an interesting sensory system. We have all had the experience of a particular smell evoking a flood of memories, often associated with strong emotional overtones. This simply demonstrates the extensive connections that the olfactory system has with components of the limbic system and therefore with other parts of the brain.

CLINICAL ASPECT

One form of epilepsy often has a significant olfactory aura (which precedes the seizure itself). The odour is usually atypical in nature such as the smell of burnt rubber. In such cases, the "trigger" area is often the orbitofrontal cortex, or an areas known as the preuncal gyrus of the temporal lobe. This particular form of epilepsy has unfortunately been called "uncinate fits." The name is derived from a significant association bundle, the uncinate bundle (an association bundle), which interconnects this part of the frontal lobe and the anterior parts of the temporal lobe where olfactory connections are located (discussed with Figure 9.6B).

ADDITIONAL DETAIL: DIAGONAL BAND

This obscure fiber bundle and nuclei associated with it (see Figure 10.5A) are additional olfactory connections, some of which interconnect the amygdala with the septal region (see also Figure 10.5B).

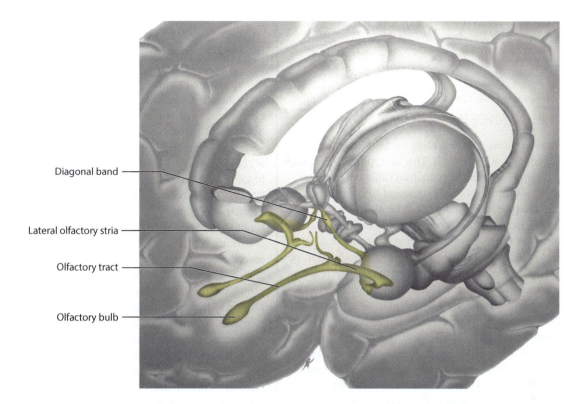

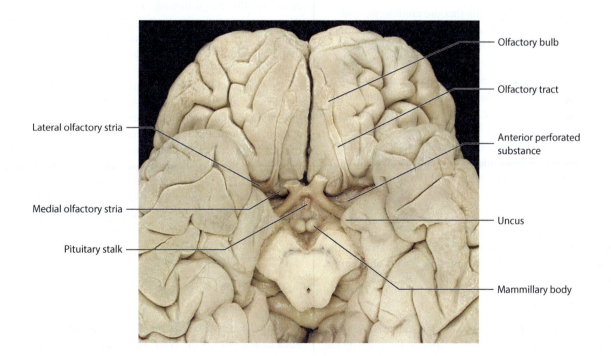

FIGURE 10.4: Olfactory System—Sense of Smell (illustration and photograph)

FIGURE 10.5A—BASAL FOREBRAIN 1

BASAL NUCLEUS (WITH T1 MAGNETIC RESONANCE IMAGING SCANS)

The basal forebrain is shown, using the same diagram of the limbic system (see Figure 9.2). This area, previously called the *substantia innominata*, contains a variety of neurons.

The **basal forebrain** area is located below the anterior commissure (which is used as a landmark; see also Figure 10.5B) and lateral to the hypothalamus. The anterior commissure is an older commissure than the corpus callosum (see Figure 2.2A and Figure 2.2B). It is considered to be the commissure of the limbic system that connects structures in the anterior temporal area. On the gross brain, this region can be found by viewing the inferior surface of the brain where the olfactory tract ends and divides into medial and lateral stria (see Figure 1.5 and Figure 1.6, as well as Figure 10.4). As discussed with Figure 10.4, this particular spot is the location where the striate arteries, penetrate the brain substance (see Figure 8.6), and it is called the anterior perforated space or substance. The basal forebrain region is found "above" this area (shown in Figure 10.5B).

The basal forebrain contains a group of diverse structures:

- Clusters of large cells that are cholinergic and which have been collectively called the *basal nucleus* (of Meynert)—discussed with this illustration.
- The ventral portions of the putamen and globus pallidus, namely, the ventral striatum, the *nucleus accumbens*, and the ventral pallidum—to be discussed with Figure 10.5B and Figure 10.7.
- Groups of cells that are continuous with the amygdala, now called the extended amygdala— to be discussed as Additional Detail.

BASAL NUCLEUS

These rather large neurons are found in clusters throughout this region. These cells project to widespread areas of the prefrontal cortex and provide that area with cholinergic innervation.

T1 MAGNETIC RESONANCE IMAGING SCANS

The small locator illustration shows the somewhat oblique coronal section through the frontal lobes (as indicated by the size and shape of the lateral ventricles seen in the T1-weighted magnetic resonance image scans underneath the locator illustration). The arching white bundle is the **anterior commissure**, connecting the anterior temporal areas, known to contain limbic structures. The basal nucleus of the forebrain is the area located below the anterior commissure; this area is labeled in the adjacent higher-magnification view.

CLINICAL ASPECT

In 1982, Bartus et al. (Reference: see Annotated Bibliography) published a report showing an association in animals and humans between cholinergic dysfunction and memory deficits, leading to "The Cholinergic Hypothesis of Geriatric Memory Dysfunction". Subsequent reports (e.g., Francis et al, 1999—Reference: see Annotated Bibliography) indicated that this was accompanied by a loss of these cholinergic cells in the basal forebrain. Many investigators thought that the "cause" of Alzheimer's disease had been uncovered (i.e., cellular degeneration of a unique group of cells and a neurotransmitter deficit). (The model for this way of thinking is Parkinson's disease.) This was followed immediately by several therapeutic trials using medication to boost the acetylcholine levels of the brain by selectively inhibiting pseudocholinesterase enzymes in the brain. These medications may lead to some short term improvement of mental functions but do not influence or slow down the underlying neurodegenerative process of Alzheimer's disease.

It is currently thought that cortical degeneration is the primary event in Alzheimer's disease, often starting in the parietal areas of the brain. We now know that several other neurotransmitters are depleted in the cortex in Alzheimer's disease. This information would lead us to postulate that the loss of the target neurons in the prefrontal cortex, the site of termination for the cholinergic neurons, would be followed, or accompanied, by the degeneration of the cholinergic cells of the basal forebrain. In addition, there is the hippocampal degeneration that goes along with the memory loss (discussed with Figure 9.4 and Figure 9.5A).

Notwithstanding this current state of our knowledge, therapeutic intervention to boost the cholinergic levels of the brain is currently considered a valid therapeutic approach, particularly in the early stages of this tragic human disease. New drugs that maintain or boost the level of acetylcholine in the brain are currently undergoing evaluation. The reports have shown in some patients, a temporary improvement, or at least a stabilization of the decline, in both memory and cognitive functions for a period of weeks or sometimes months.

ADDITIONAL DETAIL

THE EXTENDED AMYGDALA

A group of cells extends medially from the amygdaloid nucleus and follows the ventral pathway (the ventral amygdalofugal pathway; see Figure 9.6B and Figure 10.1B) through this basal forebrain region. These neurons receive a variety of inputs from the limbic cortical areas and from other parts of the amygdala. The output projects to the hypothalamus and to autonomic-related areas of the brainstem, thereby influencing neuroendocrine and autonomic, and perhaps somatomotor activities.

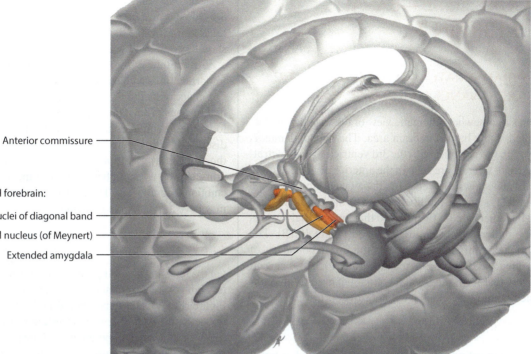

Anterior commissure

Basal forebrain:

Nuclei of diagonal band

Basal nucleus (of Meynert)

Extended amygdala

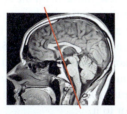

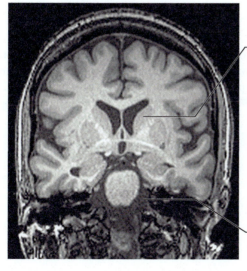

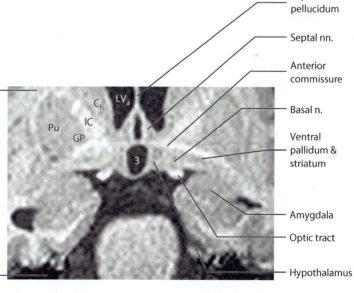

Septum
pellucidum

Septal nn.

Anterior
commissure

Basal n.

Ventral
pallidum &
striatum

Amygdala

Optic tract

Hypothalamus

C$_h$ = Caudate (head)
IC = Internal capsule
Pu = Putamen
GP = Globus pallidus

LV$_a$ = Lateral ventricle
      (anterior horn)
3 = 3rd ventricle

FIGURE 10.5A: Basal Forebrain 1—Basal Nucleus (with T1 magnetic resonance imaging scans)

FIGURE 10.5B—BASAL FOREBRAIN 2

VENTRAL STRIATUM (WITH T1 MAGNETIC RESONANCE IMAGING SCANS)

This is a somewhat schematic view of the various "nuclei" located in the basal forebrain area. The hypothalamus is shown in the midline, with the 3rd ventricle. The penetrating striate arteries are seen in the anterior perforated area. This view shows the ventral pathway emerging from the amygdala; some of the fibers going to the hypothalamus, and the others are on their way to the dorsomedial nucleus of the thalamus (see Figure 9.6B and Figure 10.1B). The anterior commissure demarcates the upper boundary of this area (see Figure 9.1A and Figure 10.5A). The cell clusters that form the basal (cholinergic) nucleus are contained within this area but are not fully portrayed.

VENTRAL STRIATUM AND VENTRAL PALLIDUM

The lowermost portions of the putamen and globus pallidus are found in the basal forebrain area; here they are referred to as the ventral striatum and ventral pallidum (see Figure 2.5B and Figure 2.7).

The ventral part of the striatum—the **nucleus accumbens**—receives input from limbic cortical areas, as well as a dopaminergic pathway from a group of dopamine-containing cells in the midbrain (the ventral tegmental area; see Figure 10.7 and Figure 10.8). The information is then relayed to the ventral pallidum (shown on the right side of the illustration; both parts of the globus pallidus are seen on the left side of the diagram). This area has a significant projection to the dorsomedial nucleus of the thalamus (and hence to prefrontal cortex).

The overall organization is therefore quite similar to that of the dorsal parts of the basal ganglia, although the sites of relay and termination are different. Just as the amygdala is now considered a limbic nucleus, many investigators now argue that the ventral striatum and pallidum should be included with the limbic system.

In summary, the region of the basal forebrain has important links with other parts of the limbic system. There is a major output to the prefrontal cortex, via the dorsomedial nucleus of the thalamus, which is considered by some to be the forebrain component of the limbic system (to be discussed with Figure 10.8). The basal forebrain is thus thought to have a strong influence on "drives" and emotions, as well as on higher cognitive functions that have an emotional component. The cholinergic neurons in this area may have a critical role in memory.

T1 MAGNETIC RESONANCE IMAGING SCANS

On the left side, the oblique coronal cut (shown in the locator image) is taken somewhat anterior to the one in Figure 10.5A. The axial (horizontal) cut on the locator image on the right side is more inferior (lower) than those used for the caudate-putamen and internal capsule (see Figure 2.10A and Figure 2.10B). Both are T1-weighted images.

On the coronal view, more so on the right side of the image, one can see the continuity between the head of the caudate nucleus and the putamen; this is the location of the nucleus accumbens (see Figure 2.5B and Figure 2.7). On the axial (horizontal) view, the cut is at the level of the anterior commissure as it crosses the midline (see the images with Figure 10.5A); the nucleus accumbens is immediately anterior to it.

ADDITIONAL DETAIL

The diagonal band (of Broca) connects the medial septal nucleus with the basal nucleus. Some of the nuclei located along its path may be cholinergic (see also Figure 10.4).

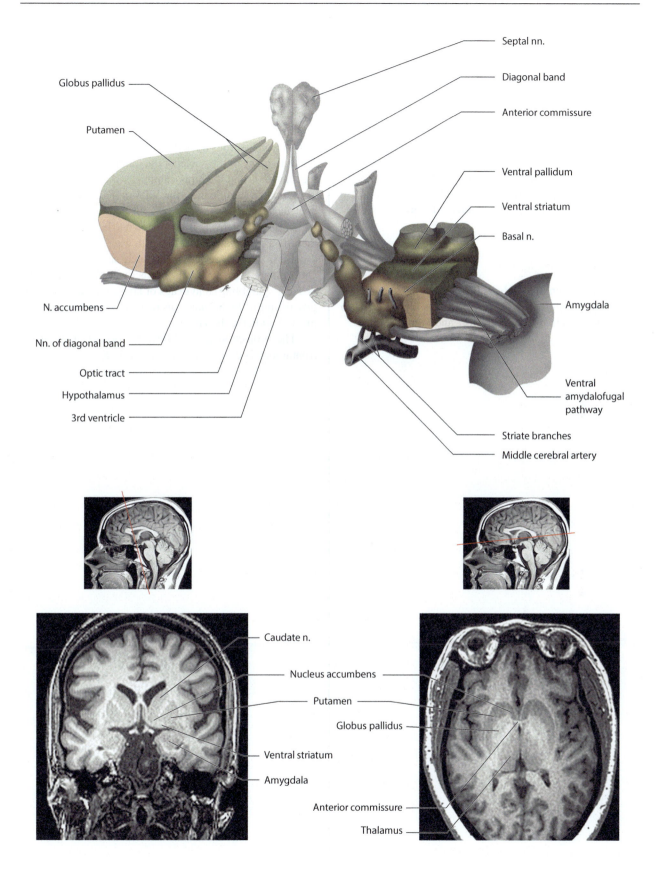

FIGURE 10.5B: Basal Forebrain 2—Ventral Striatum (with T1 magnetic resonance imaging scans)

FIGURE 10.6—VENTRAL TEGMENTAL AREA (VTA)

LIMBIC MIDBRAIN (PHOTOGRAPHIC VIEWS WITH OVERLAY)

There is a second set of dopamine neurons in the midbrain area, located ventrally, called the **ventral tegmental area**, simply called by its abbreviation, **VTA**. These dopaminergic neurons are connected with the limbic system.

UPPER ILLUSTRATION

The VTA is seen in a mid-sagittal view in the upper illustration (see Figure 1.7). This is the view that is used to show its projections (in Figure 10.7 and Figure 10.8).

LOWER ILLUSTRATION

A more anatomical view of its location is shown in the lower illustration, at the level of the upper midbrain (see Appendix Figure A.3). Note that at this level the substantia nigra is also present (review its connections; see Figure 5.14). The VTA neurons have a very different projection from neurons from the substantia nigra. The projection is to many of the nuclei of the limbic system, including the amygdala and the hippocampus.

Of much significance, the VTA projects heavily to the nucleus accumbens (discussed with Figure 10.7), and this dopaminergic pathway is now known to be activated when there is a positive association with a pleasurable sensory events in our everyday life, such as food and sex. Unfortunately, the system is also activated by illicit activities (e.g., gambling) and drugs of abuse that are known to give people a "high," hence its involvement in the development of addictive behavior.

The projection to the "limbic" cortical areas of the frontal lobe is described with Figure 10.8.

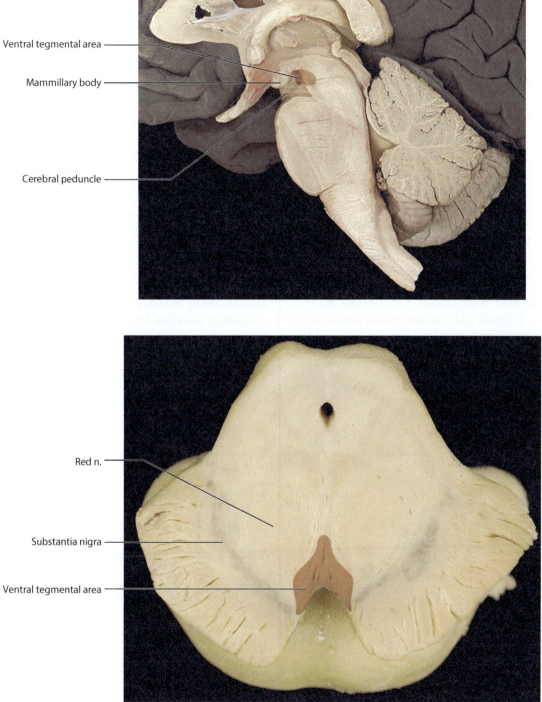

Ventral tegmental area

Mammillary body

Cerebral peduncle

Red n.

Substantia nigra

Ventral tegmental area

FIGURE 10.6: Ventral Tegmental Area (VTA)—Limbic Midbrain (photographic views with overlay)

FIGURE 10.7—VTA PROJECTION 1

NUCLEUS ACCUMBENS, AMYGDALA, HIPPOCAMPUS (SEMISCHEMATIC)

This illustration was created by using the drawing of the basal ganglia in Figure 2.7 (the middle one), adding the hippocampus and then positioning the structures within the brain, using the mid-sagittal perspective.

There is another nucleus now recognized as integral to the reward system of the brain and the development of addictions—the **nucleus accumbens**. This nucleus is located along with but inferior (ventral) to the caudate nucleus (see Figure 2.5B and Figure 2.7).

The nucleus accumbens belongs with the basal ganglia according to definition, but its connections have little to do with the caudate and the putamen, the neostriatum. From the neuroanatomical perspective (see Figure 2.5B and Figure 2.7), the caudate and putamen form the dorsal striatum. The nucleus accumbens, because it is located "below" the head of the caudate, then becomes the **ventral striatum** (which adds another layer of complexity to an already confusing terminology (see also Figure 10.5B).

This group of cells contains neurons that are part of the basal ganglia and other, possibly limbic neurons. Its functional contribution is still being clarified, although it now seems certain that this neural area becomes activated in situations that involve reward, by integrating certain cognitive aspects of the situation with the emotional component.

This nucleus receives a strong dopaminergic projection from the ventral tegmental area (VTA), which also projects to the hippocampus and the amygdala as shown.

The nucleus accumbens projects to the thalamus to the two nuclei that are associated with the limbic system, the anterior and dorsomedial nuclei (see Figure 6.13, as well as Figure 10.1A and Figure 10.1B). These nuclei project to the limbic cingulate and prefrontal cortex, further described and discussed with Figure 10.8. As such, they are part of the so-called "extended limbic lobe"—the limbic forebrain (discussed in the introduction to the Limbic System and with the Limbic system "synthesis" at the end of this section).

CLINICAL ASPECT

There is strong evidence that the nucleus accumbens is involved in addiction behavior in animals and likely in humans (discussed also with Figure 10.6).

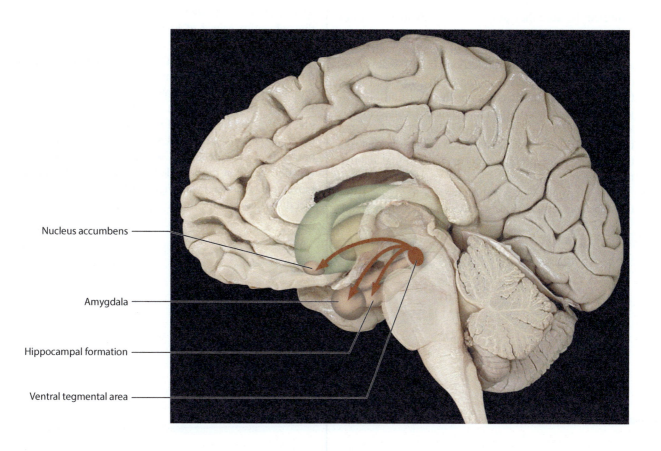

Nucleus accumbens

Amygdala

Hippocampal formation

Ventral tegmental area

FIGURE 10.7: VTA Projection 1—Nucleus Accumbens, Amygdala, Hippocampus (semischematic)

FIGURE 10.8—LIMBIC PROJECTION 2

LIMBIC PREFRONTAL CORTEX (PHOTOGRAPHIC VIEW WITH OVERLAY)

The dopaminergic ventral tegmental area (VTA) neurons project to select areas of the prefrontal cortex, namely, the medial and orbital portions (see Figure 6.13). This selective projection is quite different from the noradrenergic and serotonergic projections from other nuclei of the brainstem because they project extensively to all areas of the cortex.

The medial and orbital areas of the prefrontal cortex also receive the thalamic projections from the dorsomedial nucleus, (discussed with Figure 10.7 and see Figure 6.13). Because the nucleus accumbens is part of this projection and because the VTA-accumbens connection is known to be involved in addictive behavior (as discussed in Figure 10.7), it seems not unreasonable to ask what role these areas of the prefrontal cortex have in maladaptive human behaviors.

The prefrontal cortex, returning now to Figure 1.1 of this atlas, has several aspects: the dorsolateral, the orbital, and the medial portions (see Figure 1.3, Figure 1.4, Figure 1.5, Figure 1.6, and Figure 1.7). Many studies now implicate the dorsolateral prefrontal cortex with "decision making" and use the model of the CEO (the chief executive officer), the so-called rational dispassionate chess player (discussed with Figure 6.13).

Assuming that this is correct, what then are the roles of the other areas of the prefrontal cortex? It is possible to trace many of the limbic connections, as part of the limbic "loop," to the orbital and medial regions of the frontal lobe.

The following hypothesis is proposed: Decision making is rarely (if ever) without an "emotional" component, a "gut" feeling, an intuition. This limbic input may provide this subjective aspect of how we decide on a course of action.

One can speculate that it is the medial prefrontal area that acts as the mediator between the limbic input to the orbital areas of the frontal lobe (possibly via the uncinate fasciculus, see Figure 9.6B) and the executive functions (the CEO functions) located in the dorsolateral aspect of the frontal lobe (see Figure 4.5). One can recall the modus operandi of the cerebellum, where there is a comparator function between the intended movement and the actual movement (discussed with Figure 5.17). This notion is further discussed with the limbic "synthesis".

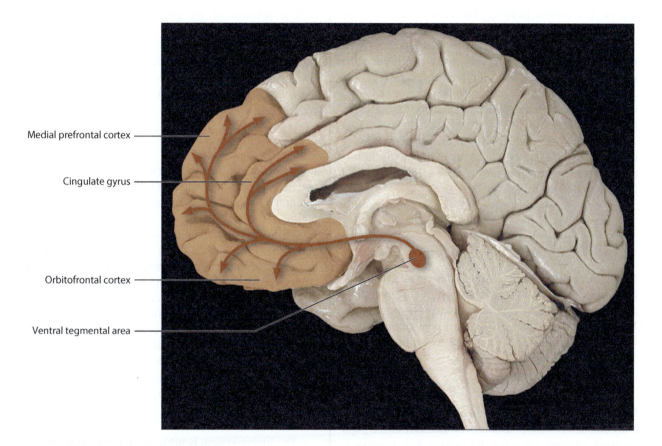

Medial prefrontal cortex

Cingulate gyrus

Orbitofrontal cortex

Ventral tegmental area

FIGURE 10.8: VTA Projection 2—Limbic Prefrontal Cortex (photographic view with overlay)

LIMBIC SYSTEM: SYNTHESIS

After studying the structures and connections of the limbic system in some detail, a synthesis of the anatomical information with the notion of an "emotional" part of the brain seems appropriate. It is not easy to understand how the limbic system is responsible for the reactions required by the definition of "emotion" proposed in the introduction to Section 4.

The key structures of the limbic system are the limbic lobe (the cortical regions, including the hippocampal formation), the amygdala, the hypothalamus, and the septal region. The limbic pathways interconnect these limbic areas (e.g., the Papez circuit). In many ways, it seems that the limbic structures communicate with each other. What is not clear is how activity in these structures influences the rest of the brain. How does the limbic system influence changes in the physiological systems (endocrine and autonomic); motor activity (behavior); and the mental state (psychological reactions)?

The following discussion is presented as a way of understanding the outcome or output of limbic function—the categories of responses are the same as those discussed in the Introduction to Section 4.

PHYSIOLOGICAL RESPONSES

ENDOCRINE AND "HOMEOSTATIC" RESPONSES

Endocrine and hormonal changes, as regulated by the hypothalamus, are part of the physiological responses to emotional states, both acute and chronic. The work of Dr. Hans Selye, for example, has shown how chronic stress has a detrimental effect on our body and mind (and perhaps also on our brain).

AUTONOMIC RESPONSES

Many parasympathetic and sympathetic responses accompany emotional states, including the diameter of the pupil (in states of fear), salivation, respiration, blood pressure, pulse, and various gastrointestinal functions. These responses are controlled in part by the hypothalamus and by the limbic connections in the midbrain and medulla.

BEHAVIORAL RESPONSES

The physiological adjustments often involve complex motor actions—consider, for example, the motor activities associated with thirst, temperature regulation, and satisfying other basic drives. The amygdala and hypothalamus are likely involved in the motor patterns associated with these basic drives.

Limbic activity involves areas of the midbrain reticular formation and other brainstem nuclei in specific ways. The best examples are perhaps the facial expressions associated with emotions, the responses to pain that are generated in part in the brainstem, and the basic "fight or flight" response to emergency situations. All these activate a considerable number of motor circuits. The ventral parts of the basal ganglia and various cortical areas are likely the areas of the central nervous system involved with the motor activities associated with emotional reactions.

PSYCHOLOGICAL REACTIONS

Neocortical areas that are involved in limbic function include portions of the prefrontal cortex (orbital and medial), the cingulate gyrus, and the parahippocampal gyrus. Activities in these limbic cortices (and the associated thalamic nuclei) are clearly candidates for the psychological (mental) reactions of emotion. These reactions influence behavior and may or may not enter consciousness.

SUMMARY

In summary, the limbic system has many connections outside itself through which it influences the endocrine, autonomic, motor, and psychological functions of the brain.

The older cortical regions of the hippocampal formation seem to have an additional function related to the formation of new episodic memories, specifically related to events and factual information. Why this is so and how this evolved are matters of speculation.

As discussed with Figure 10.8, the limbic input to the medial and orbital prefrontal cortex may have an influence on decision-making. Should this or a similar hypothesis be further developed, the implication is that all (almost all?) decisions—whether made by women or by men—have an emotional aspect or component. Perhaps, having acknowledged this, we ought to learn to pay more attention to our limbic "voice."

The limbic system is intricate and intriguing, providing a window into human behavior beyond our sensory and motor activities. It is not always clear what each part contributes to the overall functional system. In addition, some of the pathways are obscure and perhaps confusing. Nevertheless, they are part of the neuroanatomical framework for a discussion of the contribution of the limbic system to the function of the organism. It is also interesting to speculate that the elaboration of limbic functions is closely associated with the development of self-awareness, consciousness of the self.

On a final note, one can only wish that the basic activities of the limbic system that are involved in preservation of the self and species can be controlled and tamed by higher-order cortical influences, leading humankind to a more human and hopefully a more humane future.

APPENDIX: NEUROLOGICAL NEUROANATOMY

The Appendix presents the histological details of the human brainstem, and the spinal cord, with the various tracts and nuclei corresponding to the material in other sections of this atlas.

BRAINSTEM ORGANIZATION

One can approach the description of the cross-sections (axial) through the brainstem in a systematic manner. This is sometimes referred to as the floor plan of the brainstem (to be described in more detail with Figure A.2):

- **Ventral** or basal portion: The most anterior portion of each level of the brainstem contains some representation of the descending cortical fibers, specifically the cortico-bulbar, corticopontine, and cortico-spinal pathways (see Figure 5.9 and Figure 5.10).
- Central portion: The central portion of the brainstem is called the **tegmentum**. This area contains all the ascending tracts and the remaining descending tracts, as well as almost all the cranial nerve nuclei and other special nuclei including the red nucleus and the inferior olive. The reticular formation occupies the core region of the tegmentum (see Figure 3.6A and Figure 3.6B).
- **Ventricle**: The ventricular system with cerebrospinal fluid is found throughout the brainstem (see Figure 3.1, Figure 3.2, and Figure 3.3). The brainstem level can often be identified according to the ventricular system that passes through this region, namely, the aqueduct in the midbrain region and the 4th ventricle lower down.
- **Dorsal or roof**: The 4th ventricle separates the pons and medulla from the cerebellum, which is situated behind (above) the roof of the ventricle. In the midbrain region, the colliculi are located behind (dorsal to) the aqueduct of the midbrain (discussed later; see also Figure 1.9 and Figure 3.3).

BLOOD SUPPLY

The vertebro-basilar system supplies the brainstem in the following pattern (see Figure 8.1 and Figure 8.5). Penetrating branches from the **basilar** artery supply nuclei and tracts that are adjacent to the midline; these are called the **paramedian** branches. The lateral territory of the brainstem, both tracts and nuclei, is supplied by one of the cerebellar **circumferential** arteries—posterior inferior, anterior inferior, and superior (see Figure 8.1).

HISTOLOGICAL STAINING

Various histological stains are available that can show features of different normal and abnormal components of tissue. For the nervous system, there are many older stains and an ever-increasing number of newer stains using specific antibody markers, often tagged with fluorescent dyes. In general, the standard stains include those for:

- Cellular components, the cell bodies of neurons and glia (and cells lining blood vessels); these are general stains such as hematoxylin and eosin (H & E).
- The neurons, particularly the dendritic tree (including dendritic spines) and often the axons; the best known of these is the Golgi stain.
- Axonal fibers, either normal or degenerating.
- Glial elements (normal or reactive astrocytes).
- Myelin (normal or degenerating myelin).

The stain used for the histological sections in this atlas combines a cellular stain with a myelin stain; the combined stain is officially known as the Kluver-Barrera stain. Because the myelinated fibers are often compacted in certain areas, these tend to stand out clearly. The cellular neuronal areas are usually lightly stained because the cells are more dispersed, but the cell bodies can be visualized at higher magnification.

PLAN OF STUDY

- A schematic of each section is presented in the upper figure, and the corresponding histological section of the human brainstem is presented in the lower figure.
- The various nuclei of the brainstem have been colored differently (as per the color coding in the User's Guide), consistent with the color used in the tracts in this atlas (see Section 2). This visual cataloguing is maintained uniformly throughout the brainstem cross-sections.

The brainstem is described starting from the midbrain downward through to the medulla for two reasons:

1. This order follows the numbering of the cranial nerves, from midbrain downward.
2. This is the sequence that has been described for the fibers descending from the cortex.

Others may prefer to start the description of the cross-sections from the medulla upward.

Note to the Learner: The presentation of this material is the same on the accompanying Web site, with the added feature that the structure to be identified in the section is highlighted—using roll-over mouse animation—in both the schematic and histological section, at the same time. It is suggested that the student review these cross-sections by using the text together with the Web site. The histological images of the brainstem will be more understandable after this combined approach.

FIGURE A.1—BRAINSTEM HISTOLOGY A

VENTRAL VIEW (SCHEMATIC)

Study of the brainstem is continued by examining its histological neuroanatomy through a series of cross-sections. Because it is well beyond the scope of the non-specialist to know all the details, certain salient points have been selected, namely:

- The cranial nerve nuclei.
- The ascending and descending tracts.
- Certain brainstem nuclei that belong to the reticular formation.
- Other select special nuclei.

Because the focus is on the cranial nerves, only a limited number of cross-sections are studied. This diagram shows the ventral view of the brainstem, with the attached cranial nerves; the sensory cranial nerve nuclei are shown on the left side (see Figure 3.4), and the motor nuclei are shown on the right side (see Figure 3.5). The lines indicate the sections that are depicted in the series to follow.

There are *eight* cross-sections that will be studied through the three parts of the brainstem; a photographic view of each part of the brainstem was shown in Section 2 (see Chapter 4).

- Two through the midbrain (see photographic view, Figure 4.2C).
  - o Cranial nerve (CN) III—upper midbrain (superior colliculus level).
  - o CN IV—lower midbrain (inferior colliculus level).
- Three through the pons (see photographic view, Figure 4.2B).
  - o Uppermost pons (level for a special nucleus at this level).

- o CN V—mid-pons (through the principal sensory and motor nuclei).
- o CN VI, VII, and part of VIII—lowermost pons.

- Three through the medulla (see photographic view, Figure 4.2A).
  - o CN VIII (some parts)—uppermost medulla.
  - o CN IX, X, and XII—mid-medullary level.
  - o Lowermost medulla, with some special nuclei.

Two important points should be noted for this section:

1. Small images of the brainstem, both the ventral view (this illustration) and the sagittal view (see Appendix Figure A.2) are shown with each cross-sectional level with the plane of the cross-section indicated in both.
2. Three cross-sectional levels of the brainstem are the ones shown alongside the pathways in Section 2 (Functional Systems) of the atlas— upper midbrain, mid-pons, and mid-medulla (see Figure 4.6).

CLINICAL ASPECT

As has been indicated, the attachment of the cranial nerves to the brainstem is one of the keys to being able to understand this part of the brain (see Figure 1.8 and Figure 3.1). Wherever one sees a cranial nerve attached to the brainstem, one knows that its nucleus (or one of its nuclei) will be located at that level (see Figure 3.4 and Figure 3.5). Therefore, if one visually recalls or "memorizes" the attachment of the cranial nerves, one has a key to understanding the brainstem. In the clinical setting, knowledge of which cranial nerve or nerves is or are involved is usually the main clue to localize a lesion in the brainstem.

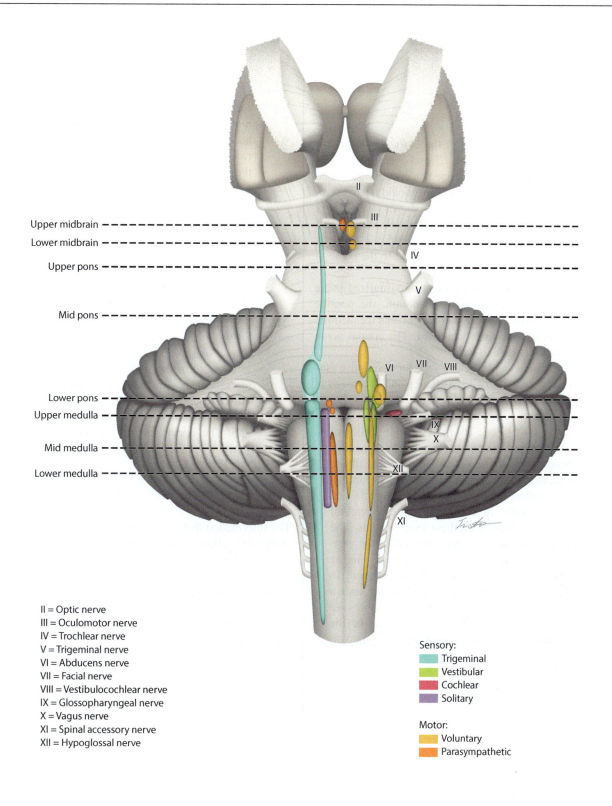

Upper midbrain
Lower midbrain
Upper pons
Mid pons
Lower pons
Upper medulla
Mid medulla
Lower medulla

II = Optic nerve
III = Oculomotor nerve
IV = Trochlear nerve
V = Trigeminal nerve
VI = Abducens nerve
VII = Facial nerve
VIII = Vestibulocochlear nerve
IX = Glossopharyngeal nerve
X = Vagus nerve
XI = Spinal accessory nerve
XII = Hypoglossal nerve

Sensory:
Trigeminal
Vestibular
Cochlear
Solitary

Motor:
Voluntary
Parasympathetic

FIGURE A.1: Brainstem Histology A—Ventral View (schematic)

FIGURE A.2—BRAINSTEM HISTOLOGY B

SAGITTAL VIEW (SCHEMATIC)

This is a schematic drawing of the brainstem seen in a mid-sagittal view (see Figure 1.7 and Figure 3.2). This view is presented because it is one that is commonly used to portray the brainstem. The student should try to correlate this view with the ventral view shown in Appendix Figure A.1, the previous diagram. This schematic is also shown in each of the cross-section diagrams, with the exact level indicated, to orient the student to the plane of section through the brainstem.

The shape of the brainstem can be visualized using this sagittal view, including the midbrain, the pontine nuclei that form the "bulging" of the pons, and the medulla which tapers to form the spinal cord. The cranial nerve attachments are shown as well.

The brainstem can be be analyzed for descriptive purposes as having what is called a "floor plan." It consists of four parts:

- **Ventral or basal**: The most anterior portion of each area of the brainstem contains some representation of the descending cortical fibers, specifically the cortico-bulbar, cortico-pontine, and cortico-spinal pathways (see Figure 5.9 and Figure 5.10). In the midbrain, the cerebral peduncles include all these axon systems. The cortico-bulbar fibers are given off to the various brainstem and cranial nerve nuclei. In the pons, the cortico-pontine fibers terminate in the pontine nuclei, which form the bulge known as the pons proper; the cortico-spinal fibers are dispersed among the pontine nuclei. In the medulla, the cortico-spinal fibers regroup to form the pyramids. The medulla ends at the point where these fibers decussate (see Figure 1.8, Figure 5.9, and Figure 6.12).

- Central: The central portion of the brainstem is called the **tegmentum**. This area contains cranial nerve nuclei CN III to CN XII and other nuclei including the red nucleus and the inferior olive, as well as all the ascending tracts and the remaining descending tracts. The reticular formation occupies the core region of the tegmentum (see Figure 3.6A and Figure 3.6B).

- Cerebrospinal fluid: The **ventricular system** is found throughout the brainstem (see Figure 3.1, Figure 3.2, Figure 3.3, and Figure 7.8). The brainstem level can often be identified according to the ventricular system that passes through this region, namely, the aqueduct in the midbrain region and the 4th ventricle lower down.

- Dorsal or roof: The four colliculi, which collectively form the **tectum**, are located behind (dorsal to) the aqueduct of the midbrain. The 4th ventricle separates the pons and medulla from the cerebellum (see Figure 1.9 and Figure 3.2). The upper part of the roof of the 4th ventricle is called the superior medullary velum (see Figure 3.3).

CLINICAL ASPECT

The information that is presented in this series should be sufficient to allow a student to recognize the clinical signs that would accompany a lesion at a particular level, particularly as it involves the cranial nerves. Such lesions would also interrupt the ascending or descending tracts, or both, and this information would assist in localizing the lesion. Specific lesions are discussed with the cross-sectional levels.

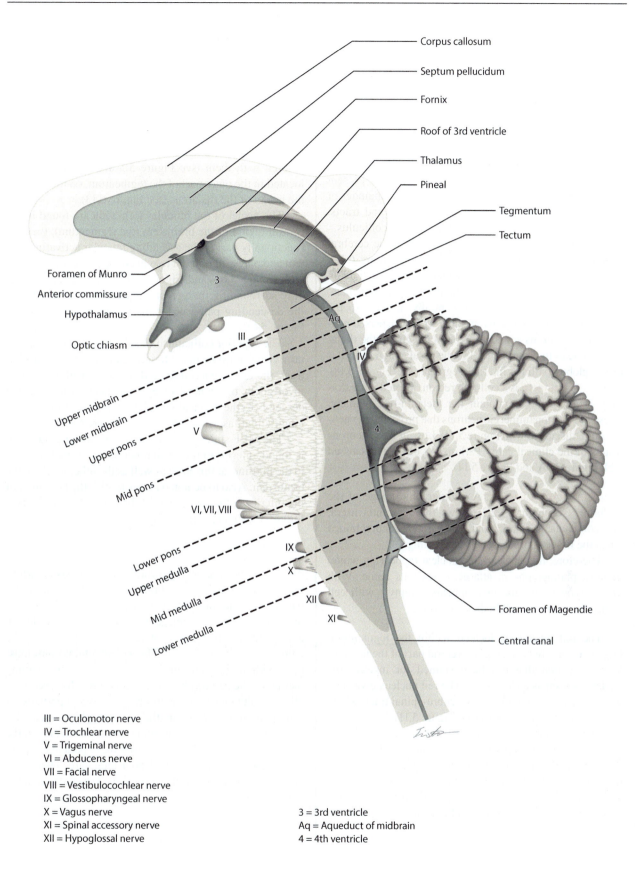

Corpus callosum
Septum pellucidum
Fornix
Roof of 3rd ventricle
Thalamus
Pineal
Tegmentum
Tectum

Foramen of Munro
Anterior commissure
Hypothalamus
Optic chiasm

3

Aq

III

IV

Upper midbrain
Lower midbrain
Upper pons

V

Mid pons

VI, VII, VIII

4

IX

Lower pons
Upper medulla

X

Mid medulla

XII

Lower medulla

XI

Foramen of Magendie
Central canal

III = Oculomotor nerve
IV = Trochlear nerve
V = Trigeminal nerve
VI = Abducens nerve
VII = Facial nerve
VIII = Vestibulocochlear nerve
IX = Glossopharyngeal nerve
X = Vagus nerve
XI = Spinal accessory nerve
XII = Hypoglossal nerve

3 = 3rd ventricle
Aq = Aqueduct of midbrain
4 = 4th ventricle

FIGURE A.2: Brainstem Histology B—Sagittal View (schematic)

FIGURE A.3—UPPER MIDBRAIN: CROSS-SECTION (SCHEMATIC AND HISTOLOGY)

The identifying features of this cross-section of the midbrain include the cerebral peduncle ventrally, with the substantia nigra posterior to it. The aqueduct is present surrounded by the periaqueductal gray. The remainder of the midbrain is the tegmentum, with nuclei and tracts. Dorsally, behind the aqueduct, is the superior colliculus.

The fiber systems are segregated within the cerebral peduncles (see Figure 5.9, Figure 5.10, and Figure 6.12). The substantia nigra consists in fact of two functionally distinct parts—the pars compacta and the pars reticulata. The **pars reticulata** lies more ventrally adjacent to the cerebral peduncle and contains some widely dispersed neurons; these neurons connect the basal ganglia to the thalamus as one of the output nuclei of the basal ganglia (similar to the globus pallidus internal segment; see Figure 5.18). The **pars compacta** is a cell-rich region, located more dorsally, whose neurons contain the melanin-like pigment. These are the dopaminergic neurons that project to the neostriatum (discussed with Figure 5.14). Loss of these neurons results in the clinical entity Parkinson's disease (discussed with Figure 2.5A and Figure 5.14).

Note to the Learner: A photographic image of the midbrain (see Figure 4.2C) shows the substantia nigra, pars compacta, as a dark band. The pigment is not retained when the tissue is processed for sectioning.

Therefore, this nuclear area is clear (appearing white) in most photographs in atlases, despite its name. With myelin-type stains, the area appears "empty"; with cell stains, the neuronal cell bodies are visible.

The red nucleus is located within the tegmentum; large neurons are typical of the ventral part of the nucleus. With a section that has been stained for myelin, the nucleus is seen as a clear zone. The red nucleus gives origin to a descending pathway, the rubro-spinal tract, which is involved in motor control (see Figure 5.11).

The oculomotor nucleus (CN III) is quite large and occupies the region in front of the periaqueductal gray, near the midline; this identifies the level as upper midbrain with the superior colliculus. These motor neurons are quite large and are easily recognizable. The parasympathetic portion of this nucleus is incorporated within it and is known as the Edinger-Westphal (E-W) nucleus (see Figure 3.5). The fibers of CN III pass anteriorly through the medial portion of the red nucleus and exit between the cerebral peduncles, in the interpeduncular fossa (see Figure 1.8).

The ascending (sensory) tracts present in the midbrain are a continuation of those present throughout the brainstem. The medial lemniscus, the ascending trigeminal pathway, and the fibers of the anterolateral system incorporated with them (see Figure 5.5 and Figure 6.11) are located in the outer part of the tegmentum, on their way to the nuclei of the thalamus (see Figure 6.13).

The nuclei of the reticular formation are found in the central region of the brainstem (the tegmentum); they are functionally part of the ascending reticular activating system and play a significant role in consciousness (discussed with Figure 3.6A and Figure 3.6B). The periaqueductal gray surrounding the cerebral aqueduct is involved with the descending pathway for the modulation of pain (see Figure 5.6).

The superior colliculus is a subcortical center for certain visual movements. These nuclei give rise to a fiber tract, the tecto-spinal tract, a descending pathway that is involved in the control of eye and neck movements; it descends to the cervical spinal cord as part of the medial longitudinal fasciculus (MLF) (see Figure 6.9).

The MLF stains heavily with a myelin-type stain and is found anterior to the cranial nerve motor nucleus, next to the midline, at this level as well as the other levels of the brainstem. Also to be noted at this level is the brachium of the inferior colliculus, a part of the auditory pathway (see Figure 3.3, Figure 6.1, and Figure 6.2).

CLINICAL ASPECT

A specific lesion involving thrombosis of the basilar artery may destroy much of the brainstem, yet leave the inner part of the midbrain intact. Few people actually survive this cerebrovascular damage, but those who do are left in a suspended (rather tragic) state of living, known by the name **locked-in syndrome**. The patient retains consciousness, with intellectual functions generally intact, meaning that he or she can think and feel as before. However, usually, all voluntary movements are gone, except perhaps for some eye movements (in the vertical plane—see also Figure 6.8), or occasionally some small movements in the hands and fingers. This means that these patients require a respirator to breathe and 24-hour total care. There may also be a loss of all sensations, or some sensation from the body may be retained.

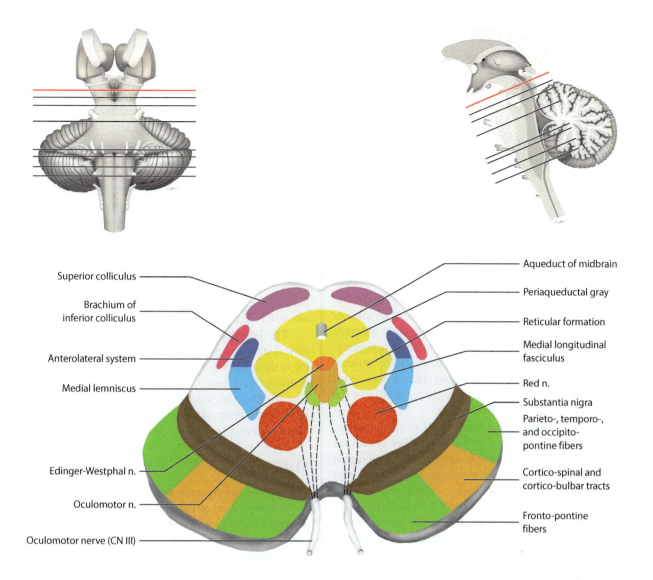

Superior colliculus

Brachium of
inferior colliculus

Anterolateral system

Medial lemniscus

Edinger-Westphal n.

Oculomotor n.

Oculomotor nerve (CN III)

Aqueduct of midbrain

Periaqueductal gray

Reticular formation

Medial longitudinal
fasciculus

Red n.

Substantia nigra

Parieto-, temporo-,
and occipito-
pontine fibers

Cortico-spinal and
cortico-bulbar tracts

Fronto-pontine
fibers

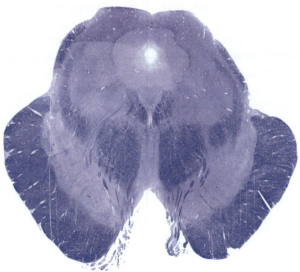

FIGURE A.3: Upper Midbrain: Cross-Section (schematic and histology)

FIGURE A.4—LOWER MIDBRAIN: CROSS-SECTION (SCHEMATIC AND HISTOLOGY)

This cross-section includes the cerebral peduncles, still located anteriorly, and the substantia nigra located immediately behind these fibers. The unique feature in the lower midbrain is the decussation (crossing) of the superior cerebellar peduncles that occupies the central area of the section; this identifies the section as the inferior collicular level. Posteriorly, the aqueduct is surrounded by the periaqueductal gray, and behind the aqueduct is the inferior colliculus. Often, the cross-section at this level includes some of the pontine nuclei (as seen in the histological section in the lower part of the figure). Therefore, one may see a somewhat confusing mixture of structures.

The arrangement of the fibers in the cerebral peduncle is the same as found in the upper midbrain. The tegmentum contains the ascending tracts, the medial lemniscus, the trigeminal pathway, and the anterolateral fibers (system), which are situated together at the outer edge of the lower midbrain (see Figure 6.11).

In sections through the lower levels of the midbrain, there is a brief appearance of a massive fiber system (as seen with a myelin-type stain) occupying the central region of the lower midbrain. These fibers are the continuation of the superior cerebellar peduncles, which are crossing (decussating) at this level (see Figure 3.3 and Figure 6.11). The fibers are coming from the deep cerebellar nuclei (the intracerebellar nuclei), mainly the dentate nucleus, and are headed for the ventral lateral nucleus of the thalamus and then on to the motor cortex (discussed with Figure 5.18). Some of the fibers that come from the intermediate deep cerebellar nucleus synapse in the red nucleus.

The nuclei of the reticular formation found in the central region (the tegmentum) at this level are functionally part of the ascending reticular activating system (ARAS)

and play a significant role in consciousness (see Figure 3.6A and Figure 3.6B). Between the cerebral peduncles is a small nucleus, the interpeduncular nucleus, which belongs with the limbic system. The periaqueductal gray surrounding the aqueduct of the midbrain is involved with pain and also with the descending pathway for the modulation of pain (see Figure 5.6).

The nucleus of CN IV, the trochlear nucleus, is located in front of the periaqueductal gray, next to the midline. Because it supplies only one extra-ocular muscle, it is a smaller nucleus than the oculomotor nucleus. CN IV heads dorsally and exits from the brainstem below the inferior colliculus (see Figure 6.12), on the posterior aspect of the brainstem. The medial longitudinal fasciculus (MLF) lies just anterior to the trochlear nucleus. Some unusually large round cells are often seen at the edges of the periaqueductal gray; these cells are part of the **mesencephalic nucleus** of the trigeminal nerve, CN V (see Figure 3.4).

The lateral lemniscus, the ascending auditory pathway, is still present at this level and its fibers are terminating in the inferior colliculus, a relay nucleus in the auditory pathway (see Figure 6.1 and Figure 6.2). After synapsing here, the fibers are relayed to the medial geniculate nucleus via the brachium of the inferior colliculus, seen at the upper midbrain level (see Appendix Figure A.3).

CLINICAL ASPECT

The presence of the pain and temperature fibers that are found at this level at the outer edge of the midbrain has prompted the possibility, in very select cases, to sever the sensory ascending pathways surgically at this level. This highly dangerous neurosurgical procedure would be done particularly for patients with cancer who are suffering from intractable pain. Nowadays, it would be considered only as a measure of last resort. Pain control is currently managed through the use of drugs, either as part of palliative care or in the setting of a pain "clinic," accompanied by other measures.

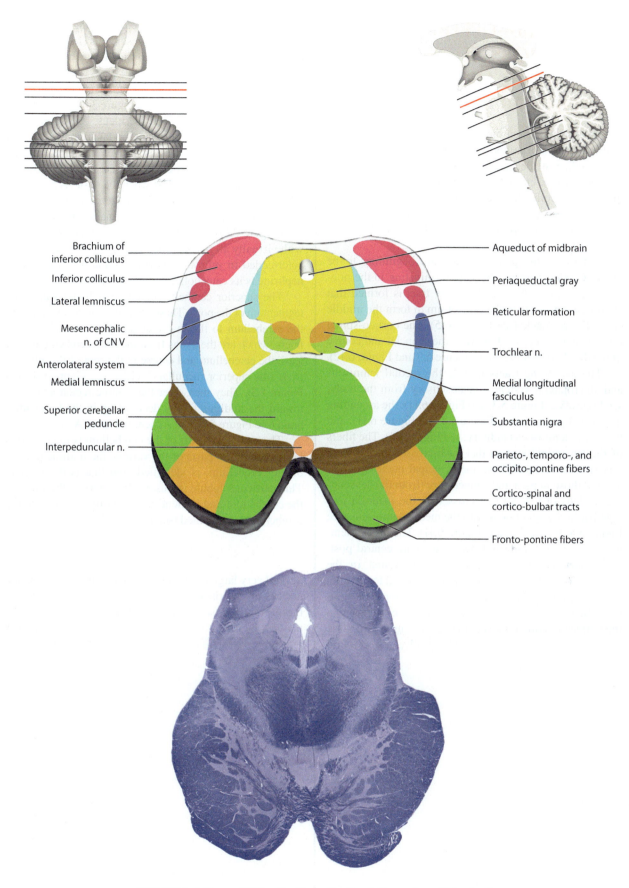

Brachium of inferior colliculus

Inferior colliculus

Lateral lemniscus

Mesencephalic n. of CN V

Anterolateral system

Medial lemniscus

Superior cerebellar peduncle

Interpeduncular n.

Aqueduct of midbrain

Periaqueductal gray

Reticular formation

Trochlear n.

Medial longitudinal fasciculus

Substantia nigra

Parieto-, temporo-, and occipito-pontine fibers

Cortico-spinal and cortico-bulbar tracts

Fronto-pontine fibers

FIGURE A.4: Lower Midbrain: Cross-Section (schematic and histology)

FIGURE A.5—UPPERMOST PONS: CROSS-SECTION (SCHEMATIC AND HISTOLOGY)

This level is presented mainly to allow an understanding of the transition of midbrain to pons. This particular section is taken at the uppermost pontine level, where the trochlear nerve, cranial nerve IV (CN IV), exits (below the inferior colliculus; see Figure 1.9, Figure 3.3, and Figure 5.17). This is the only cranial nerve that exits posteriorly; its fibers cross (decussate) before exiting (see Figure 6.12).

Anteriorly, the pontine nuclei are beginning to be found. Cortico-pontine fibers are terminating in the pontine nuclei. From these cells, a new tract is formed that crosses and projects to the cerebellum to form the middle cerebellar peduncle (see Figure 5.15). The cortico-spinal fibers become dispersed between these nuclei and course in bundles between them (see Figure 5.9 and Figure 6.12).

The ascending tracts include the medial lemniscus and anterolateral system (somatosensory from the body; see Figure 5.2, Figure 5.3, and Figure 6.11), the ascending trigeminal pathway (see Figure 5.4 and Figure 6.11), and the lateral lemniscus (auditory; see Figure 6.1). The fibers of the trigeminal system that have crossed in the pons (discriminative touch from the principal nucleus of CN V), and those of pain and temperature (from the descending nucleus of CN V) that crossed in the medulla join together in the upper pons with the medial lemniscus (see Figure 5.4, Figure 5.5, and Figure 6.11). The medial lemniscus is located midway between its more central position inferiorly and the lateral position found in the midbrain (see Figure 6.11). In sections stained for myelin, it has a typical "comma-shaped" configuration. The auditory fibers are located dorsally, just before terminating in the inferior colliculus in the lower midbrain (see Figure 6.2 and Figure 6.11). Centrally, the cerebral

aqueduct is beginning to enlarge, becoming the 4th ventricle. The medial longitudinal fasciculus (MLF) is found in its typical location ventral to the 4th ventricle, next to the midline.

The nuclei of the reticular formation are located in the tegmentum (see Figure 3.6A and Figure 3.6B). The special nucleus at this level, the locus ceruleus, is located in the dorsal part of the tegmentum not too far from the edges of the 4th ventricle. The nucleus derives its name from its bluish color in fresh specimens (see Figure 4.2B). As explained, the pigment is lost when the tissue is processed for histological examination. The locus ceruleus is usually considered part of the reticular formation (as discussed with Figure 3.6B) because of its widespread connections with virtually all parts of the brain. (It has therefore been color-coded in yellow.) It is also unique because norepinephrine is its catecholamine neurotransmitter substance.

The superior cerebellar peduncle is found within the tegmentum of the pons. These fibers carry information from the cerebellum to the thalamus and the red nucleus. The fibers, which are the axons from the deep cerebellar nuclei, leave the cerebellum and course in the roof of the 4th ventricle (the superior medullary velum; see Figure 1.10 and Figure 5.3). They then enter the pontine region and move toward the midline, finally decussating in the lower midbrain (see Figure 5.17 and Appendix Figure A.4).

The uppermost part of the cerebellum is found at this level. One of the parts of the vermis, the midline portion of the cerebellum, is identified, and that is the lingula. This particular lobule is a useful landmark in the study of the cerebellum and is identified when the anatomy of the cerebellum is explained (see Figure 3.7).

ADDITIONAL DETAIL

Several very large neurons belonging to the mesencephalic nucleus of the trigeminal may be found near the edges of the 4th ventricle (see Figure 3.4). This small cluster of cells may not be found in every cross-section of this particular region.

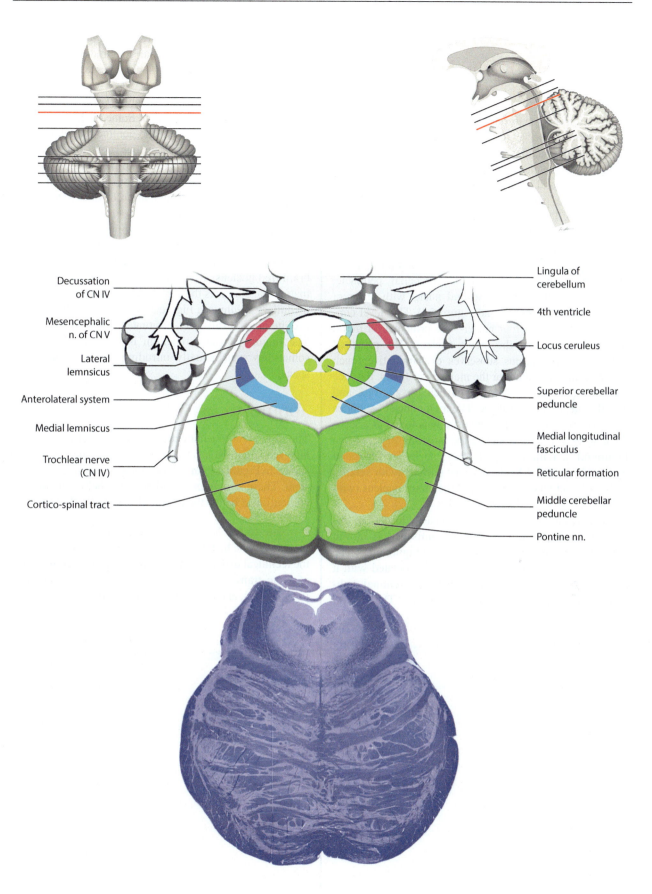

Decussation of CN IV

Mesencephalic n. of CN V

Lateral lemniscus

Anterolateral system

Medial lemniscus

Trochlear nerve (CN IV)

Cortico-spinal tract

Lingula of cerebellum

4th ventricle

Locus ceruleus

Superior cerebellar peduncle

Medial longitudinal fasciculus

Reticular formation

Middle cerebellar peduncle

Pontine nn.

FIGURE A.5: Uppermost Pons: Cross-Section (schematic and histology)

FIGURE A.6—MID-PONS: CROSS-SECTION (SCHEMATIC AND HISTOLOGY)

This section is taken through the level of the attachment of the trigeminal nerve (CN V). Anteriorly, the pontine nuclei and the bundles of cortico-spinal fibers are easily recognized. The pontine cells (nuclei) and their axons, which cross and then become the middle cerebellar peduncle, are particularly numerous at this level (see Figure 5.15). The cortico-spinal fibers are seen as distinct bundles that are widely dispersed among the pontine nuclei at this level (see Figure 5.9 and Figure 6.12).

The trigeminal nerve enters and exits the brainstem along the course of the middle cerebellar peduncle. Cranial nerve (CN) V has several nuclei with different functions (see Figure 3.4 and Figure 5.4). This level contains only two of its four nuclei—the principal (or main) sensory nucleus and the motor nucleus. The principal (main) sensory nucleus subserves discriminative touch sensation and accounts for the majority of fibers; the face area is extensively innervated, particularly the lips, and also the surface of the tongue. The motor nucleus supplies the muscles of mastication and usually is found as a separable nerve as it exits alongside the large sensory root (see Figure 6.12). Within the pons, these nuclei are separated by the fibers of CN V; the sensory nucleus (with smaller cells) is found more laterally, and the motor nucleus (with larger cells) is located more medially.

The ascending fiber systems are easily located at this cross-sectional level. The medial lemniscus has moved away from the midline as it ascends (see Figure 6.11). The anterolateral fiber system has become associated with it by this level. In addition, the ascending trigeminal pathway joins with the medial lemniscus. The lateral lemniscus is seen as a distinct tract, lying just lateral to the medial lemniscus. The medial longitudinal fasciculus (MLF) is found in its typical location anterior to the 4th ventricle.

The core area of the tegmentum is occupied by the nuclei of the reticular formation. Some of the nuclei here are called the oral portion of the pontine reticular formation (see Figure 3.6B). This "nucleus" contributes fibers to a descending medial reticulo-spinal tract that is involved in the indirect voluntary pathway for motor control, and it plays a major role in the regulation of muscle tone (discussed with Figure 5.12B).

The 4th ventricle has become quite wide at this level. The superior cerebellar peduncles are found at its edges, exiting from the cerebellum and heading toward the midbrain (red nucleus) and thalamus. The thin sheet of white matter that connects these peduncles is called the superior medullary velum (see Figure 3.3). The cerebellum, which is quite large at this level, is situated behind the ventricle. The lingula of the cerebellum is again labeled and is sometimes seen actually intruding into the ventricular space.

ADDITIONAL DETAIL

The superior cerebellar peduncles and the superior medullary velum can be located in a specimen (such as the one shown in Figure 1.9) in a dorsal view of the isolated brainstem. These structures would be found below the inferior colliculi, just below the exiting fibers of CN IV dorsally.

Note to the Learner: The cerebellum is usually not included in the histological sections of the pons because of the technical difficulty of sectioning such a large fragment of tissue, transferring the section through the various staining solutions, and mounting the section on large slides.

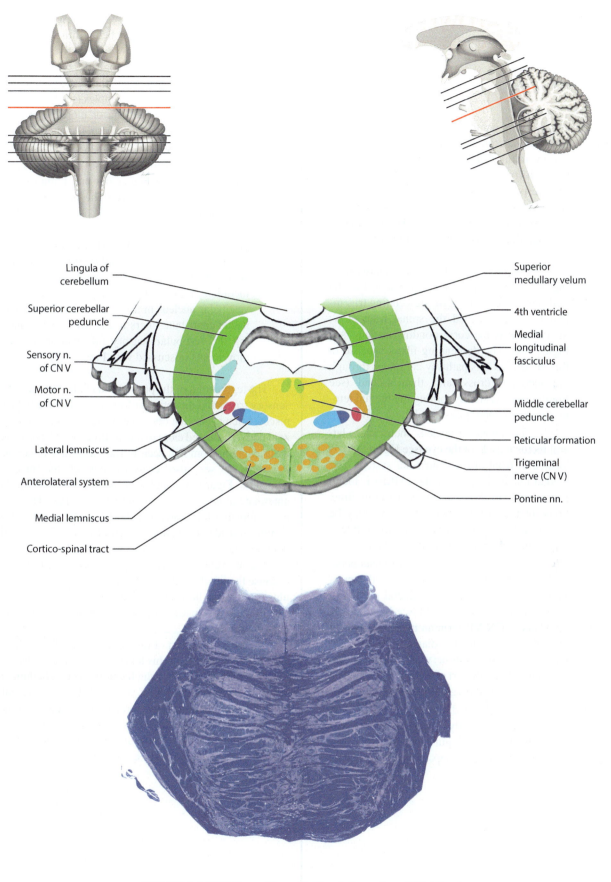

FIGURE A.6: Mid-Pons: Cross-Section (schematic and histology)

FIGURE A.7—LOWERMOST PONS: CROSS-SECTION (SCHEMATIC AND HISTOLOGY)

This section is very complex because of the number of nuclei related to the cranial nerves located in the tegmental portion, including CN V, VI, VII, and VIII. Some of the tracts are shifting in position or are forming, or both. Anteriorly, the pontine nuclei have all but disappeared, and the fibers of the cortico-spinal tract are regrouping into a more compact bundle that will become the pyramids in the medulla (see Appendix Figure A.8).

- **CN V**: The fibers of the trigeminal nerve carrying pain and temperature, that entered at the mid-pontine level, form the descending trigeminal tract, also called the spinal tract of V; medial to it is the corresponding nucleus (see Figure 3.4). The descending fibers synapse in this nucleus as this pathway continues through the medulla, cross, and then ascend (see Figure 5.4), eventually joining the medial lemniscus in the upper pons (see Figure 5.5).

- **CN VI**: The abducens nucleus, motor to the lateral rectus muscle of the eye (see Figure 3.5), is located in front of the ventricular system. The medial longitudinal fasciculus (MLF) is found just anterior to these nuclei, near the midline. Some of the exiting fibers of CN VI may be seen (see below) as the nerve emerges anteriorly, at the junction of pons and medulla.

- **CN VII**: The motor neurons of the facial nerve nucleus, supplying the muscles of facial expression, are located in the ventrolateral portion of the tegmentum. As explained (see Figure 6.12), the fibers of CN VII form an internal loop over the abducens nucleus. The diagram is drawn as if the whole course of this nerve is present in a single section, but only part of this nerve is found on an actual section through this level of the pons.

- **CN VIII**—cochlear division: CN VIII enters the brainstem slightly lower, at the ponto-cerebellar angle (see Figure 1.8 and Figure 3.1). The auditory fibers synapse in the dorsal and ventral cochlear nuclei, which is seen in the medulla in a section just below this level (see also Figure A.8). The two distinctive parts of this nerve at this histological level are the crossing fibers that form the trapezoid body and the superior olivary complex (see Figure 6.1 and Figure 6.9). After one or more synapses, the fibers then ascend and form the lateral lemniscus, which actually commences at this level.

- **CN VIII**—vestibular division: Of the four vestibular nuclei (see Figure 6.8 and Figure 6.9), three are found at this level. The lateral vestibular nucleus, with its giant cells, is located at the lateral edge of the 4th ventricle; this nucleus gives rise to the lateral vestibulo-spinal tract (see Figure 5.13). The medial vestibular nucleus is also present at this level, an extension of the medullary region. There is also a small superior vestibular nucleus in this region. The latter two nuclei contribute fibers to the MLF, thus relating the vestibular sensory information to eye movements (discussed with Figure 6.9).

The tegmentum of the pons also includes the ascending sensory tracts and the reticular formation. The medial lemniscus, often somewhat obscured by the fibers of the trapezoid body, is situated close to the midline, but it has changed its orientation from that seen in the medullary region (see Figure 6.11; see also cross-sections of the medulla; Appendix Figure A.9 and Appendix Figure A.10). The anterolateral system is too small to be identified. The nuclei of the reticular formation include the caudal portion of the pontine reticular formation, which also contributes to the pontine reticulo-spinal tract (see Figure 5.12A).

The 4th ventricle is very large but often seems smaller because the lobule of the cerebellar vermis called the nodulus (part of the flocculonodular lobe; refer to Figure 3.7) impinges on its space. The MLF is found anterior to it, near the midline.

The lowermost part of the middle cerebellar peduncle can still be identified at this level. Also present is the inferior cerebellar peduncle, which enters the cerebellum at a lower level (see Figure 1.8); it is found more internally within the cerebellum. The intracerebellar (deep cerebellar) nuclei are also found at this cross-sectional level and are located within the white matter of the cerebellum (discussed with Figure 3.8 and Figure 5.16).

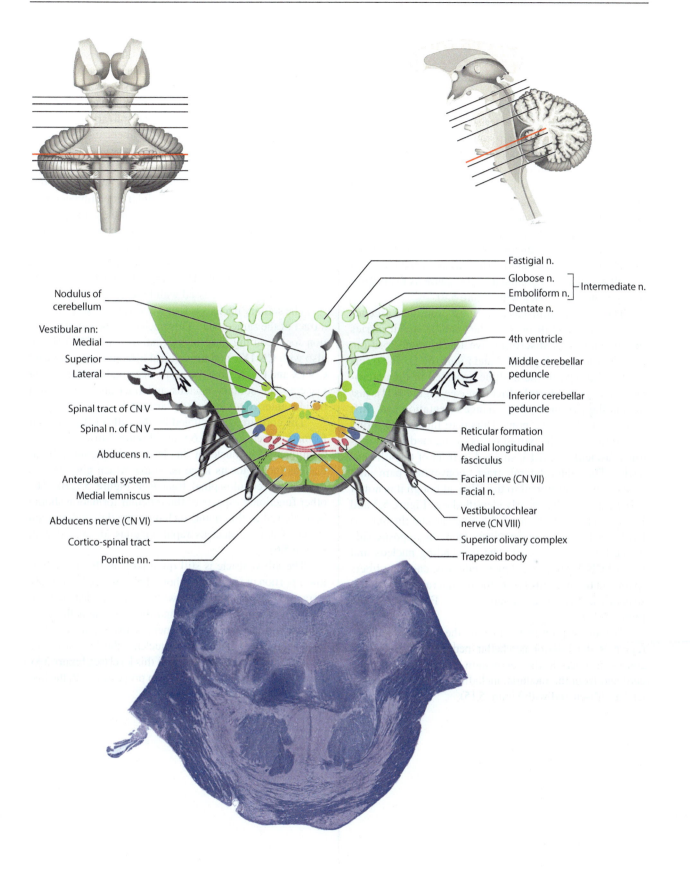

Fastigial n.
Globose n.
Emboliform n. ⎱ Intermediate n.
Dentate n.

Nodulus of cerebellum

Vestibular nn:
Medial
Superior
Lateral

4th ventricle

Middle cerebellar peduncle

Inferior cerebellar peduncle

Spinal tract of CN V
Spinal n. of CN V
Abducens n.
Anterolateral system
Medial lemniscus
Abducens nerve (CN VI)
Cortico-spinal tract
Pontine nn.

Reticular formation
Medial longitudinal fasciculus
Facial nerve (CN VII)
Facial n.
Vestibulocochlear nerve (CN VIII)
Superior olivary complex
Trapezoid body

FIGURE A.7: Lowermost Pons: Cross-Section (schematic and histology)

FIGURE A.8—UPPER MEDULLA: CROSS-SECTION (SCHEMATIC AND HISTOLOGY)

This section has the characteristic features of the medullary region, namely, the pyramids anteriorly with the inferior olivary nucleus situated just laterally and behind.

The cortico-spinal voluntary motor fibers from areas 4 and 6 go through the white matter of the hemispheres, funnel via the internal capsule (posterior limb), continue through the cerebral peduncles of the midbrain and the pontine region, and emerge as a distinct bundle in the medulla within the pyramids. The cortico-spinal tract is in fact often called the pyramidal tract because its fibers form the pyramids (discussed with Figure 5.9).

The medial lemniscus is the most prominent ascending (sensory) tract throughout the medulla, and it carries the modalities of discriminative touch, joint position, and vibration (see Figure 5.2 and Figure 6.11). The tracts are located next to the midline, oriented in the antero-posterior (ventrodorsal) direction (see Figure 6.11), just behind the pyramids; they change orientation and shift more laterally in the pons. Dorsal to them, also along the midline, are the paired tracts of the medial longitudinal fasciculus (MLF), situated in front of the 4th ventricle. The anterolateral tract, conveying pain and temperature, lies dorsal to the olive, although it is not of sufficient size to be clearly identified (see Figure 5.3 and Figure 6.11). Both the medial lemniscus and the anterolateral system are carrying fibers from the opposite side of the body at this level. The descending nucleus and tract of CN V are present more laterally, carrying fibers (pain and temperature) from the ipsilateral face and oral structures, before decussating (see Figure 5.4 and Figure 6.11).

The other prominent tract in the upper medullary region is the inferior cerebellar peduncle. This tract is conveying fibers to the cerebellum, both from the spinal cord and from the medulla, including the inferior olivary nucleus (discussed with Figure 5.15).

Cranial nerve VIII enters the medulla at its uppermost level, at the cerebello-pontine angle, and passes over the inferior cerebellar peduncle. The nerve has two nuclei along its course, the ventral and dorsal cochlear nuclei (see Figure 3.4). The auditory fibers synapse in these nuclei and then go on to the superior olivary complex in the lower pons region. The crossing fibers are seen in the lowermost pontine region as the trapezoid body (see Figure 6.1 and Figure 6.11).

The vestibular part of CN VIII is represented at this level by two nuclei, the medial and inferior vestibular nuclei (see Figure 6.8). Both these nuclei lie in the same position as the vestibular nuclei in the pontine section, adjacent to the lateral edge of the 4th ventricle. The inferior vestibular nucleus is rather distinct because of the many axon bundles that course through it. The vestibular nuclei contribute fibers to the MLF (discussed with Figure 6.9).

The solitary nucleus is found at this level, surrounding a tract of the same name. This nucleus is the synaptic station for incoming taste fibers (mainly with CN VII, but also with CN IX), and for visceral afferents entering with CN IX and CN X from the gastrointestinal tract and other viscera. The solitary nucleus and tract are situated just beside (anterior to) the vestibular nuclei.

The core area is occupied by the cells of the reticular formation (see Figure 3.6A and Figure 3.6B). The most prominent of its nuclei at this level is the gigantocellular nucleus (noted for its large neurons), which gives rise to the lateral reticulo-spinal tract (see Figure 5.12B). The other functional aspects of the reticular formation should be reviewed at this point, including the descending pain system from the nucleus raphe magnus (discussed with Figure 5.6).

The 4th ventricle is still quite large at this level. The lower portion of its roof has choroid plexus (see Figure 7.8); a fragment of this is present with the histological section, although the roof is torn. Behind the ventricle is the cerebellum, with the vermis (midline) portion and the cerebellar hemispheres. The dentate nucleus, the largest of the intracerebellar nuclei, is present at this level (see Figure 3.8). Again, the cerebellum has not been processed with the histological specimen.

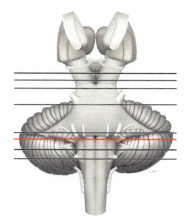

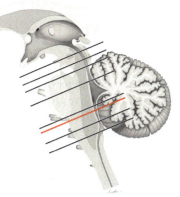

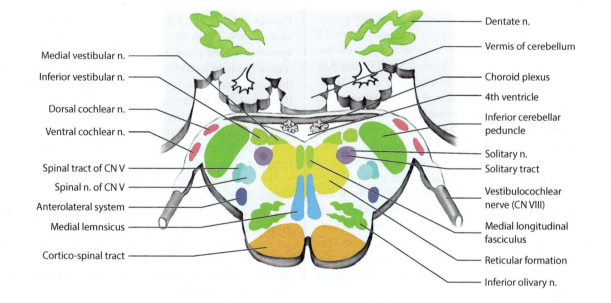

Medial vestibular n.

Inferior vestibular n.

Dorsal cochlear n.

Ventral cochlear n.

Spinal tract of CN V

Spinal n. of CN V

Anterolateral system

Medial lemnsicus

Cortico-spinal tract

Dentate n.

Vermis of cerebellum

Choroid plexus

4th ventricle

Inferior cerebellar peduncle

Solitary n.

Solitary tract

Vestibulocochlear nerve (CN VIII)

Medial longitudinal fasciculus

Reticular formation

Inferior olivary n.

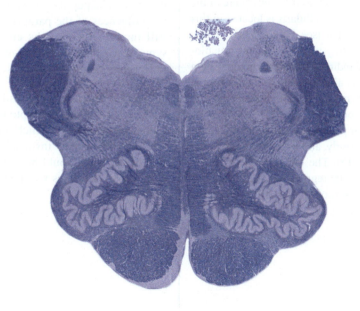

FIGURE A.8: Upper Medulla: Cross-Section (schematic and histology)

FIGURE A.9—MID-MEDULLA: CROSS-SECTION (SCHEMATIC AND HISTOLOGY)

This is a classic level for descriptive purposes. The pyramids and inferior olive are easily recognized anteriorly.

The medial lemniscus occupies the area between the olives, on either side of the midline (see Figure 6.11). The medial longitudinal fasciculus (MLF) lies behind (dorsal) to the medial lemniscus, also situated adjacent to the midline. The fibers of the anterolateral system are situated dorsal to the olive. The descending nucleus and tract of the trigeminal system have the same location as seen previously in the lateral aspect of the tegmentum.

The hypoglossal nucleus (CN XII) is found near the midline and in front of the ventricle; its fibers exit anteriorly, between the pyramid and the olive (see Figure 1.8 and Figure 3.1). CN IX and CN X are attached at the lateral aspect of the medulla (see Figure 1.8 and Figure 3.1). Their efferent fibers are derived from two nuclei (indicated by the dashed lines): the dorsal motor nucleus of the vagus (CN X), which is parasympathetic, and the nucleus ambiguus, which is motor to the muscles of the pharynx and larynx (see Figure 3.5). The dorsal motor nucleus lies adjacent to the 4th ventricle just lateral to the nucleus of CN XII. The nucleus ambiguus lies dorsal to the olivary nucleus; in a single cross-section only a few cells of this nucleus are usually seen, thus making its identification difficult (i.e., "ambiguous") in actual sections. The taste and visceral afferents that are carried in these nerves synapse in the solitary nucleus, which is located in the posterior aspect of the tegmentum, surrounded by the tract of the same name.

The reticular formation occupies the central core of the tegmentum; the nucleus gigantocellularis is located in this part of the reticular formation (see Figure 3.6B). These cells give rise to a descending tract, the lateral reticulospinal tract, as part of the indirect voluntary motor system (see Figure 5.12B); there is also a strong influence on the excitability of the lower motor neuron, influencing the stretch reflex and muscle tone.

The inferior cerebellar peduncle is found at the lateral edge of this section, posteriorly, and it carries fibers to the cerebellum (see Figure 5.15). The 4th ventricle is still a rather large space, behind the tegmentum, with the choroid plexus attached to its roof in this area; often the ventricle appears "open," likely because this thin tissue has been torn. The other possibility is that there is no cerebellar tissue posteriorly because the section is below the level of the cerebellum (see the sagittal schematic accompanying this figure).

CLINICAL ASPECT

Vascular lesions in this area of the brainstem are not uncommon. The midline area is supplied by the paramedian branches from the vertebral artery (see Figure 8.1). The structures included in this territory are the corticospinal fibers, the medial lemniscus, and the hypoglossal nucleus.

The lateral portion is supplied by the posterior inferior cerebellar artery, a branch of the vertebral artery (see Figures 8.1, Figure 8.2, and Figure 8.5), called by its abbreviation **PICA** by neuroradiologists. This artery is apparently quite prone to infarction, for some unknown reason (perhaps its tortuosity). Included in its territory are the cranial nerve nuclei and fibers of CN IX and CN X, the descending trigeminal nucleus and tract, fibers of the anterolateral system, and the solitary nucleus and tract, as well as descending autonomic fibers. The inferior cerebellar peduncle or vestibular nuclei, or both, may also be involved. The whole clinical picture is called the **lateral medullary syndrome** (of Wallenberg; also discussed with Figure 6.11).

Interruption of the descending autonomic fibers gives rise to a clinical condition called **Horner's syndrome**. In this syndrome (also discussed with Figure 6.7), there is loss of the autonomic sympathetic supply to one side of the face, ipsilaterally. This leads to drooping of the upper eyelid (ptosis), dry skin, and constriction of the pupil. The pupillary change results from the competing influences of the parasympathetic fibers, which are still intact. Other lesions elsewhere that interrupt the sympathetic fibers in their long course can also give rise to Horner's syndrome.

Note to the Learner: It is instructive for a student to work out the clinical symptoms of both these vascular lesions and to indicate which function is lost with each of the tracts or nucleus involved in the lesion, and which side of the body would be affected. A clinical problem with these syndromes will be included with the Clinical Cases.

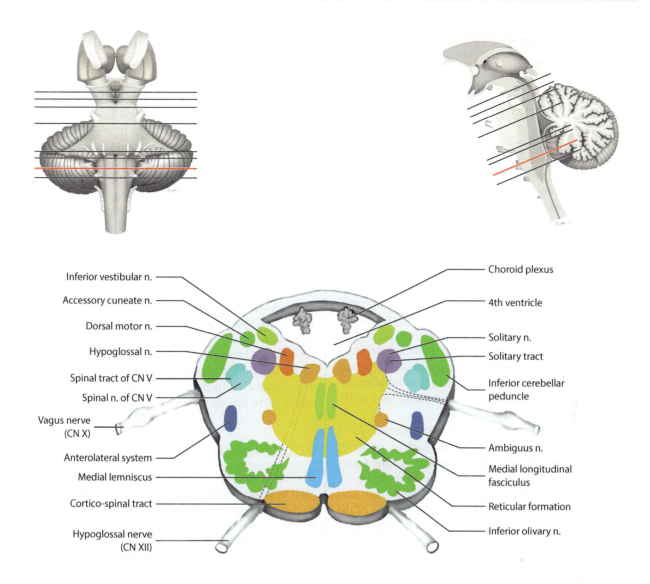

Inferior vestibular n.

Accessory cuneate n.

Dorsal motor n.

Hypoglossal n.

Spinal tract of CN V

Spinal n. of CN V

Vagus nerve (CN X)

Anterolateral system

Medial lemniscus

Cortico-spinal tract

Hypoglossal nerve (CN XII)

Choroid plexus

4th ventricle

Solitary n.

Solitary tract

Inferior cerebellar peduncle

Ambiguus n.

Medial longitudinal fasciculus

Reticular formation

Inferior olivary n.

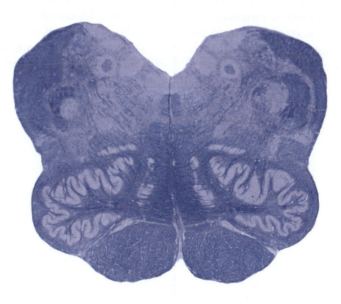

FIGURE A.9: Mid-Medulla: Cross-Section (schematic and histology)

FIGURE A.10—LOWER MEDULLA: CROSS-SECTION (SCHEMATIC AND HISTOLOGY)

The medulla seems significantly smaller at this level, approaching the size of the spinal cord below. The section is still easily recognized as medullary because of the presence of the pyramids anteriorly (the cortico-spinal tract) and the adjacent inferior olivary nucleus.

The tegmentum contains the cranial nerve nuclei, the reticular formation, and the other tracts. The nuclei of cranial nerve (CN) X and CN XII, as well as the descending nucleus and tract of CN V, are present as before (as in the mid-medullary section; see Appendix Figure A.9). The medial longitudinal fasciculus (MLF) and anterolateral fibers are also in the same position. The solitary tract and nucleus are still found in the same location. The internal arcuate fibers are present at this level; these are the fibers from the nuclei gracilis and cuneatus that cross (decussate) to form the medial lemniscus (see later). These fibers usually obscure visualization of the nucleus ambiguus. Finally, the reticular formation is still present.

The dorsal aspect of the medullary tegmentum is occupied by two large nuclei—the nucleus cuneatus (cuneate nucleus) laterally and the nucleus gracilis (gracile nucleus) more medially. These nuclei are found on the dorsal aspect of the medulla (see Figure 1.9 and Figure 6.11). These nuclei are the synaptic stations of the tracts of the same name that have ascended the spinal cord in the dorsal column (see Figure 5.2, Figure 6.10, and Appendix Figure A.11). The gracilis is mainly for the lower limb and lower body; the cuneatus carries information from the upper body and upper limb. The fibers relay in these nuclei and then move through the medulla anteriorly as the internal arcuate fibers, cross (decussate), and form the medial lemniscus on the opposite side (see Figure 6.11). At this level, the medial lemniscus is situated between the olivary nuclei and dorsal to the pyramids and is oriented anteroposteriorly.

Posteriorly, the 4th ventricle is tapering down in size, giving a "V-shaped" appearance to the dorsal aspect of the medulla (see Figure 3.1 and Figure 3.3). It is usual for the ventricle roof to be absent at this level. This is likely accounted for by the presence of the foramen of Magendie, through which the cerebrospinal fluid escapes from the ventricular system into the subarachnoid space (see the sagittal schematic accompanying this figure; also see Figure 3.2 and Figure 7.8). Posterior to this area is the cerebello-medullary cistern, otherwise known as the cisterna magna (see the T1 magnetic resonance scan in Figure 3.2, and also Figure 7.8).

One special nucleus is found in the "floor" of the ventricle at this level (see Figure 3.3 but not indicated in the illustration), the **area postrema**. This forms a little bulge that can be appreciated on some sections. The nucleus is part of the system that controls vomiting, and it is often referred to as the vomiting "center." This region lacks a blood-brain barrier, thus allowing this particular nucleus to be "exposed" directly to whatever is circulating in the bloodstream. It likely connects with the nuclei of the vagus nerve that are involved in the act of vomiting.

ADDITIONAL DETAIL

The accessory cuneate nucleus is found at this level, as well as at the mid-medullary level. This nucleus is a relay for some of the cerebellar afferents from the upper extremity (see Figure 5.15 and Figure 6.10). The fibers then go to the cerebellum via the inferior cerebellar peduncle. The inferior cerebellar peduncle has not yet been formed at this level.

Cross-sections through the lowermost part of the medulla may include the decussating cortico-spinal fibers (i.e., the pyramidal decussation; see Figure 6.11); this would therefore alter significantly the appearance of the structures in the actual section.

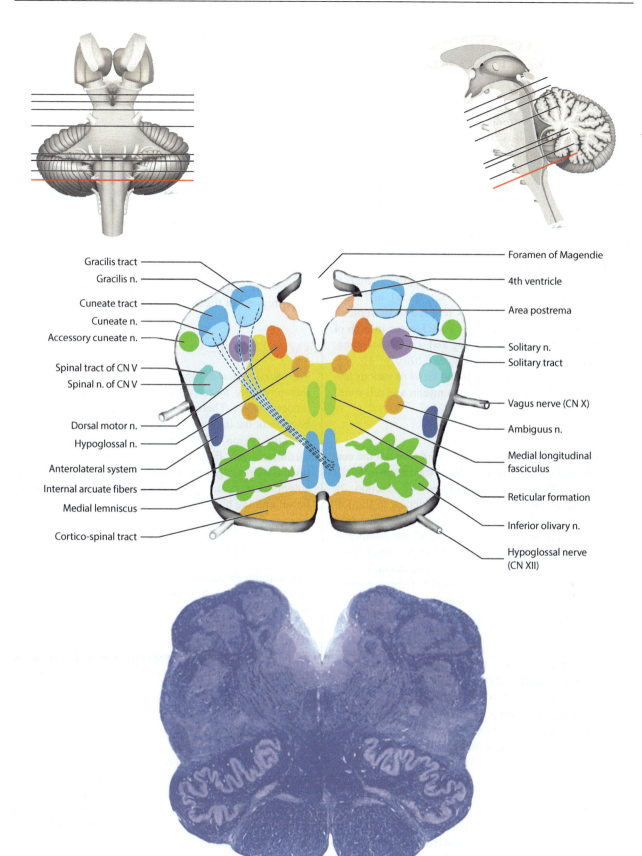

Gracilis tract
Gracilis n.
Cuneate tract
Cuneate n.
Accessory cuneate n.
Spinal tract of CN V
Spinal n. of CN V
Dorsal motor n.
Hypoglossal n.
Anterolateral system
Internal arcuate fibers
Medial lemniscus
Cortico-spinal tract

Foramen of Magendie
4th ventricle
Area postrema
Solitary n.
Solitary tract
Vagus nerve (CN X)
Ambiguus n.
Medial longitudinal fasciculus
Reticular formation
Inferior olivary n.
Hypoglossal nerve (CN XII)

FIGURE A.10: Lower Medulla: Cross-Section (schematic and histology)

FIGURE A.11—SPINAL CORD: CROSS-SECTIONAL VIEWS (HISTOLOGICAL VIEWS)

The spinal cord is introduced in Section 1 of the atlas, the external views in Chapter 1 (see Figure 1.10 and Figure 1.11). The organization of the nervous tissue in the cord has the gray matter inside, in a typical "butterfly" or "H-shaped" configuration, with the white matter surrounding it (see Figure 3.9). The functional aspects of the spinal cord are presented in Section 2, including the nuclei and connections for the afferent fibers (sensory; see Figure 5.1), and the efferent circuits with some reflexes (motor; see Figure 5.7).

The white matter surrounding the gray matter is divided by it into three areas—dorsal, lateral, and anterior. These zones are sometimes referred to as funiculi (singular, funiculus). Various tracts are located in each of these three zones, some ascending and some descending, which have been reviewed (see Figure 6.10).

The following are cross-sectional views of various levels of the spinal cord, stained with a myelin and cell stain.

CERVICAL LEVEL—C8

This is a cross-section of the spinal cord through the cervical enlargement. This level has been used in many of the illustrations of the various pathways (in Section 2). Because the cervical enlargement contributes to the formation of the brachial plexus to the upper limb, the gray matter ventrally is very large because of the number of neurons involved in the innervation of the upper limb, particularly the muscles of the hand. The dorsal horn is similarly large because of the number of afferents coming from the skin of the fingers and hand.

The white matter is comparatively larger at this level because:

- All the ascending tracts are present and are carrying information from the lower parts of the body, as well as the upper limb.
- All the descending tracts are fully represented because many of the fibers terminate in the cervical region of the spinal cord. In fact, some of them do not descend to lower levels.

THORACIC LEVEL—T6

The thoracic region of the spinal cord presents an altered morphology because of the decrease in the amount of gray matter. There are fewer muscles and less dense innervation of the skin in the thoracic region. The gray matter has, in addition, a lateral horn, which represents the sympathetic preganglionic neurons (see Figure 3.9). The lateral horn is present from T1 to L2.

LUMBAR LEVEL—L3

This cross-sectional level of the spinal cord has been used in the various illustrations of the pathways in Section 2 of the atlas. This cross-section is similar in appearance to the cervical section because both are innervating the limbs. There is, however, proportionately less white matter at the lumbar level. The descending tracts are smaller because many of the fibers have terminated at higher levels. The ascending tracts are smaller because they are conveying information only from the lower regions of the body.

SACRAL LEVEL—S3 (NOT SHOWN)

The sacral region of the spinal cord is the smallest and is therefore easy to recognize. The white matter is quite reduced in size. There is still a fair amount of gray matter because of the innervation of the pelvic musculature.

This region of the spinal cord, roughly the conus medullaris (see Figure 1.10 and Figure 1.11), also contains the preganglionic parasympathetic neurons of the autonomic nervous system. These neurons innervate the bowel and the bladder.

BLOOD SUPPLY

The anterior spinal artery, the main blood supply to the spinal cord, comes from branches from each of the vertebral arteries that join (see Figure 8.8); it descends in the midline (see Figure 1.10 and Figure 8.7) and supplies the ventral horn and the anterior and lateral group of tracts, including the lateral cortico-spinal pathway. The posterior spinal arteries supply the dorsal horn and the dorsal columns (see Figure 8.7).

CLINICAL ASPECT

It is known that this blood supply is marginal, particularly in the lower thoracic region (see Figure 8.8).

The student is encouraged to work out the clinical symptoms of lesions of the spinal cord at various levels. Various examples of lesions will be included with the Clinical Cases.

Cervical level

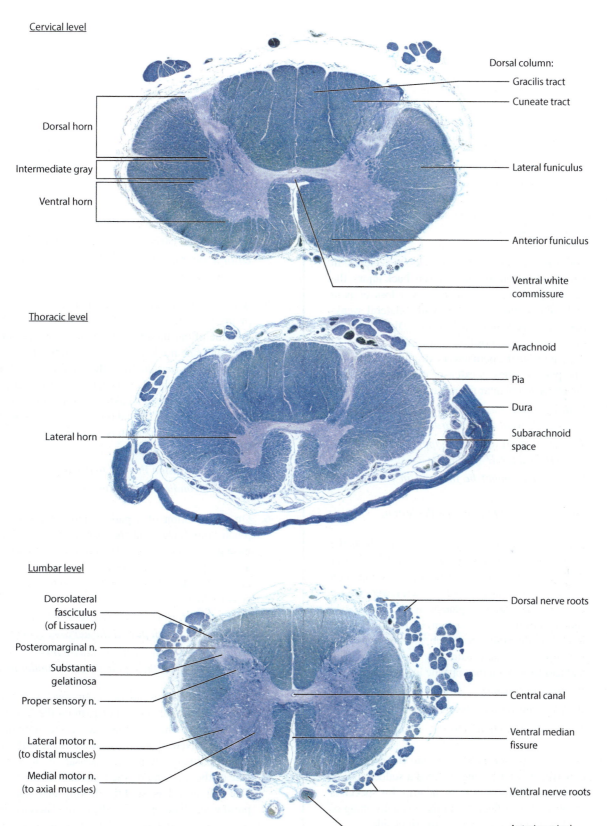

Dorsal column:
Gracilis tract
Cuneate tract

Dorsal horn

Intermediate gray

Ventral horn

Lateral funiculus

Anterior funiculus

Ventral white commissure

Thoracic level

Lateral horn

Arachnoid

Pia

Dura

Subarachnoid space

Lumbar level

Dorsolateral fasciculus (of Lissauer)

Posteromarginal n.

Substantia gelatinosa

Proper sensory n.

Lateral motor n. (to distal muscles)

Medial motor n. (to axial muscles)

Dorsal nerve roots

Central canal

Ventral median fissure

Ventral nerve roots

Anterior spinal artery

FIGURE A.11: Spinal Cord: Cross-Sectional Views (histological views)

CLINICAL CASES

The following clinical problems are designed to emphasize the importance of a knowledge of neuroanatomy as the foundation for clinical neurology. They have been created in consultation with practicing neurologists (in adult and pediatric practice).

Note to the Learner: Answers and explanations to the problems are found on the Atlas 3 Web site (www.atlasbrain.com).

Additional problems may be added on the Web site—please check periodically.

1. Following a serious motor vehicle accident in which the person suffered a severe back injury, the neurological examination reveals a loss of pain and temperature sensations with bilateral weakness and hyperreflexia below T12. Discriminative touch sensation, position sense and a sense of "vibration" in the both lower extremities are intact. The plantar reflex is extensor (upgoing) which is the abnormal plantar response.

   Which pathway has been damaged and at approximately what (spinal cord) level?

   This level corresponds to approximately which vertebral level?

   Which artery might be involved in causing this lesion?

   Draw a diagram to illustrate the lesion.

2. Following a serious bicycle injury in which the rider (who was wearing a helmet) was thrown over the handlebars, the neurological examination reveals a loss of movement in the left lower extremity, with an extensor plantar reflex (previously called a positive Babinski sign) on that side. There is a loss of discriminative touch, position sense, and vibration on the same extremity and a loss of pain and temperature on the opposite lower extremity.

   Where is the lesion?

   Which pathways have been affected?

   Draw a diagram to illustrate the lesion.

3. An elderly person is brought into the emergency department after having suffered a sudden loss of movement of his right face and arm; his speech is incoherent (non-fluent), and there is a drooping of the lower part of the face on the right side.

   What is the nature of the lesion?

   Is it likely to be vascular in origin, and if so, which cerebral blood vessel is involved—and on which side?

   What would be revealed by examining the sensory system, the visual field, the motor system, and the reflexes?

4. What would be the visual field loss with a lesion in the following locations:
   a. The optic nerve on the right side.
   b. The optic chiasm.
   c. The optic radiation in the left temporal lobe.
   d. The visual (striate) cortex on the right side.

   Draw a diagram of the visual field to illustrate the loss of vision for each of these lesions.

5. A young person visits her family doctor and complains that she has developed a drooping of the left side of her mouth over the past few days and that she can no longer play her tuba. Physical examination reveals that she could no longer wrinkle the forehead on the left side, close her left eyelid and there is a weakness on the left side of the cheek and lower facial muscles. The rest of the neurological examination is normal.

   Which nerve is affected?

   Where is the lesion thought to occur?

   Does this condition have a name?

6. On examination of a patient and asking her to touch your finger and then touch the tip of her nose as rapidly as possible, you note that there is a "past pointing" as the movement approaches either end. There is no motor weakness and no sensory loss.

   This type of tremor is usually attributed to a lesion in which part of the nervous system?

   Explain the course of the pathway involved and which part of the motor system is modulated.

7. An older adult presents in the emergency department with a sudden onset of pain on the left side of the face and hiccups that have lasted more than 24 hours. Neurological examination reveals the loss of pain and temperature sensation on the left side of the face and on the right side of the body. The left eyelid is slightly droopy and the left pupil is small compared to the right. There is also some loss of balance and unsteadiness of gait (ataxia).

   Which pathways have been affected? On which side?

   Where is the location of the lesion?

Does this syndrome have a special name?

Draw a diagram of the location of the lesion.

How do you explain the sensory loss on the right side of the body?

8. An older man shuffles into your office with his wife, and you notice immediately that he has a shaky continuous movement of his right hand that consists of a coarse "pill-rolling" action.

   What would you look for on neurological examination?

   What is the name of this condition?

   Which part of the nervous system is involved?

   What else would you look for or ask about?

   Explain what is known of the mechanism of this disease.

   What does the acronym TRAP stand for?

9. A young person is seen in the emergency department following a severe head injury in a car accident. The person in non-responsive. Examination of the pupils reveals that the pupil on the right is dilated and does not respond to light when light is shone on the eye and the right eye is abducted and deflected slightly downward. There is an extensor plantar response (previously called a positive Babinski sign) on the left side.

   What is happening inside the head?

   Which pathways have been disrupted?

   Draw a diagram of the connections involved in the pupillary light reflex and eye movement abnormality.

10. A 50-year-old man comes into your office and complains of a recurrent headache that is not responding to over-the-counter medication. He also says that he has noted that he is having more and more difficulty hearing, particularly on the left side, and has noted "ringing" (tinnitus) in that ear occasionally.

    On examination, the Weber test lateralizes to the right and there is a positive Rinne test on both sides.

    What lesion would you be thinking of ruling out as soon as possible?

    What "test" would you order?

    Where, neuroanatomically, may the lesion be located?

11. At a 6-week follow-up examination of a newborn, you note that the measurement of the head is at the upper end of normal and that the anterior fontanelle seems to be "bulging" slightly.

    What condition would you be concerned about?

    What is the problem named?

    Draw a diagram of the circulation of the cerebrospinal fluid.

12. A 55-year-old man comes into your office and complains of diffuse muscle pain. He is taking statin medication as prescribed to control a higher than recommended cholesterol level for several months. He also says that he has noted that he is having more difficulty climbing stairs. Examination reveals weakness, and pain on palpation in all four limbs. His sensory exam is normal, his reflexes decreased throughout.

    What lesion would you be thinking of ruling out as soon as possible?

    What "test" would you order?

    Where, neuroanatomically, may the lesion be located?

13. You receive a telephone call from the emergency department that one of your patients, 66 years old and known to be overweight and hypertensive, has been brought in after a sudden episode of one-sided weakness. There is hyperreflexia and an extensor plantar response on the left, as well as weakness of the left face, arm and leg; sensation is reported to be normal in the face, arm, trunk, and leg.

    Which pathway has been interrupted?

    Draw a diagram of the pathway.

    What is the likely location of the lesion?

14. A teen-aged male patient is brought in from the ski hill by the paramedics. The patient is on a board with sandbags on either side of his head. He is unconscious and is reported to have a head injury after being found snowboarding "off-piste"; he was not wearing a helmet. The injury is said to have occurred about 2 hours ago. There is a bruise over the left temple. His left pupil is dilated and the left eye is abducted and deflected slightly downward. The plantar response is normal on the left side and extensor on the right side.

    What lesion would you be concerned about?

    Is this an emergency?

    Which artery has been traumatized?

    Describe the anatomical information necessary to understand this lesion.

15. You are a second-year medical student and are told about a friend of your parents who is now recovering in the hospital after an operation for a sudden severe headache. Your parents were present when the person complained that he felt like a "thunderclap" hit his head. There was no mention of trauma or of any previously known condition.

What explanations would you suggest as possibilities?

What questions might you ask the patient about the onset of the headache which might be useful?

On second thought, do you think that you know enough neuroanatomy and have a sufficient understanding of clinical neurology to offer an explanation?

GLOSSARY

Note to the Learner: This glossary contains neuronatomical terms, as well as terms commonly used clinically to describe neurological symptoms and physical findings of a neurological examination; few clinical syndromes are included.

Abducens nerve Sixth cranial nerve (CN VI); to lateral rectus muscle for abduction of the eye.

Accessory nerve Eleventh cranial nerve (CN XI)—see Spinal accessory nerve.

Afferent Conduction toward the central nervous system; usually means sensory.

Agnosia Loss of ability to recognize the significance of sensory stimuli (tactile, auditory, visual), even though the primary sensory systems are intact.

Agonist A muscle that performs a certain movement of the joint; the opposing muscle is called the antagonist.

Agraphia Inability to write because of a lesion of higher brain centers, even though muscle strength and coordination are preserved.

Akinesia Absence or loss of motor function; lack of spontaneous movement; difficulty in initiating movement (as in Parkinson's disease).

Alexia Loss of ability to grasp the meaning of written words; inability to read because of a central lesion; word blindness.

Allocortex The phylogenetically older cerebral cortex, consisting of less than six layers; includes paleocortex (e.g., subicular region = three to five layers) and archicortex (e.g., hippocampus proper and dentate = three layers).

Alpha motor neuron Another name for the anterior (ventral) horn cell, also called the lower motor neuron.

Ammon horn The hippocampus proper, which has an outline in cross-section suggestive of a ram's horn; also called the cornu ammonis (CA).

Amygdala Amygdaloid nucleus or body in the temporal lobe of the cerebral hemisphere; a nucleus of the limbic system.

Angiogram Display of blood vessels for diagnostic purposes, using x-rays, magnetic resonance imaging, or computed tomography, usually by using contrast medium injected into the vascular system.

Anopia A defect in the visual field (e.g., hemianopia—loss of one half of the visual field; quadrantanopia—loss of one fourth of the visual field).

Antagonist A muscle that opposes or resists the action of another muscle, which is called the agonist.

Antidromic Relating to the propagation of an impulse along an axon in a direction that is the reverse of the normal or usual direction.

Aphasia An acquired disruption or disorder of language, specifically a deficit of expression using speech or of comprehending spoken or written language; global aphasia is a severe form affecting all language areas.

Apoptosis Programmed cell death, either genetically determined or following an insult or injury to the cell.

Apraxia Loss of ability to carry out purposeful or skilled movements despite the preservation of power, sensation, and coordination.

Arachnoid The middle meningeal layer, forming the outer boundary of the subarachnoid space.

Archicerebellum A phylogenetically old part of the cerebellum that functions in the maintenance of equilibrium; anatomically, the flocculonodular lobe.

Archicortex Three-layered cortex included in the limbic system; located mainly in the hippocampus proper and dentate gyrus of the temporal lobe.

Area postrema An area involved in vomiting; located in the caudal part of the floor of the fourth ventricle, with no blood-brain-barrier.

Areflexia Loss of reflex as tested using the myotatic, stretch, deep tendon reflex.

Ascending tract Central sensory pathway (e.g., from spinal cord to brainstem, cerebellum, or thalamus).

Association fibers Fibers connecting parts of the cerebral hemisphere, on the same side.

Astereognosis Loss of ability to recognize the nature of objects or to appreciate their shape by touching or feeling them.

Astrocyte A type of neuroglial cell with metabolic and structural functions; reacts to injury of the central nervous system by forming a gliotic "scar."

Asynergy Disturbance of the proper sequencing in the contraction of muscles, at the proper moment, and of the proper degree, so that an action is not executed smoothly or accurately.

Ataxia A loss of coordination of voluntary movements; often associated with cerebellar dysfunction.

Athetosis Slow writhing movements of the limbs, especially of the hands, not under voluntary control, caused by degenerative changes in the striatum.

Autonomic Pertaining to autonomic nervous system; usually taken to mean the efferent or motor innervation of viscera (smooth muscle and glands).

Autonomic nervous system (ANS) Visceral innervation; sympathetic and parasympathetic divisions.

Axon Efferent process of a neuron that conducts impulses to other neurons or to muscle fibers (striated and smooth) and gland cells.

Babinski response Babinski reflex is not correct; stroking the outer border of the sole of the foot in an adult normally results in a plantar (downgoing) of the toes; the abnormal response consists of an upgoing of the first toe and a fanning of the other toes, indicating a lesion of the pyramidal (cortico-spinal) tract. The term **extensor plantar response** is currently preferred.

Basal ganglia (nuclei) Central nervous system nuclei involved in motor control—the caudate, putamen, and globus pallidus (the lentiform nucleus), including, functionally, the subthalamus and the substantia nigra.

Basilar artery The major artery supplying the brainstem and cerebellum, formed by the two vertebral arteries.

Brachium A large bundle of fibers connecting one part with another (e.g., brachium associated with the inferior and superior colliculi of the midbrain).

Bradykinesia Abnormally slow initiation of voluntary movements (usually seen in Parkinson's disease).

Brainstem Includes the medulla, pons, and midbrain.

Brodmann areas Numerical subdivisions of the cerebral cortex on the basis of histological differences among different functional areas (e.g., area 4 = motor cortex; area 17 = primary visual area).

Bulb Referred at one time to the medulla, but in the context of "cortico-bulbar tract" refers to the whole brainstem, in which the motor nuclei of cranial nerves and other nuclei are located.

Carotid siphon Hairpin bend of the internal carotid artery within the skull.

Cauda equina "Horse's tail"—the lower lumbar, sacral, and coccygeal spinal nerve roots within the subarachnoid space of the lumbar (cerebrospinal fluid) cistern.

Caudal Toward the tail, or hindmost part of neuraxis.

Caudate nucleus Part of the neostriatum—consists of a head, body, and tail (which extends into the temporal lobe).

Central nervous system (CNS) Brain (cerebral hemispheres), including diencephalon, cerebellum, brainstem, and spinal cord.

Cerebellar peduncles Inferior, middle, and superior; fiber tracts linking the cerebellum and brainstem.

Cerebellum The little brain; an older part of the brain with motor functions, dorsal to the brainstem, situated in the posterior cranial fossa.

Cerebral aqueduct (of Sylvius) Aqueduct of the midbrain; passageway carrying cerebrospinal fluid through the midbrain, as part of the ventricular system.

Cerebral peduncle Descending cortical fibers in the "basal" (ventral) portion of the midbrain; sometimes includes the substantia nigra (located immediately behind).

Cerebrospinal fluid (CSF) Fluid in the ventricles and in the subarachnoid space and cisterns.

Cerebrum Includes the cerebral hemispheres and diencephalon but not the brainstem and cerebellum.

Cervical Referring to the neck region; the part of the spinal cord that supplies the structures of the neck; C1 to C7 vertebrae; C1 to C8 spinal segments.

Chorda tympani Part of the seventh cranial nerve (CN VII) (see Facial nerve); carrying taste from the anterior two-thirds of the tongue and parasympathetic innervation to glands.

Chorea A motor disorder characterized by abnormal, irregular, spasmodic, jerky, uncontrollable movements of the limbs or facial muscles, thought to be caused by degenerative changes in the basal ganglia.

Choroid A delicate membrane; choroid plexuses are found in the ventricles of the brain.

Choroid plexus Vascular structure consisting of pia with blood vessels, with a surface layer of ependymal cells; responsible for the production of cerebrospinal fluid.

Cingulum A bundle of association fibers in the white matter under the cortex of the cingulate gyrus; part of the Papez (limbic) circuit.

Circle of Willis Anastomosis between the internal carotid and basilar arteries, located at the base of the brain, surrounding the pituitary gland.

Cistern(a) Expanded portion of the subarachnoid space containing cerebrospinal fluid (e.g., cisterna magna [cerebello-medullary cistern], lumbar cistern).

Claustrum A thin sheet of gray matter, of unknown function, situated between the lentiform nucleus and the insula.

Clonus Abnormal sustained series of contractions and relaxations following stretch of the muscle; usually elicited in the ankle joint; present following lesions of the descending motor pathways and associated with spasticity.

CNS Abbreviation for central nervous system.

Colliculus A small elevation; superior and inferior colliculi comprising the tectum of the midbrain; also facial colliculus in the floor of the fourth ventricle.

Commissure A group of nerve fibers in the central nervous system connecting structures on one side to the other across the midline (e.g., corpus callosum of the cerebral hemispheres; anterior commissure).

Conjugate eye movement Coordinated movement of both eyes together, so that the image falls on corresponding points of both retinas.

Consensual reflex Light reflex; refers to the bilateral response of the pupil after shining a light in one eye.

Contralateral On the opposite side (e.g., contralateral to a lesion).

Corona radiata Fibers radiating from the internal capsule to various parts of the cerebral cortex—a term often used by neuroradiologists.

Corpus callosum The main (largest) neocortical commissure of the cerebral hemispheres.

Corpus striatum Caudate, putamen, and globus pallidus, the nuclei inside the cerebral hemispheres, with motor function; the basal ganglia.

Cortex Layers of gray matter (neurons and neuropil) on the surface of the cerebral hemispheres (mostly six layers) and cerebellum (three layers).

Cortico-bulbar fibers Descending fibers connecting motor cortex with motor cranial nerve nuclei and other nuclei of brainstem (including reticular formation).

Corticofugal fibers Axons carrying impulses away from the cerebral cortex.

Corticopetal fibers Axons carrying impulses toward the cerebral cortex.

Cortico-spinal tract Descending tract, from motor cortex to anterior (ventral) horn cells of the spinal cord (sometimes direct); also called the pyramidal tract.

Cranial nerve nuclei Collections of cells in the brainstem giving rise to or receiving fibers from cranial nerves (CN III to XII); may be sensory, motor, or autonomic.

Cranial nerves Twelve pairs of nerves arising from the brain and innervating structures of the head and neck (CN II is actually a central nervous system tract).

CSF Cerebrospinal fluid, in the ventricles and subarachnoid space (and cisterns).

CT or CAT scan Computerized (axial) tomography Computed tomography (computerized axial tomography); a diagnostic imaging technique that uses x-rays and computer reconstruction of the brain.

Cuneatus (cuneate) Sensory tract (fasciculus cuneatus) of the dorsal column of spinal cord, from the upper limbs and body; cuneate nucleus of the medulla.

Decerebrate posturing (rigidity) Characterized by extension of the upper and lower limbs; lesion at the brainstem level between the vestibular nuclei and the red nucleus.

Decorticate posturing (rigidity) Characterized by extension of the lower limbs and flexion of the upper; lesion is located above the level of the red nucleus.

Decussation The point of crossing of central nervous system tracts (e.g., decussations of the pyramidal [cortico-spinal] tract, medial lemnisci, and superior cerebellar peduncles).

Dementia Progressive brain disorder that gradually destroys a person's memory, starting with short-term memory, and loss of intellectual ability (e.g., the ability to learn, reason, make judgments, and communicate), and finally, inability to carry out normal activities of daily living; usually affects people of advancing age.

Dendrite Receptive process of a neuron; usually several processes emerge from the cell body, each of which branches in a characteristic pattern.

Dendritic spine Cytoplasmic excrescence of a dendrite and the site of an excitatory synapse.

Dentate (toothed or notched) Dentate nucleus of the cerebellum (intracerebellar nucleus); dentate gyrus of the hippocampal formation.

Dermatome A patch of skin innervated by a single spinal cord segment (e.g., T1 supplies the skin of the inner aspect of the upper arm; T10 supplies umbilical region).

Descending tract Central motor pathway (e.g., from cortex to brainstem or spinal cord).

Diencephalon Consisting of the thalamus, epithalamus (pineal), subthalamus, and hypothalamus.

Diplopia Double vision; a single object is seen as two objects.

Dominant hemisphere The hemisphere responsible for language; this is the left hemisphere in about 85% to 90% of people (including left-handed individuals).

Dorsal column Fasciculus gracilis and fasciculus cuneatus of the spinal cord, pathways (tracts) for discriminative touch, conscious proprioception, and vibration.

Dorsal root Afferent sensory component of a spinal nerve, located in the subarachnoid (cerebrospinal fluid) space.

Dorsal root ganglion (DRG) A group of peripheral neurons along the dorsal root whose axons carry afferent information from the periphery; their central process enters the spinal cord.

Dura Dura mater, the thick external layer of the meninges (brain and spinal cord).

Dural venous sinuses Large venous channels for draining blood from the brain; located within dura of the meninges.

Dysarthria Difficulty with the articulation of words.

Dyskinesia Purposeless movements of the limbs or trunk, usually caused by a lesion of the basal ganglia; also difficulty in performing voluntary movements.

Dysmetria Disturbance of the ability to control the range of movement in muscular action, causing undershooting or overshooting of the target (usually associated with cerebellar lesions).

Dysphagia Difficulty with swallowing.

Dyspraxia Impaired ability to perform a voluntary act previously well performed, with intact movement, coordination, and sensation.

Efferent Away from the central nervous system; usually means motor to muscles.

Emboliform Emboliform nucleus of the cerebellum, one of the intracerebellar (deep cerebellar) nuclei; with globose nucleus forms the interposed nucleus.

Entorhinal Associated with olfaction (smell); the entorhinal area is the anterior part of the parahippocampal gyrus, adjacent to the uncus.

Ependyma Epithelium lining of ventricles of the brain and central canal of spinal cord; specialized tight junctions at the site of the choroid plexus.

Extensor plantar response This is the abnormal plantar reflex indicating a lesion of the pyramidal (cortico-spinal) tract, including an upgoing of the first toe and a fanning of the other toes.

Extrapyramidal system An older clinically used term, usually intended to include the basal ganglia portion of the motor systems and not the pyramidal (cortico-spinal) motor system.

Facial nerve Seventh cranial nerve (CN VII); motor to muscles of facial expression; carries taste from the anterior two thirds of the tongue; also parasympathetic to two salivary glands, lacrimal and nasal glands (see also Chorda tympani).

Falx Dural partition in the midline of the cranial cavity; the large falx cerebri between the cerebral hemispheres, and the small falx cerebelli.

Fascicle A small bundle of nerve fibers.

Fasciculus A large tract or bundle of nerve fibers.

Fasciculus cuneatus Part of the dorsal column of the spinal cord; ascending tract for discriminative touch, conscious proprioception, and vibration from the upper body and upper limb.

Fasciculus gracilis Part of the dorsal column of the spinal cord; ascending tract for discriminative touch, conscious proprioception, and vibration from the lower body and lower limb.

Fastigial nucleus One of the deep cerebellar (intracerebellar) nuclei.

Fiber Synonymous with an axon (either peripheral or central).

Flaccid paralysis Muscle paralysis with hypotonia caused by a lower motor neuron lesion.

Flocculus Lateral part of flocculonodular lobe of cerebellum (vestibulocerebellum).

Folium (plural folia) A flat leaf-like fold of the cerebellar cortex.

Foramen An opening, aperture, between spaces containing cerebrospinal fluid (e.g., Monro, between the lateral ventricles and the third ventricle; Magendie, between the fourth ventricle and the cisterna magna; Luschka, the lateral foramen of the fourth ventricle).

Forebrain Anterior division of embryonic brain; cerebrum and diencephalon.

Fornix The efferent (non-cortical) tract of the hippocampal formation, arching over the thalamus and terminating in the mammillary nucleus of the hypothalamus and in the septal region.

Fourth ventricle Cavity between the brainstem and the cerebellum and containing cerebrospinal fluid.

Funiculus A large aggregation of white matter in the spinal cord, may contain several tracts.

Ganglion (plural ganglia) A collection of nerve cells in the peripheral nervous system—dorsal root ganglion (DRG) and sympathetic ganglion; also inappropriately used for certain regions of gray matter in the brain (i.e., basal ganglia).

Geniculate bodies Specific relay nuclei of thalamus—medial (auditory) and lateral (visual).

Genu Knee or bend; middle portion of the internal capsule; genu of the facial nerve.

Glial cell Also called neuroglial cell; supporting cells in the central nervous system—astrocyte, oligodendrocyte, and ependymal—also microglia.

Globus pallidus Efferent part of basal ganglia; part of the lentiform nucleus with the putamen; located medially.

Glossopharyngeal nerve Ninth cranial nerve (CN IX); motor to muscles of swallowing and carries taste from the posterior one third of the tongue; nerve for the gag reflex.

Gracilis (gracile) Sensory tract (fasciculus gracilis) of the dorsal column of the spinal cord; nucleus gracilis of the medulla.

Gray matter Nervous tissue, mainly nerve cell bodies and adjacent neuropil; looks "grayish" after fixation in formalin.

Gyrus (plural gyri) A convolution or fold of the cerebral hemisphere; includes cortex and white matter.

Habenula A nucleus of the limbic system, adjacent to the posterior end of the roof of the third ventricle (part of the epithalamus).

Hemiballismus Violent jerking or flinging movements of one limb, not under voluntary control, caused by a lesion of subthalamic nucleus.

Hemiparesis Muscular weakness affecting one side of the body.

Hemiplegia Paralysis of one side of the body.

Herniation Bulging or expansion of the tissue beyond its normal boundary.

Heteronymous hemianopia Loss of different halves of the visual field of both eyes, as defined by projection to the visual cortex of both sides; bitemporal for the temporal halves and binasal for the nasal halves.

Hindbrain Posterior division of the embryonic brain; includes the pons, medulla, and cerebellum (located in the posterior cranial fossa).

Hippocampus or hippocampus "proper" Part of the limbic system; a cortical area "buried" within the medial temporal lobe and consisting of phylogenetically old (three-layered) cortex; protrudes into the floor of the inferior horn of the lateral ventricle.

Homonymous hemianopia Loss of the same visual field in both eyes (i.e., left or right) as defined by the projection to the visual cortex on one side—involving the nasal half of the visual field in one eye and the temporal half in the other eye; also quadrantanopia.

Horner syndrome Miosis (constriction of the pupil), anhidrosis (dry skin with no sweat), and ptosis (drooping of the upper eyelid) resulting from a lesion of the sympathetic pathway to the head.

Hydrocephalus Enlargement of the ventricles, usually from excessive accumulation of cerebrospinal fluid within the ventricles (e.g., obstruction).

Hypoglossal nerve Twelfth cranial nerve (CN XII); motor to muscles of the tongue.

Hypokinesia Markedly diminished movements (spontaneous).

Hyporeflexia/hyperreflexia Decrease (hyporeflexia) or increase (hyperreflexia) of the stretch (deep tendon) reflex.

Hypothalamus A region of the diencephalon that serves as the main controlling center of the autonomic nervous system and is involved in several limbic circuits; also regulates the pituitary gland.

Hypotonia/hypertonia Decrease (hypotonia) or increase (hypertonia) of the tone of muscles, manifested by decreased or increased resistance to passive movements.

Infarction Local death of an area of tissue caused by loss of its blood supply.

Infundibulum (funnel) Infundibular stem of the posterior pituitary (neurohypophysis).

Innervation Nerve supply, sensory and/or motor.

Insula (island) Cerebral cortical area not visible from the outside view and situated at the bottom of the lateral fissure (also called the island of Reil).

Internal capsule White matter between the lentiform nucleus and head of caudate nucleus and the thalamus; consists of an anterior limb, genu, and posterior limb.

Ipsilateral On the same side of the body (e.g., ipsilateral to a lesion).

Ischemia A condition in which an area is not receiving an adequate blood supply.

Ischemic penumbra A region adjacent to or surrounding an area of infarcted brain tissue that is not receiving sufficient blood; the neurons may still be viable.

Kinesthesia The conscious sense of position and movement.

Lacune A pathological small "hole" remaining after an infarct in the internal capsule; also an irregularly shaped venous "lake" or channel draining into the superior sagittal sinus.

Lateral ventricle Cerebrospinal fluid cavity in each cerebral hemisphere; consists of an anterior horn, body, atrium (or trigone), posterior horn, and inferior (temporal) horn.

Lemniscus A specific pathway in the central nervous system (medial lemniscus for discriminative touch, conscious proprioception, and vibration; lateral lemniscus for audition).

Lentiform Lens-shaped; lentiform nucleus, a part of the corpus striatum; also called lenticular nucleus; composed of the putamen (laterally) and globus pallidus.

Leptomeninges Arachnoid and pia mater, part of the meninges.

Lesion Any injury or damage to tissue (e.g., vascular, traumatic).

Limbic system Part of the brain associated with emotional behavior.

Locus ceruleus A small nucleus located in the uppermost pons on each side of the fourth ventricle; contains melanin-like pigment, visible as a dark-bluish area in freshly sectioned brain.

Lower motor neuron Anterior horn cell of the spinal cord and its axon; also the cells in the motor cranial nerve nuclei of the brainstem; called the alpha motor neuron; its loss leads to atrophy of the muscle and weakness, with hypotonia and hyporeflexia; also fasciculations are to be noted.

Mammillary Mammillary bodies; nuclei of the hypothalamus that are seen as small swellings on the ventral surface of diencephalon (also spelled mamillary).

Massa intermedia A bridge of gray matter connecting the thalami of the two sides across third ventricle; present in 70% of human brains (also called the inter-thalamic adhesion).

Medial lemniscus Brainstem portion of the sensory pathway for discriminative touch, conscious proprioception, and vibration, formed after synapse (relay) in the nucleus gracilis and nucleus cuneatus.

Medial longitudinal fasciculus (MLF) A tract throughout the brainstem and upper cervical spinal cord that interconnects visual and vestibular input with other nuclei controlling movements of the eyes and the head and neck.

Medulla Caudal portion of the brainstem; may also refer to the spinal cord as in a lesion within (intramedullary) or outside (extramedullary) the spinal cord.

Meninges Covering layers of the central nervous system (dura, arachnoid, and pia).

Mesencephalon The midbrain (upper part of the brainstem).

Microglia The "scavenger" cells of the central nervous system (i.e., macrophages); considered by some as one of the neuroglia.

Midbrain Part of the brainstem; also known as the mesencephalon (the middle division of the embryonic brain).

Motor Associated with movement or response.

Motor unit A lower motor neuron, its axon, and the muscle fibers that it innervates.

MRI (NMR) Magnetic resonance imaging (nuclear magnetic resonance), a diagnostic imaging technique that uses an extremely strong magnet, not x-rays.

Muscle spindle Specialized receptor within voluntary muscles that detects muscle length; necessary for the stretch/myotatic reflex (deep tendon reflex); contains muscle fibers within itself capable of adjusting the sensitivity of the receptor.

Myelin Proteolipid layers surrounding nerve fibers, formed in segments, which is important for rapid (saltatory) nerve conduction.

Myelin sheath Covering of a nerve fiber, formed and maintained by the oligodendrocyte in the central nervous system and the Schwann cell in the peripheral nervous system; interrupted by nodes of Ranvier.

Myelopathy Generic term for disease affecting the spinal cord.

Myopathy Generic term for muscle disease.

Myotatic reflex Stretch reflex, also called deep tendon reflex (DTR); elicited by stretching the muscle; causes a reflex contraction of the same muscle; monosynaptic (also spelled myotactic reflex).

Myotome Muscle groups innervated by a single spinal cord segment; in fact, usually two adjacent segments are involved (e.g., biceps, C5 and C6).

Neocerebellum The phylogenetically newest part of the cerebellum, present in mammals and especially well developed in humans; involved in coordinating precise voluntary movements and also in motor planning.

Neocortex The phylogenetically newest part of the cerebral cortex, consisting of six layers (and sublayers); characteristic of mammals and constituting most of the cerebral cortex in humans.

Neostriatum The phylogenetically newer part of the basal ganglia, consisting of the caudate nucleus and putamen; also called the striatum.

Nerve fiber Axonal cell process, plus myelin sheath, if present.

Neuralgia Pain—severe, shooting, "electrical," along the distribution of a peripheral nerve (spinal or cranial).

Neuraxis The straight longitudinal axis of the embryonic or primitive neural tube, bent in later evolution and development.

Neuroglia Accessory or interstitial cells of the central nervous system; they include astrocytes, oligodendrocytes, ependymal cells, and microglial cells.

Neuron The basic structural unit of the nervous system, consisting of the nerve cell body and its processes—dendrites and axon.

Neuropathy Disorder of one or more peripheral nerves.

Neuropil An area between nerve cells consisting of a complex arrangement of nerve cell processes, including axon terminals, dendrites, and synapses.

Nociception Refers to an injurious stimulus causing a neuronal response; may or may not be associated with the sensation of pain.

Node of Ranvier Gap in the myelin sheath between two successive internodes; necessary for saltatory (rapid) conduction.

Nucleus (plural nuclei) An aggregation of neurons within the central nervous system; in histology, the nucleus of a cell.

Nystagmus An involuntary oscillation of the eye(s), slow in one direction and rapid in the other; named for the direction of the quick movement.

Oculomotor nerve Third cranial nerve (CN III); motor to most muscles of the eye.

Olfactory nerve First cranial nerve (CN I); special sense of smell.

Oligodendrocyte A neuroglial cell, forms and maintains the myelin sheath in the central nervous system; each cell is responsible for several internodes on different axons.

Optic chiasm(a) Partial crossing of optic nerves—nasal half of retina representing the temporal visual fields—after which the optic tracts are formed.

Optic disc Area of the retina where the optic nerve exits; also the site for the central retinal artery and vein; devoid of receptors, hence the blind spot.

Optic nerve Second cranial nerve (CN II); special sense of vision; actually a tract of the central nervous system, from the ganglion cells of the retina until the optic chiasm.

Paleocortex The phylogenetically older cerebral cortex consisting of three to five layers.

Papilledema Edema of the optic disc, visualized with an ophthalmoscope (also called a choked disc); usually a sign of abnormal increased intracranial pressure.

Paralysis Complete loss of muscular action.

Paraplegia Paralysis of both legs and lower part of trunk.

Paresis Muscle weakness or partial paralysis.

Paresthesia Spontaneous abnormal sensation (e.g., tingling; pins and needles).

Pathway A chain of functionally related neurons (nuclei) and their axons, making a connection between one region of the central nervous system and another; a tract (e.g., visual pathway, dorsal column-medial lemniscus sensory pathway).

Peduncle A thick stalk or stem; a bundle of nerve fibers (cerebral peduncle of the midbrain; also three cerebellar peduncles—superior, middle, and inferior).

Perikaryon The cytoplasm surrounding the nucleus of a cell; sometimes refers to the cell body of a neuron.

Peripheral nervous system (PNS) Nerve roots, peripheral nerves, and ganglia outside the central nervous system (motor, sensory, and autonomic).

PET Positron emission tomography; a technique used to visualize areas of the living brain that become "activated" under certain task conditions; uses very short-acting biologically active radioactive compounds.

Pia (mater) The thin innermost layer of the meninges, attached to the surface of the brain and spinal cord; forms the inner boundary of the subarachnoid space.

Plexus An interweaving arrangement of vessels or nerves.

Pons (bridge) The middle section of the brainstem that lies between the medulla and the midbrain; appears to constitute a bridge between the two hemispheres of the cerebellum.

Projection fibers Bidirectional fibers connecting the cerebral cortex with structures below, including basal ganglia, thalamus, brainstem, and spinal cord.

Proprioception The sense of body position (conscious or unconscious).

Proprioceptor One of the specialized sensory endings in muscles, tendons, and joints; provides information concerning movement and position of body parts (proprioception).

Prosody Vocal tone, inflection, and melody accompanying speech.

Ptosis Drooping of the upper eyelid.

Pulvinar The posterior nucleus of the thalamus; functionally, involved with vision.

Putamen The larger (lateral) part of the lentiform nucleus, with the globus pallidus; part of the neostriatum with the caudate nucleus.

Pyramidal system Named because the cortico-spinal tracts occupy pyramid-shaped areas on the ventral aspect of the medulla; may include cortico-bulbar fibers; the term pyramidal tract refers specifically to the corticospinal tract.

Quadrigeminal Referring to the four colliculi of the midbrain; also called the tectum.

Quadriplegia Paralysis affecting the four limbs (also called tetraplegia).

Radicular Refers to a nerve root (motor or sensory).

Ramus (plural rami) The division of the mixed spinal nerve (containing sensory, motor, and autonomic fibers) into anterior and posterior.

Raphe An anatomical structure in the midline; in the brainstem, several nuclei of the reticular formation are in the midline of the medulla, pons, and midbrain (these nuclei use serotonin as the neurotransmitter).

Red nucleus Nucleus in the midbrain (reddish color in a fresh specimen).

Reflex Involuntary movement of a fixed nature in response to a stimulus.

Reflex arc Consisting of an afferent fiber, a central connection, a motor neuron, and its efferent axon leading to a muscle movement.

Reticular Pertaining to or resembling a net—reticular formation of brainstem.

Reticular formation Diffuse nervous tissue, nuclei, and connections, in brainstem; quite old phylogenetically.

Rhinencephalon In humans, refers to structures related to the olfactory system.

Rigidity Abnormal muscle stiffness (increased tone) with increased resistance to passive movement of both agonists and antagonists (e.g., flexors and extensors), usually seen in Parkinson's disease; velocity independent.

Root The peripheral nerves—sensory (afferent, dorsal) and motor (efferent, ventral)—as they emerge from the spinal cord and are found in the subarachnoid space.

Rostral Toward the nose, or the most anterior end of the neuraxis.

Rubro Red; pertaining to the red nucleus, as in rubrospinal tract and cortico-rubral fibers.

Saccadic To jerk; extremely quick movements, normally of both eyes together (conjugate movement), in changing the direction of gaze.

Schwann cell Neuroglial cell of the peripheral nervous system responsible for formation and maintenance of myelin; there is one Schwann cell for each internode of myelin.

Secretomotor Parasympathetic motor nerve supply to a gland.

Sensory Afferent; to do with receiving information, from the skin, the muscles, the external environment, or internal organs.

Septal region An area below the anterior end of the corpus callosum on the medial aspect of the frontal lobe that includes cortex and the septal nuclei.

Septum pellucidum A double membrane of connective tissue separating the anterior horns of the lateral ventricles, situated in the median plane.

Somatic Used in neurology to denote the body, exclusive of the viscera (as in somatic afferent neurons from the skin and body wall); the word soma is also used to refer to the cell body of a neuron.

Somatic senses Touch (discriminative and crude), pain, temperature, proprioception, and the "sense of vibration."

Somatotopic The orderly representation of the body parts in central nervous system pathways, nuclei, thalamus, and cortex; topographical representation.

Somesthetic Consciousness of having a body; somesthetic senses are the general senses of touch, pain, temperature, position, movement, and "vibration."

Spasticity Velocity-dependent increased tone and increased resistance to passive stretch of the antigravity muscles; in humans, flexors of the upper limb and extensors of the lower limb; usually accompanied by hyperreflexia.

Special senses Sight (vision), hearing (audition), balance (vestibular), taste (gustatory), and smell (olfactory).

Spinal accessory nerve Eleventh cranial nerve (CN XI); refers usually to the part of the nerve that originates in the upper spinal cord (C1 to C5) and innervates the muscles of the neck, the sternomastoid and trapezius muscles.

Spinal shock Complete "shut down" of all spinal cord activity (in humans) following an acute complete lesion of the cord (e.g., severed cord after a diving or motor vehicle accident); usually up to 2 to 3 weeks in duration.

Spino-cerebellar tracts Ascending tracts of the spinal cord, anterior and posterior, for "unconscious" proprioception to the cerebellum.

Spino-thalamic tracts Ascending tracts of the spinal cord for pain and temperature (lateral) and non-discriminative or light touch and pressure (anterior).

Split brain A brain in which the corpus callosum has been severed in the midline, usually as a therapeutic measure for intractable epilepsy.

Stereognosis The recognition of an object using the tactile senses and also central processing, involving association areas especially in the parietal lobe.

Strabismus A squint; lack of conjugate fixation of the eyes; may be constant or variable.

Stria A slender strand of fibers (e.g., stria terminalis from amygdala).

Striatum The phylogenetically more recent part of the basal ganglia (neostriatum) consisting of the caudate nucleus and the putamen (lateral portion of the lentiform nucleus).

Stroke A sudden severe attack of the central nervous system; usually refers to a sudden focal loss of neurological function caused by death of neural tissue; mostly resulting from a vascular lesion, either infarct (embolus, occlusion) or hemorrhage.

Subarachnoid space Space between the arachnoid and pia mater and containing cerebrospinal fluid.

Subcortical Not in the cerebral cortex (i.e., at a functionally or evolutionary "lower" level in the central nervous system); usually refers to the white matter of the cerebral hemispheres, and also may include the basal ganglia.

Subicular region Part of hippocampal formation; transitional cortex (three to five layers) between that of the hippocampus proper and the parahippocampal gyrus.

Substantia gelatinosa A nucleus of the gray matter of the dorsal (sensory) horn of the spinal cord composed of small neurons; receives pain and temperature afferents.

Substantia nigra A flattened nucleus in the midbrain with motor functions—consisting of two parts: the pars compacta with melanin pigment in the neurons (the dopamine neurons, which degenerate in Parkinson's disease) and the pars reticulata, which is an output nucleus of the basal ganglia.

Subthalamus Region of the diencephalon beneath the thalamus that contains fiber tracts and the subthalamic nucleus; part of the functional basal ganglia.

Sulcus (plural sulci) Groove between adjacent gyri of the cerebral cortex; a deep sulcus may be called a fissure.

Synapse Area of structural and functional specialization between neurons where transmission occurs (excitatory, inhibitory, or modulation), using neurotransmitter substances (e.g., glutamate, gamma-aminobutyric acid [GABA]); similarly at the neuromuscular junction (using acetylcholine).

Syringomyelia A pathological condition characterized by expansion of the central canal of the spinal cord with destruction of nervous tissue around the cavity.

Tectum The "roof" of the midbrain (behind the aqueduct) consisting of the paired superior and inferior colliculi; also called the quadrigeminal plate.

Tegmentum The "core area" of the brainstem, between the ventricle (or aqueduct) and the cortico-spinal tract; contains the reticular formation, cranial nerve and other nuclei, and various tracts.

Telencephalon Rostral part of embryonic forebrain; primarily the cerebral hemispheres of the adult brain.

Tentorium The tentorium cerebelli is a sheet of dura between the occipital lobes of the cerebral hemispheres and the cerebellum; its hiatus or notch is the opening for the brainstem—at the level of the midbrain.

Thalamus A major portion of the diencephalon with sensory, motor, and integrative functions; consists of several nuclei with connections to areas of the cerebral cortex.

Third ventricle Midline ventricle at the level of the diencephalon (between the thalamus of each side) and containing cerebrospinal fluid.

Tic Brief, repeated, stereotyped, semipurposeful muscle contraction; not under voluntary control, although it may be suppressed for a limited time.

Tinnitus Persistent ringing or buzzing sound in one or both ears.

Tomography Radiological images, done sectionally, including computed tomography and magnetic resonance imaging.

Tone Referring to muscle, its firmness, and elasticity—normal tone, hypertonia, hypotonia—elicited by passive movement and also assessed by palpation.

Tract A bundle of nerve fibers within the central nervous system, with a common origin and termination (e.g., optic tract, cortico-spinal tract).

Transient ischemic attack (TIA) A nonpermanent focal deficit, caused by a vascular event; by definition, usually reversible within a few hours, up to a maximum of 24 hours.

Trapezoid body Transverse crossing fibers of the auditory pathway situated in the ventral portion of the tegmentum of the lower pons.

Tremor Oscillating, "rhythmic" movements of the hands, limbs, head, or voice; intention (kinetic) tremor of the limb commonly seen with cerebellar lesions; tremor at rest commonly associated with Parkinson disease.

Trigeminal nerve Fifth cranial nerve (CN V); major sensory nerve of the head (face, eye, tongue, nose, sinuses); also supplies muscles of mastication.

Trochlear nerve Fourth cranial nerve (CN IV); motor to the superior oblique eye muscle.

Two-point discrimination Recognition of the simultaneous application of two points close together on the skin; distance varies with the area of the body (compare fingertip to back).

Uncus An area of cortex—the medial protrusion of the rostral (anterior) part of the parahippocampal gyrus of the temporal lobe; the amygdala is situated deep to this area; important clinically as in uncal herniation.

Upper motor neuron Neuron located in the motor cortex or other motor areas of the cerebral cortex or in the brainstem—giving rise to a descending tract to lower motor neurons in the brainstem (for cranial nerves) or spinal cord (for body and limbs).

Upper motor neuron lesion A lesion of the brain (cortex, white matter of hemisphere), brainstem, or spinal cord interrupting descending motor influences to the lower motor neurons of the brainstem or spinal cord and characterized by weakness, spasticity, hyperreflexia, and often clonus; usually accompanied by an extensor plantar response.

Vagus Tenth cranial nerve (CN X); supplies motor fibers to the larynx; the major parasympathetic nerve to organs of the thorax and abdomen.

Velum A membranous structure; the superior medullary velum forms the roof of the fourth ventricle.

Ventricles Cerebrospinal fluid–filled cavities inside the brain.

Vermis Unpaired midline portion of the cerebellum between the hemispheres.

Vertigo Abnormal sense of spinning, whirling, or motion, either of the self or of one's environment.

Vestibulocochlear Eighth cranial nerve (CN VIII); special senses of hearing and balance (acoustic nerve is not really correct).

White matter Nervous tissue of the central nervous system that is made up of nerve fibers (axons), some of which are myelinated; appears "whitish" after fixation in formalin.

ANNOTATED BIBLIOGRAPHY

This is a select list of references with some commentary to help the learner choose additional learning resources about the structure, function, and diseases of the human brain.

The perspective is for medical students and practitioners not involved with neurology, as well as those in related fields in the allied health professions. The listing includes text and atlases and some Web sites; more recent publications (since 2000) have been preferentially selected.

Note to the Learner: This listing has been updated to June 2014.

NEUROANATOMICAL TEXTS

Afifi, A.K., and Bergman, R.A. *Functional Neuroanatomy Text and Atlas,* 2nd ed., 2005, Lange Medical Books, McGraw-Hill, New York.

This is a neuroanatomical text with the addition of functional information on clinical syndromes. A chapter on normal features is followed by a chapter on clinical syndromes (e.g., of the cerebellum). The book is richly illustrated (in two colors) using semi-anatomical diagrams and magnetic resonance imaging (MRI) scans. Each chapter has key points at the beginning and terminology for that chapter at the end. It is a pleasant book visually and quite readable. There is an atlas of the central nervous system (CNS) at the end, but it is not in color, and also several brain MRI scans.

Arslan, O. *Neuroanatomical Basis of Clinical Neurology,* 2nd ed., 2014, CRC Press, Boca Raton, FL.

This book has been thoroughly and extensively revised since its first publication (in 2001), including the illustrations and the text. The orientation remains the same—to bridge the gap between the basic neuroanatomy and neurological diseases. As a neuroanatomy text book, it is quite detailed in describing the structures and connections of the central nervous system, with plentiful illustrations. The clinical aspects are added in separate "boxes," sometimes accompanied by an illustration; the disease entities are not case-based.

Carpenter, M.B. *Core Text of Neuroanatomy,* 4th ed., 1991, Williams & Wilkins, Baltimore.

This "classic" textbook by a highly respected author presents a detailed description of the nervous system from the perspective of a neuroanatomist. A more complete version is also available as a reference text—*Carpenter's Human Neuroanatomy,* 9th ed., 1995, now with A. Parent as the author.

FitzGerald, M.J.T., Gruener, G., and Mtui, E., eds. *Clinical Neuroanatomy and Neuroscience,* 6th ed., 2012, Saunders Elsevier, Edinburgh.

These are the same co-authors who edited the fifth edition of the FitzGerald book. That edition added several chapters, including electrical events, transmitters and receptors (and modulators), electrodiagnostic examination, electroencephalography and evoked potentials, with modifications of some of the illustrations and some additional ones. No major change was noted in this sixth edition. The online version has related videos, some of which are clinically related.

Haines, D.E., ed. *Fundamental Neuroscience for Basic and Clinical Applications,* 4th ed., 2013, Saunders Elsevier, Philadelphia.

This multi-edited large text, with many color illustrations, is an excellent reference book, mainly for neuroanatomical detail. The title of the previous (third) edition was changed to *Fundamental Neuroscience for Basic and Clinical Applications,* reflecting a modification of the approach to the subject matter. The size of the book was also increased. Further enhancements have been done for the fourth edition, but no major changes were noted. The related chapter videos were not clinically related.

Kandel, E.R., ed. *Principles of Neural Science,* 5th ed., 2013, McGraw-Hill Medical, New York.

This multi-edited thorough textbook presents a physiological depiction of the nervous system, with experimental details and information from animal studies. It is suitable as a reference book and for graduate students. It is extensively updated since the last edition (in 2000), with new information available from molecular biology and genetics and on diseases of the nervous system. This edition explores new approaches for the understanding of the behavior of systems and cognition.

Kiernan, J.A., and Rajakumar, N. *Barr's The Human Nervous System: An Anatomical Viewpoint,* 10th ed., 2014, Wolters Kluwer/Lippincott Williams & Wilkins, Baltimore.

This newest edition of Barr's book presents a change of formatting, but this is still a "classic" neuroanatomical textbook. Most of the diagrams are the same, now with added color, and the clinical notes remain. The chapter on imaging has been updated with more examples. The chapters on major systems are presented as previously. The book is clearly written and clearly presented, and it includes a glossary. There is no longer a CD-ROM, but purchasing the book provides access to the publisher's Web site with many learning resources.

Kolb, B., and. Whishaw, I.Q., eds. *Fundamentals of Human Neuropsychology,* 6th ed., 2009, Worth Publishers, New York.

This is a classic in the field and highly recommended for a good understanding of the human brain in action. Topics discussed include memory, attention, language, and the limbic system.

Martin, J.H. *Neuroanatomy: Text and Atlas,* 4th ed., 2012, McGraw-Hill, New York.

This is a very complete text with a neuroanatomical perspective. The material is clearly presented, with explanations of how systems function. This edition has undergone a major revision and presents a major upgrade of all the illustrations to full color. The chapter on the development of the CNS has been dropped, and the somatic sensory section has been expanded. Each chapter begins with a clinical case scenario and ends with summary and study (multiple-choice) questions (with answers). A detailed atlas section is included at the end, as well as a glossary of terms.

Nieuwenhuys, R., Voogd, J., and van Huijzen, C., eds. *The Human Central Nervous System,* 4th ed., 2008, Springer, Berlin.

This edition is totally different from previous editions—now it is a thick neuroanatomy textbook, with many of the original illustrations (1981, see the listing under Neuroanatomical Atlases) and many more new illustrations (with color). Section I provides an orientation, Section II is descriptive, and Section III describes functional systems.

Nolte, J. *The Human Brain,* 6th ed., 2009, Mosby Elsevier, St. Louis.

This is a new edition of an excellent neuroscience text, with anatomical and functional (physiological) information on the nervous system, complemented by clinically relevant material. The textbook includes scores of illustrations in full color, stained brainstem and spinal cord cross-sections, and three-dimensional brain reconstructions by John Sundsten. A glossary has been added. The book is now thicker, in part because of an increase in the font size of the text (making the text easier to read), with added material to reflect new knowledge. Existing illustrations have been enhanced, with new illustrations such as diffusion tensor imaging of tracts. This is a "Student Consult" book with access to the online material available with the purchase of the book.

Steward, O. *Functional Neuroscience,* 2000, Springer, Berlin.

According to the author, this is a book for medical students that blends the physiological systems approach with the structural aspects. The emphasis is on the "processing" of information, for example, in the visual system. Chapters at the end discuss arousal, attention, consciousness, and sleep. It is nicely formatted and readable.

Williams, P., and Warwick, R. *Functional Neuroanatomy of Man,* 1975, Saunders, Philadelphia.

This is the "neuro" section from *Gray's Anatomy.* Although somewhat dated, there is excellent reference material on the CNS, as well as the nerves and autonomic parts of the peripheral nervous system. The limbic system and its development are also well described.

Wilson-Pauwels, L., Akesson, E.J., and Stewart, P.A. *Cranial Nerves: Anatomy and Clinical Comments,* 1988, B.C. Decker, Toronto.

This is a handy resource on the cranial nerves, with some very nice illustrations. It is relatively complete and easy to follow.

NEUROANATOMICAL ATLASES

Crossman, A.R., and Neary, D. *Neuroanatomy: An Illustrated Colour Text,* 4th ed., 2010, Elsevier Churchill Livingstone, Edinburgh.

This book format is part of a series of "Illustrated Colour Texts" (the companion neurology text is by Fuller and Manford; see Clinical Texts). The text is accompanied by many illustrations, including diagrams and photographs of the brain. This book may be a helpful addendum for review purposes to medical students, and it may be useful for health sciences students learning about the nervous system for the first time.

DeArmond, S.J., Fusco, M.M., and Dewey, M.M. *Structure of the Human Brain: A Photographic Atlas,* 3rd ed., 1989, Oxford University Press, Oxford.

This is an excellent and classic reference to the neuroanatomy of the human CNS. It has no explanatory text and no color.

England, M.A., and Wakely, J. *Color Atlas of the Brain and Spinal Cord,* 2nd ed., 2006, Mosby Elsevier, St. Louis.

This is a very well-illustrated atlas, with most of the photographs and sections in color. It has little in the way of explanatory text. This atlas has been updated with enhancement of the colors, the addition of new illustrations (new histological stains, functional MRI), and a larger, clearer format.

Felten, D. L., and Shetty, A.N., eds. *Netter's Atlas of Neuroscience,* 2nd ed., 2010, Saunders Elsevier, Philadelphia.

The familiar illustrations of Netter on the nervous system have been collected into a single atlas, with limited commentary. The diagrams are extensively labeled. The co-authorship has been changed to Anil N. Shetty for this edition. Both peripheral and autonomic nervous systems are included. There has been a major reorganization of the material into chapters, with text boxes added for clinical points. New material has been included, particularly imaging (computed tomography, MRI, positron emission tomography, and diffusion tensor imaging). Related videos in the online version often have clinical relevance.

Fix, J.D. *Atlas of the Human Brain and Spinal Cord*, 2nd ed., 2008, Jones and Bartlett Publishers, Boston.

This atlas is designed to provide a photographic survey of the central nervous system, both gross and microscopic views. The sections are from the special Yakovlev Collection of the Armed Forces Institute of Pathology in Washington, D.C. The illustrations are shown on one side with the structures numbered and the numbers with the names of the structures are found on the opposite page. No text is provided. There is some additional clinical material including lesions, tumors and other diseases.

Haines, D. *Neuroanatomy: An Atlas of Structures, Sections and Systems,* 8th ed., 2012, Wolters Kluwer/Lippincott, Williams & Wilkins, Baltimore.

In addition to enhancements of the illustrations (e.g., color), new or additional neuroimaging has been added. The section on functional systems (pathways, tracts) has been significantly modified, with the addition of images and information regarding lesions. There is an expanded description of the cranial nerves. This edition does not have a CD-ROM, but access to "thePoint," the publisher's online resource is available with purchase of the book. Some United States Medical Licensing Examination (USMLE) questions and answers are still in the text, but all are available online.

Netter, F.H. *The CIBA Collection of Medical Illustrations,* Volume 1, Part 1, 1983, CIBA, Summit, NJ.

This is a classic. It has excellent illustrations of the nervous system, as well as of the skull, the autonomic and peripheral nervous systems, and embryology. The text is interesting but may be dated. The *Netter Collection of Medical Illustrations,* 2nd ed., 2013, has been published by Saunders Elsevier—Volume 7, The Nervous System: Part I, The Brain, and Part II, Spinal Cord and Peripheral Nervous System, edited by H. Royden Jones, T.M. Burns, M.J. Aminoff, and S.C. Pomeroy.

The editors have combined Netter's classic medical illustrations with a new text including basic science and clinical information "discussing the anatomy, physiology, pathology and clinical presentation of many neurological disorders." Also see *Netter's Atlas of Neuroscience,* by Felton and Shetty (listed earlier in this section of the bibliography), as well as *Netter's Neurology,* edited by Royden (listed under Clinical Texts in this bibliography).

Nieuwenhuys, R., Voogd, J., and van Huijzen, C., eds. *The Human Central Nervous System: A Synopsis and Atlas,* 2nd ed., 1981, Springer, Berlin.

Unique three-dimensional drawings of the CNS and its pathways are presented, in tones of gray. These diagrams are extensively labeled, with no explanatory text.

Nolte, J., and Angevine, J.B., eds. *The Human Brain in Photographs and Diagrams,* 4th ed., 2013, Elsevier Saunders, Philadelphia.

This is a well-illustrated color atlas with neuroradiology. Functional systems are drawn onto the brain sections with emphasis on the neuroanatomy; the accompanying text is quite detailed. The third edition (2007) increased the number of illustrations, with color enhancements of the existing illustrations as well as improved MRI images. The labeling of the illustrations was modified to be more selective for students. The format size is now larger, and standard binding is used (not spiral binding). The glossary now includes small images. All the illustrations in the text are on the CD-ROM included with the book. Excellent three-dimensional brain reconstructions by J.W. Sundsten are still included.

Woolsey, T.A., Hanaway, J., and Gado, M.H., eds. *The Brain Atlas: A Visual Guide to the Human Central Nervous System,* 3rd ed., 2008, John Wiley and Sons, Hoboken, NJ.

This atlas presents a complete pictorial presentation of the human brain, with labeled illustrations, some including color, and radiographic material. Parts III and IV consist of histological sections of the hemispheres, brainstem, spinal cord, and limbic structures. Part V presents the pathways, accompanied by some explanatory text.

CLINICAL TEXTS

Asbury, A.K., McKhann, G.M., McDonald, W.I., Goodsby, P.J., and McArthur, J.C. *Diseases of the Nervous System: Clinical Neurobiology,* 3rd ed., 2002, Cambridge University Press, Cambridge.

This is a complete neurology text, in two volumes, on all aspects of basic and clinical neurology and the therapeutic approach to diseases of the nervous system.

Donaghy, M., ed. *Brain's Diseases of the Nervous System,* 12th ed., 2009, Oxford University Press, Oxford.

A very trusted source of information about clinical diseases and their treatment.

Greenberg, D.A., Aminoff, M.J., and Simon, R.P., eds. *Clinical Neurology,* 8th ed., 2012, McGraw-Hill, New York.

If a student wishes to consult a clinical book for a quick look at a disease or syndrome, then this is a suitable book of the survey type. Clinical findings are given, and investigative studies are included, as well as treatment. The illustrations are adequate (in two colors), and there are many tables with classifications and causes.

Fuller, G., and Manford, M., eds. *Neurology: An Illustrated Colour Text,* 3rd ed., 2010, Elsevier Churchill Livingstone, Edinburgh.

A concise explanation of select clinical entities is presented, with many illustrations (in full color); it is not a comprehensive textbook. The large format and presentation make this an appealing but limited book. At the end there are several case histories and the answers. This new edition includes "newer investigations and treatments" and a brief section on sleep, but it is otherwise unchanged from the previous (2006) edition.

Kasper, D.L., Braunwald, E., Fauci, A.S., Hauser, S.L., Longo, D.L., and Jameson, J.L., eds., *Harrison's Principles of Internal Medicine,* 18th ed., 2012, McGraw-Hill, New York.

Harrison's Online, Volume 1096-7133, McGraw-Hill, New York.

Harrison's is a trusted, authoritative source of information, with few illustrations. Part 2 in Section 3 (Volume I) has chapters on the presentation of disease; Part 15 (Volume II) is on all neurological disorders of the CNS, nerve and muscle diseases, and mental disorders. The online version of *Harrison's* has updates, search capability, practice guidelines, and online lectures and reviews, as well as illustrations.

Hendelman, W.J., Humphreys, P., and Skinner, C. *The Integrated Nervous System: A Systematic and Diagnostic Approach*, 2010, CRC Press, Taylor and Francis Group, Boca Raton, FL.

This is a clinical case-based textbook which is co-authored by Dr. W. Hendelman (the author of this *Atlas of Functional Neuroanatomy*) with two clinical neurologists, Dr. P. Humphreys (pediatric) and Dr. C. Skinner (adult). The textbook is accompanied by an extensive Web site with over 40 clinical cases, including radiographs and other clinical tests. Its goal is to teach non-neurologists (medical students and others) how to localize a neurological problem and how to determine the most likely etiology of the disease process.

Ropper, A.H., and Martin, A. S. *Adams and Victor's Principles of Neurology,* 9th ed., 2009, McGraw-Hill, New York.

This is a comprehensive neurology text—devoted partly to cardinal manifestations of neurological diseases and partly to major categories of diseases.

Rowland, L.P., Pedley, T.A., and Merritt, H.H., eds. *Merritt's Neurology,* 12th ed., 2010, Lippincott Williams & Wilkins, Philadelphia.

This is a well-known, complete, and trustworthy neurology textbook, now edited by L.P. Rowland.

Royden, H.R., Jr., editor-in-chief. *Netter's Neurology,* 2nd ed., 2012, Elsevier Saunders, Philadelphia.

Netter's neurological illustrations have been collected in one textbook, with the addition of Netter-style clinical pictures; these add an interesting dimension to the descriptive text. There is broad coverage of many disease states, although not in depth, with clinical scenarios in each chapter. It is now available with a CD-ROM. The Preface to this edition states: "Every chapter in this second edition has been carefully reviewed and in most instances significantly rewritten." Also, new vignettes have been added as well as new MRI images. The number of chapters has been reduced from 108 to 76, indicating a significant review of the book. The online edition of the book has many related videos that are clinically relevant. The look and feel of the text remain "Netter."

PEDIATRIC NEUROLOGY

Fenichel, G.M. *Clinical Pediatric Neurology: A Signs and Symptoms Approach,* 6th ed., 2009, Saunders Elsevier, Philadelphia.

This book, by a highly experienced pediatric neurologist, is recommended for medical students and other novices as a basic text with a clinical approach that uses signs and symptoms.

NEUROPATHOLOGY

Kumar, V., Abbas, A.K., Fausto, N. and Aster, J., eds. *Robbins and Cotran Pathologic Basis of Disease,* 8th ed., 2010, Saunders Elsevier, Philadelphia.

This is a complete source of information on all aspects of pathology for learners, including neuropathology. Purchase of the book includes a CD-ROM with interactive clinical cases and access to the Web site and other learning resources.

Kumar, V., Abbas, A.K., and Aster, J.C., eds. *Robbins Basic Pathology,* 9th ed., 2013, Elsevier Saunders, Philadelphia.

This book is not as complete as the other text by the same authors (previous entry).

CHAPTER 10 REFERENCES

Bartus, R.T., Dean III, R.L., Beer, B., and Lippa, A.S. The cholinergic hypothesis of geriatric memory dysfunction. *Science.* 1982;217(4558):408–417.

Francis, P.T., Palmer, A.P., Snape, M., and Wilcock, G.W. The cholinergic hypothesis of Alzheimer's disease: a review of progress. *Journal of Neurology, Neurosurgery and Psychiatry.* 1999;66:137–147.

Olds, J. and Milner, P. Positive reinforcement produced by electrical stimulation of septal area and other regions of rat brain. *Journal of Comparative and Physiological Psychology.* 1954;47(6):419–427.

WEB SITES

Web sites should be recommended to students only *after* the sites have been critically evaluated by the teaching faculty. If keeping up with various teaching texts is difficult, a critical evaluation of the various Web resources is an impossible task for any one person. This is indeed a task to be shared with colleagues and perhaps by a consortium of teachers and students.

Additional sources of reliable information on diseases are usually available on the disease-specific Web site maintained by an organization, usually with clear explanatory text on the disease and often accompanied by excellent illustrations.

The following sites have been visited by the author, and several of them are gateways to other sites—clearly not every one of the links has been viewed. Although some are intended for the general public, they may contain good illustrations or other links.

The usual World Wide Web precaution prevails—look carefully at who created the Web site and when.

SOCIETY FOR NEUROSCIENCE

http://web.sfn.org/

This is the official Web site for the Society for Neuroscience, a very large and vibrant organization with an annual meeting attended by more than 30,000 neuroscientists from all over the world. The Society maintains an active educational branch, which is responsible for sponsoring a Brain Awareness Week aimed at the public at large and, particularly, at students in elementary and high schools. The following are examples of their publications.

SEARCHING FOR ANSWERS: FAMILIES AND BRAIN DISORDERS

This four-part DVD shows the human face of degenerative brain diseases. Researchers tell how they are working to find treatments and cures for Huntington disease, Parkinson disease, amyotrophic lateral sclerosis (ALS), and Alzheimer disease. Patients and families describe the powerful physical, emotional, and financial impact of these devastating disorders.

BRAIN FACTS

http://www.brainfacts.org/

Brain Facts is an on-line primer on the brain and nervous system that is published by the Society for Neuroscience. It is a starting point for a general audience interested in neuroscience. The new edition updates all sections and includes new information on brain development, addiction, neurological and psychiatric illnesses, and potential therapies.

Educational resources are now available online at www.BrainFacts.org. This site includes information about neuroscience, sections on brain basics, sensing, thinking, and behaving, and diseases and disorders, as well as a section called "across the lifespan." In addition, there are entries called "discoveries" and links to "recent neuroscience in the news," as well as blogs for opinion and conversation about what is new, notable, or inspiring in neuroscience.

DIGITAL ANATOMIST PROJECT

http://www9.biostr.washington.edu/da.html

BRAIN ATLAS

The material includes two-dimensional and three-dimensional views of the brain from cadaver sections, magnetic resonance imaging scans, and computer reconstructions. Authored by John W. Sundsten.

NEUROANATOMY INTERACTIVE SYLLABUS

This syllabus uses the images in the Brain Atlas (the previous entry) and many others. It is organized into functional chapters suitable as a laboratory guide, with an instructive caption accompanying each image. It contains the following: three-dimensional computer graphic reconstructions of brain material; MRI scans; tissue sections, some enhanced with pathways; gross brain specimens and dissections; and summary drawings. Chapter titles include Topography and Development, Vessels and Ventricles, Spinal Cord, Brainstem and Cranial Nerves, Sensory and Motor Systems, Cerebellum and Basal Ganglia, Eye Movements, Hypothalamus and Limbic System, Cortical Connections, and Forebrain and MRI Scan Serial Sections. Authored by John W. Sundsten and Kathleen A. Mulligan, Digital Anatomist Project, Department of Biological Structure, University of Washington, Seattle, Washington.

BRAINSOURCE

http://www.brainsource.com/

BrainSource is an informational Web site aimed at enriching professional, practical, and responsible applications of neuropsychological and neuroscientific knowledge. The Web site is presented by neuropsychologist Dennis P. Swiercinsky.

The site includes a broad and growing collection of information and resources about the following: normal and injured brains; clinical and forensic neuropsychology; brain injury rehabilitation; creativity, memory, and other brain processes; education; brain-body health; and other topics in brain science. BrainSource is also a guide to products, books, continuing education, and Internet resources in neuroscience.

This Web site originated in 1998 for promotion of clinical services and as a portal for dissemination of certain documents useful for attorneys, insurance professionals, students, families and persons with brain injury, rehabilitation specialists, and others working in the field of brain injury. The Web site is growing to expand content to broader areas of neuropsychological application.

HARVARD: THE WHOLE BRAIN IMAGING ATLAS

http://www.med.harvard.edu/AANLIB/home.html

This is a neuroradiological resource using a variety of neuroimaging modalities. Images include the normal brain, cerebrovascular disease (stroke), as well as neoplastic, degenerative and inflammatory diseases.

The Brain from Top to Bottom

http://thebrain.mcgill.ca/

This interesting Web site is designed to let users choose the content that matches their level of knowledge. For every topic and subtopic covered on this site, you can choose from three different levels of explanation—beginner, intermediate, or advanced. The major topics include brain basics (anatomy, evolution and the brain, development and ethics, pleasure and pain, the senses and movement), a section called brain and mind (memory, emotions, language, sleep and consciousness), and a section on some brain disorders; other subject areas are under development. This site focuses on five major levels of organization—social, psychological, neurological, cellular, and molecular. On each page of this site, you can click to move among these five levels and learn what role each plays in the subject under discussion.

Note to the Learner: This site can be viewed in both English and French.

The Dana Foundation

http://www.dana.org/

The Dana Foundation is a private philanthropic organization with a special interest in brain science, immunology, and arts education. It was founded in 1950. The Dana Alliance is a nonprofit organization of more than 200 preeminent scientists dedicated to advancing education about the progress and promise of brain research.

The Brain Center of this site is a gateway to the latest research on the human brain. The Brain Information and Brain Web sections access links to validated sites related to more than 25 brain disorders.

Neuroscience For Kids

http://faculty.washington.edu/chudler/neurok.html

Neuroscience for Kids was created for all students and teachers who would like to learn about the nervous system. The site contains a wide variety of resources, including images—not only for kids. Sections include exploring the brain, Internet neuroscience resources, neuroscience in the news, and reference to books, magazines articles, and newspaper articles about the brain.

Neuroscience for Kids is maintained by Eric H. Chudler and is supported by a Science Education Partnership Award (R25 RR12312) from the National Center for Research Resources.

To receive this interesting monthly newsletter, contact: Eric H. Chudler, Ph.D. (e-mail: chudler@u.washington.edu).

TELEVISION SERIES

http://www.pbs.org/wnet/brain/index.html

The Secret Life of the Brain, a David Grubin Production, reveals the fascinating processes involved in brain development across a lifetime. This five-part series, which was shown nationally by the Public Broadcasting Service in the winter of 2002, informs viewers of exciting new information in the brain sciences, introduces the foremost researchers in the field, and uses dynamic visual imagery and compelling human stories to help a general audience understand otherwise difficult scientific concepts.

The material includes a history of the brain, three-dimensional brain anatomy, mind illusions, and scanning the brain. Episodes include The Baby's Brain, The Child's Brain, The Teenage Brain, The Adult Brain, and The Aging Brain.

The *Secret Life of the Brain* is a co-production of Thirteen/WNET New York and David Grubin Productions, © 2001 Educational Broadcasting Corporation and David Grubin Productions, Inc.

VIDEOTAPES (BY THE AUTHOR)

These edited presentations are on the skull and the brain as the material would be shown to students in the gross anatomy laboratory. They have been prepared with the same teaching orientation as this atlas and are particularly useful for self-study or small groups. Each video lesson is fully narrated and lasts for about 20 to 25 minutes. These videotapes of actual specimens are particularly useful for students who have limited or no access to brain specimens.

Note to Learner: These video lessons are now available on the atlas Web site: www.atlasbrain.com.

INTERIOR OF THE SKULL

This program includes a detailed look at the bones of the skull, the cranial fossa, and the various foramina for the cranial nerves and other structures. Included are views of the meninges and venous sinuses.

THE GROSS ANATOMY OF THE HUMAN BRAIN SERIES

Part I: The Hemispheres

This is a presentation on the hemispheres and the functional areas of the cerebral cortex, including the basal ganglia.

Part II: Diencephalon, Brainstem, and Cerebellum

This is a detailed look at the brainstem, with a focus on the cranial nerves and a functional presentation of the cerebellum.

Part III: Cerebrovascular System and Cerebrospinal Fluid

This is a presentation on cerebrovascular system and the cerebrospinal fluid.

INDEX

Pages followed by f indicate figures; those followed by t indicate tables.